THE EUROPEAN CITIES AND TECHNOLOGY READER

Cities and Technology, a series of three textbooks and three readers, explores one of the most fundamental changes in the history of human society: the transition from predominantly rural to urban ways of living. This series presents a new social history of technology, using primarily urban settings as a source of historical evidence and a focus for the interpretation of the historical relations of technology and society.

Drawing on perspectives and writings from across a number of disciplines involved in urban historical studies – including archaeology, urban history, historical geography and architectural history – the books in the series explore: how towns and cities have been shaped by the applications of a range of technologies; and how such technological applications have been influenced by their contexts, including politics, economics, culture and the natural environment. *The European Cities and Technology Reader* is designed to be used on its own or as a companion volume to the accompanying *European Cities and Technology textbook* in the same series.

The European Cities and Technology Reader is divided into three main sections presenting key readings on: Cities of the Industrial Revolution (to 1870), European Cities since 1870 and Urban Technology Transfer. Among the topics covered are: the rise of Merthyr Tydfil, Manchester and Glasgow; the similarities and differences in the urban growth and built environment of London and Paris; the rise of modern urban planning; the complicated explanation of Berlin's twentieth-century development; the intricacies of technological transfer; the maze of post-war reconstruction; the technological influence of the West in British India and Russia's urban development. Compiled as a reference source for students, this reader offers a deeper understanding of the historical role of technological change in urban development.

David Goodman is Head of the Department of the History of Science and Technology at the Open University.

The Cities and Technology series
Pre-industrial Cities and Technology
The Pre-industrial Cities and Technology Reader

European Cities and Technology: industrial to post-industrial city
The European Cities and Technology Reader: industrial to post-industrial city

American Cities and Technology: wilderness to wired city
The American Cities and Technology Reader: wilderness to wired city

THE EUROPEAN CITIES AND TECHNOLOGY READER

Industrial to post-industrial city

Edited by David Goodman

London and New York

First published 1999
by Routledge
11 New Fetter Lane, London EC4P 4EE

Simultaneously published in the USA and Canada
by Routledge
29 West 35th Street, New York, NY 10001

Typeset in Palatino by J&L Composition Ltd, Filey, North Yorkshire
Printed and bound in Great Britain by Bath Press Ltd

British Library Cataloguing in Publication Data
A catalogue record for this book is available from the British Library

Library of Congress Cataloging in Publication Data
The European cities and technology reader: industrial to post-industrial city/
edited by David Goodman.
p. cm. - (Cities and technology)
1. Technology-Social aspects-Europe-History. 2. Urbanization-Europe-History.
I. Goodman, David.
II. Series: Chant, Colin. Cities and technology.
T14.5.E955 1999
307.76′094-dc21
98-38748

ISBN 0-415-20081-4 (hbk)
ISBN 0-415-20082-2 (pbk)

CONTENTS

LIST OF FIGURES AND TABLES

Figures

Tables

INTRODUCTION

This reader is part of a series about the technological dimension of one of the most fundamental changes in the history of human society: the transition from predominantly rural to urban ways of living. It is first and foremost a series in the social history of technology; the urban setting serves above all as a repository of historical evidence with which to interpret the historical relations of technology and society. The main focus, though not an exclusive one, is on the social relations of technology as exhibited in the physical form and fabric of towns and cities.

The main aim of the series is twofold: to investigate the extent to which major changes in the physical form and fabric of towns and cities have been stimulated by technological developments (and conversely how far urban development has been constrained by the existing state of technology); and to explore within the urban setting the social origins and contexts of technology. The series draws upon a number of disciplines involved in urban historical studies (urban archaeology, urban history, urban historical geography, architectural history and environmental history) in order to correct any illusion of perspective that the greatest changes in urban form and fabric might be sufficiently explained by technological innovations. In brief, the series attempts to show not only how towns and cities have been shaped by applications of technology, but also how such applications have been influenced by, for example, politics, economics, culture and the natural environment.

The wide chronological and geographical compass of the series serves to bring out the general features of urban form which differentiate particular civilizations and economic orders. Attention to these differences shows how civilizations and societies can be characterized by their use of certain complexes of technologies and also by the peculiar political, social and economic pathways through which the potentials of these technologies are channelled and shaped. Despite its wide sweep, the series does not sacrifice depth for breadth: studies of technologies in particular urban settings form the bulk of the three collections of existing texts in the series, this being the second.

This book is designed to stand on its own, but readers might note that there is a companion volume of original essays.[1] With a general and undergraduate readership in mind, scholarly apparatus has been kept to a minimum. This means the omission of many primary and secondary source references, but these have been retained where there is direct quotation.

This collection is a product of that distinctive, collaborative unit – the Open University Course Team. Editorship therefore signifies rather less individual contribution than in other readers with a single name on the cover. Only for the first section – Cities of the Industrial

Revolution (to 1870) - has selection and editing of sources been my task alone. All the rest is due to the Course Team. And my thanks go also to Denise Hall for her invaluable administrative assistance.

Note

1 D. Goodman and C. Chant (eds), *European Cities and Technology: Industrial to Post-industrial City*, Routledge, 1999.

Part 1
CITIES OF THE INDUSTRIAL REVOLUTION (TO 1870)

Britain's Industrial Revolution has been a subject of historical controversy for much of the twentieth century. Controversy has persisted over the alleged causes of the Revolution, its dates and, more recently, its reality has been brought into question. At the time of writing, there has been a definite shift in historical opinion away from this scepticism and towards acceptance of 'Industrial Revolution' as an appropriate description of what occurred first in eighteenth-century Britain and then on the Continent. The transformation of cities is a central topic of the history of this period. The dramatic effects of industrialization on cities have frequently evoked pithy assertions: 'Cotton made Manchester!' This type of simplistic and deterministic explanation continues to creep into today's academic history writing as readily as it does into more popular forms of expression. Reading 1 in this volume shows why 'iron made Merthyr Tydfil' is an unacceptable simplification of the way in which this wholly new town sprang up in eighteenth-century Wales. As the analysis of Chris Evans reveals, the local abundance of iron-ore and rich coal-seams are only part of the story; a complex amalgam of human industrial tradition and human decisions explains the form of Merthyr – or rather, the lack of form.

The setting remains Merthyr in reading 2. Based on a study of local records, R. K. Grant presents a grim picture of sanitary conditions in the nineteenth-century town, leading to a concerted drive for improvement. Here was an urban environment of utter squalor, the inhabitants helpless victims of industrial pollution and uncontrolled sewage disposal. As elsewhere in Britain, it took the terrifying outbreaks of cholera to stir a movement for urban reform. And again here, we see how the implementation of engineering solutions was not automatic, but a process tempered by vested interests and other human influence.

New building-types were generated by the Industrial Revolution. Iron-framed warehouses and multi-storeyed factories were conspicuous features in the urban landscape of nineteenth-century Britain's industrial centres. Booming production of textiles and chemicals resulted in an unprecedented demand for storage facilities in cities and ports. Many of the products were highly inflammable and the sense of apprehension of imminent, disastrous urban conflagration in mid-Victorian Britain is nowhere more perceptible than in reading 3, written by C. F. Young, a civil engineer of the time.

One of the greatest of all changes to life in both town and country during the era of industrialization was the revolution in transport brought about by the railways. John Kellett's pioneering study of the effects of the railways on Victorian cities, first published 30 years ago, retains validity and vigour. Reading 4, from the introduction to that impressive monograph, contains wise advice for anyone seeking to understand how railways affected cities: there has to be a comparison between numerous cities so that common consequences can be discerned along with distinctive, individual urban change. There is no technological determinism to be found here, either. There was no inevitability about the onward progress of steam locomotives towards the city centres of Victorian Britain. The actual years when railways first came to particular Victorian cities, the siting of urban railway facilities and, above all, the routes of lines into cities depended on contingent local conditions, such as the accidents of land ownership and prevailing attitudes of local authorities. Kellett also questioned the widespread assumption that railways must have been the stimulus for that familiar component of modern cities, the suburb.

The real complexity of the origin of suburbs comes out clearly in reading 5, extracts from F. M. L. Thompson's masterly analysis. This distinguishes very effectively between sufficient and necessary conditions for suburban growth, and rightly insists on the importance of chronology - cheap, mass-transit did not precede Victorian suburbanization but came along after. The force of the argument destroys the whole idea of a necessary, causal connection between transport improvement and the creation of suburbs. Instead, the emphasis is put on human choice. This far-ranging discussion is full of originality and new ideas on what stimulated the generation of suburbs.

Manchester is the subject of the next two pieces. Reading 6 is an extract from Vigier's study of Liverpool and Manchester during the Industrial Revolution. First published in 1970, this book remains of value as an introductory treatment for undergraduates. Enlivened by maps and data, this text explains how an entire manufacturing region developed around Manchester. The changing relationship between the centre, Manchester, and the periphery - towns such as Bolton, Bury and Oldham - is closely examined from the viewpoints of technological innovation and regional urbanization. From around 1800 we see Manchester's function changing from a marketing to a manufacturing centre with an associated expansion of the city's built-up area. As Manchester's population swelled, efficient water-supply became a pressing requirement.

Reading 7 is from a paper by Hassan and Wilson, communicating the results of research on how this problem was confronted by the recently created city corporation. Soon after the municipalization of water services, Manchester Corporation embarked on the largest of hydraulic engineering projects, designed to bring pure spring water into the city from the outlying Pennines. The authors present the main features of the ambitious scheme, indicate the formidable engineering difficulties encountered, and assess the success of the enterprise. Their description of components of the long-distance supply, various structures that had to be built miles beyond the city, provides good examples of the way in which a large city can alter the appearance of its rural surrounds.

Glasgow, that other great centre of industrialization in northern Britain, is the focus of the next four texts. Reading 8 considers rapidly expanding Glasgow within the wider context of Scotland's marked urban development and industrial concentration in the later eighteenth and early nineteenth century. It raises the interesting question whether there was anything distinctive about Scottish cities and analyses the causes of Glasgow's soaring industrial importance. The indispensable starting point for understanding Glasgow's rise to second city of the Empire is river improvement. Readings 9 and 10, both by John Riddell, together provide by far the best available discussion of the rising pressures for improving the navigability of the Clyde and the far-reaching consequences of the various phases of engineering works achieved. The initial eighteenth-century phase of river engineering is the subject of reading 9, which vividly conveys the nature of the problems to be tackled and clearly describes the solutions adopted. Reading 10 is a much broader sweep of these changes, extending its purview to cover the nineteenth century. The result is a fascinating portrayal of the eventual transformation, by the river works, of the form and fabric of Glasgow, altering the lives of citizens and work-force, as the city was filled with harbours and docks, and Clydeside became a booming centre of shipbuilding. But another traditional function of the Clyde would have to be discontinued: the river had served as the city's principal source of drinking water. Now its waters were becoming increasingly polluted by industrialization. Reading 11 is a letter of 1852 that records the initiative of two scientists that would eventually result in a large-scale engineering project to bring pure water to the expanding city from a distant loch. As this letter shows, public health was not the only concern that motivated this ambitious plan for change.

London's role in the Industrial Revolution has been wholly reassessed by recent research. In place of the former parasitic portraits, the image of the metropolis now emerges in a much more positive light. That can be perceived in reading 12, an extract from Roy Porter's synthesis of recent revisionist research. Another historical reinterpretation of London in the era of industrialization concerns the importance of the railways for supplying the capital's millions with fresh milk. The question is not nearly as clear cut as one might at first suppose. The debate can be followed in readings 13–15.

Reading 16, on Paris, has been introduced to encourage comparison with London. As with London, historians have recently been busy embellishing the industrial image of the French capital. This text, by Barrie Ratcliffe, is an impressive example of the trend. Drawing on contemporary census sources recording the occupations of nineteenth-century Parisians, and presenting detailed statistical tables and pie-charts, the paper arrives at challenging new conclusions. But what most of all inspires the paper is the conviction that historians of industrialization have been led astray by adopting an unduly narrow definition of 'industry'. Ratcliffe insists that historians must abandon the confining criteria of heavy industry, large installations with teeming work-forces and situated on the edge of coalfields. Only then will they understand that nineteenth-century Paris was a dynamic centre of manufacturing industry.

Whatever the verdict on Paris – hive of industry or not – no one doubts that the French capital was totally transformed during the Second Empire by a combination of comprehensive town planning on an unprecedented scale and innovative technology. Pinkney's pioneering study of the creation of modern Paris first appeared in 1958, but it continues to be the most illuminating account. The extract from Pinkney's monograph, reproduced here as reading 17, conveys an accurate picture of the magnitude of the tasks undertaken in one project that was just a part of the enormous master-plan to modernize the city: the installation of an efficient sewer-network that still serves today's Paris.

I

INDUSTRY AND URBAN FORM IN EIGHTEENTH-CENTURY MERTHYR

Chris Evans

Source: Chris Evans, 'Merthyr Tydfil in the Eighteenth Century: Urban by Default?', in P. Clark and P. Corfield (eds), *Industry and Urbanization in Eighteenth-century England*, Leicester, Leicester University Press, 1994, pp. 11–16 and 18

Merthyr was an iron town. Iron alone justified the building of a town at the head of the Taff valley, on the northern-most edge of Glamorgan. Before blast furnaces began to stud the landscape around the pre-industrial village the place could lay no claim to urban credentials, and since the death of the town's smelting industry in 1930 it has been an open question as to what else can sustain an urban community in so unpromising a location. The aim of this paper is to explore the relationship between the iron industry and urban development at Merthyr. At a basic level this relationship is perfectly simple: iron manufacture, and the swarming workforce it required at the end of the eighteenth-century, made Merthyr. On the other hand, the early modern iron industry had a peculiarly non-, even anti-urban character. And the environment in which the pioneer ironmasters were to settle – upland south Wales – was entirely bereft of urban antecedents. As a result, the town of Merthyr Tydfil was one which conspicuously lacked the type of urban facilities that were commonplace in other eighteenth-century urban centres. Indeed, some contemporary observers were puzzled as to whether Merthyr could be classified as a town at all, for all its sprawl and all its people. Merthyr was assuredly a working community, several working communities in fact, but the work which was performed at Merthyr had only ambivalently urban connotations. In what follows below particular attention will be paid to the forms of work, social organization and culture which Merthyr's inhabitants inherited from the early modern iron industry. This heritage was in large measure antithetical to the emergence of a stable urban environment or a settled urban identity at Merthyr. [. . .]

The raw facts of Merthyr's industrial and urban development are easily given. The parish stretched for nearly ten miles along the valley of the Taff, divided into five hamlets to compensate for its vastness. The village of Merthyr stood at the northern end of the parish, some five hundred feet above sea level. From the narrow valley floor the ground rose steeply to a mountain plateau nearly 1,500 feet above sea level. Merthyr's agriculture was meagre, its population sparse and its hinterland threadbare; the pre-industrial village had not the least commercial or administrative importance. As is usually the case with Wales, the demographic indicators upon which English social historians would rely are absent. However, it is unlikely that the parish could claim more than seven or eight hundred inhabitants on the eve of industrialisation.

The incursion of the iron industry transformed this situation with revolutionary speed and completeness. The first furnace at Dowlais went into blast in 1760; another furnace, named in honour of the ground landlord, the Earl of Plymouth, was in operation by 1766; and another was built at Cyfarthfa in 1766–7. A fourth works was established at Penydarren in 1783. Together these works embodied the acme of iron-making technology in the late eighteenth century, smelting and refining iron with mineral fuel. Indeed, Merthyr was uniquely well-equipped to make use of the new coal technology that came into use in the 1750s. The mineral inputs demanded by coke smelting literally jutted from the ground in the north of the parish. Merthyr knew no prior tradition of iron smelting, it was true, and this may be considered an impediment to the industry's growth in the sense that precious ironmaking skills had to be engrafted from older centres of iron production in the Midlands. Yet the lack of any foundation for heavy industry was not without benefits, for it enabled Merthyr to exemplify the dynamic of combined and uneven development. The Merthyr ironmasters had no need to adapt squat, charcoal-fuelled blast furnaces to the new discipline of coke smelting. They built plant de novo, employing the latest techniques on an unprecedentedly large scale. In doing so they catapulted Merthyr to the head of the British iron industry. By the mid 1790s the Cyfarthfa ironworks of Richard Crawshay was the world's largest, and it was with justice that one visitor to the town described Merthyr as the 'metropolis of ironmasters'. A metropolis not just of ironmasters, Merthyr was a metropolis of ironworkers, miners and colliers as well. When the first census was taken in 1801 Merthyr ('which twenty years ago scarcely deserved the name of village') was found to have 7,705 inhabitants. Ten years later the population of the parish had leapt to over 11,000 – a decennial increase of 44 percent.

This concentration of production and population on a single confined spot was unheralded and without precedent in the history of the British iron industry. Indeed, it was a striking reversal of the iron trade's earlier organizational priorities. The clustering of furnaces and forges on coalfield sites such as Merthyr overturned the tendency of the old charcoal-fuelled iron trade towards the dispersal and mobility of production. As is well known, the iron industry had been, since the introduction of the blast furnace to Britain in the early sixteenth century, a voracious consumer of charcoal. This was enough to ensure the industry's banishment to locations that were remote from or inaccessible to urban consumers whose demand for domestic and industrial fuel drove up the price of charcoal. The early establishment of iron smelting in the Weald of Kent and Sussex, where the luxuriance of timber close to London was offset by the impenetrability of the woodlands, met this condition: ironmasters were relatively undisturbed by the capital's appetite for fuel. However, the expansion of iron production to meet growing consumer demand was only to be achieved by the extension of the industry to new woodland sites in south Wales, the Welsh borders, the Midlands and the West Riding of Yorkshire – a process of colonisation which was effected between the late sixteenth and early eighteenth century. The net result of this was the steady flight of the iron industry away from major centres of population towards the more distant northerly and westerly reaches of the British archipelago. This flight permitted national output of pig iron to reach an all time high in the early eighteenth century, a level which was maintained until the advent of coke in the 1750s. The British iron industry was not – as was once thought – in terminal decline before the introduction of coke smelting. On the contrary, it was busy and active. Nevertheless, it seems clear that charcoal smelting had hit a production ceiling through which it could not break. And it could sustain that level of production only by seeking out ever more recondite sources of wood. [. . .]

The centrifugal tendencies of the late charcoal era had some simple consequences for the men and women who worked in the iron industry. For one thing, their working lives were peripatetic, hastening from one site to another. Moreover, their migration within the industry was now taking place over increasingly long distances. This movement was not, however, restricted to the newer and geographically peripheral centres of iron production. There is evidence that the workforces at older, more established furnaces and forges were also subject to displacement and removal in the middle decades of the eighteenth century. Production at furnaces operated by the Knight family in the Severn valley, for example, was being staggered in an apparent effort to conserve charcoal resources. Bringewood furnace in Shropshire (now Herefordshire) was put into blast on alternate years only for the best part of the 1730s, 1740s and 1750s. The same could be said - albeit for shorter periods - of other Knight furnaces in Worcestershire, Warwickshire and Shropshire. The implications of such a strategy are clear: those in the Knight's employ had to be regularly redeployed between different sites.

It was men from this peripatetic workforce who physically built Merthyr's ironworks: veterans of the east Shropshire coalfield who came to Dowlais and men from the forges of the Stour valley to Penydarren. The construction of the Cyfarthfa works was overseen by Charles Wood, former manager of the Lowmill forge in Cumberland, whose workforce included forge carpenters and masons who had accompanied him from the far northwest, as their names (Postlethwaite, Dickinson, Craven) testified. Others were old hands who had worked along the arc of the Severn. As construction got underway at Cyfarthfa in the summer of 1766 prospective recruits began to arrive in the expectation of work, illustrating well the speed with which news and rumour spread through the iron trade. One forgeman, who had worked for Charles Wood in Cumberland, presented himself when the Cyfarthfa forge was still only half-built. He announced that another old hand from Lowmill, a hammerman who was then working at Tidnor forge near Hereford, was also about to try his luck in Wales.

In short, the establishment of an iron industry at Merthyr Tydfil, remote and undistinguished by any other form of industrial endeavour, was quite consistent with the prevailing conditions in the charcoal iron industry. It was an industry that sought seclusion, and whose workforce was highly mobile, moving from job to job through channels that were largely specific to the industry and which were not centred on the usual urban nodes. Of course, the introduction of coke smelting and coal-fired refining techniques did much to transform this. Iron production could now be concentrated on coalfield sites. Furnaces and forges could multiply without fear of exhausting fuel supplies, each works giving employment to dozens of iron workmen and still greater numbers of colliers and miners. At Merthyr, where four such works sat cheek-by-jowl, these new organizational arrangements gave rise to a settlement that was perforce urban, by virtue of its swollen population if little else. And at Merthyr there was little else. For the poverty of Merthyr's urban credentials was ramified on every side: by the paucity of an urban tradition in Wales as a whole; by the anti-urban heritage of the old charcoal iron industry; and by the very nature of the town's iron industry, wedded to capital goods production and incapable of stimulating ancillary urban functions. Some of the keynotes of contemporary urban society (diversity, internal specialisation of function, the sophistication of the means of exchange) never featured at Merthyr.

'Notwithstanding its magnitude and commercial consequence', announced one visitor in

1811, 'Merthyr Tydfil is but a village, although by courtesy it enjoys the title of town'.[1] Indeed, the under-provision of urban amenities at Merthyr was spectacular. The parish church constituted the sole public building. Street lighting was not introduced until the second half of the nineteenth century. (Parsimonious ratepayers took the view that the flare from the furnaces provided light enough.) The service economy was so stunted that a directory of 1795 amounted to nothing more than a roll-call of butchers and victuallers. The provision of retail facilities was so inadequate that as late as 1822 Merthyr had one shop for every 400 of its inhabitants, when an established commercial and ecclesiastical centre like York could boast a ratio of one per 70 inhabitants. Professional, educational and recreational facilities which would have gone unremarked in a small market town in England were absent from what was a world centre of industrial production.

Social amenities of this sort foundered because they had no clientele. Merthyr society was famously polarised. At the pinnacle of the town's shaky social structure stood a knot of ironmasters, the very wealthiest of whom, Richard Crawshay, died a millionaire. [. . .] Merthyr's proletarian base enjoyed an infinitely more modest existence, and very little of any social substance stretched between these extremes. Admittedly, within the ironworks themselves the ironmasters could call upon the services of a distinct managerial stratum. Indeed, 'management' had already – in the final decades of the charcoal era – crystallised as the function of a professional caste in the iron industry, the essential intermediary between iron capitalists and the diaspora of iron workers they set to labour. However, Merthyr's clerks and book-keepers were numerically insignificant and – for all their powers of industrial administration – never featured as the agents of a wider social or cultural hegemony in Merthyr. For one thing, they simply lacked the social eminence with which to overawe workmen who set enormous store on the respect due to those who exercised the craft skills of iron-making. [. . .]

In short, Merthyr lacked the 'middling' social groups which were to be found in such profusion in Britain's more established urban centres, including newer centres of manufacturing, even in areas of iron production. [. . .] Merthyr did have a 'sub-urban' nucleus of traders and dealers. And it was upon this manifestly inadequate foundation that a good deal of Merthyr's urban infrastructure was erected, for the ironmasters proved reluctant civic leaders.

The ironmasters' evasion is most readily apparent in the provision of housing. The last charcoal ironmasters had, as a matter of course, provided accommodation for the workers they shipped to furnace sites on the coasts of north Wales and Scotland. New charcoal furnaces were usually on greenfield sites where no form of shelter would have been available if the masters had not built it. This tradition was carried over into the early coke era. When Charles Wood began the construction of the first forge at Cyfarthfa in 1766 he prepared plans for apartments to be built over the top of the work area, providing (noisy) accommodation for eight families. This would, he thought, 'save the Erecting separate Houses for them, at much greater expense; and the convenience of every workman being near his work, will be an ease to them; & an advantage to the Masters'. [. . .]

Neverthless, the commitment of the ironmasters to housing their workforce was limited and grudging. Housing the furnace and forge workers of the old charcoal industry had not been an onerous undertaking. It was a matter of catering for a dozen households per site at most. The new coke industry made far larger demands. The Dowlais works employed '400 men and boys . . . exclusive of familys' by 1794; the larger Penydarren works gave employment, by one estimate, to over nine hundred men, women and children ('reckoning in the miners') by 1802. The ironmasters could scarcely look on the task of accommodating a working population of this magnitude with any equanimity. Indeed, they did their utmost to evade the problem. In 1798, when the workforce at Cyfarthfa must have been approaching one thousand, Richard Crawshay had no more than 70 stone-walled houses at his disposal. The Penydarren partners controlled only 49 company dwellings, the Plymouth Company a mere fifteen. [. . .] That Merthyr's ironmasters were able to avoid a building programme [. . .] was due to the presence of that group of petty entrepreneurs which Merthyr village harboured. This latter group were more than eager to meet Merthyr's housing deficit with piecemeal, speculative building. In the absence of any form of regulation, or of any systematically planned development by the iron companies, houses were thrown up without pattern or design, dotted about the fields and yards on the edge of the village, crammed into the interstices of existing courts and terraces, or stretched along the sides of the lanes and gutters about the ironworks.

The effect was disastrous. Speculative endeavour could only partly assuage ironmaster inaction. It failed to keep pace with the population that came to throng the ironworks. The outcome was chronic overcrowding and an ecological blight [. . .]

[. . .] after its first and most dynamic phase of growth the iron trade of Merthyr and its associated iron settlements immediately to the east and west acquired a self-sustaining weight: Tredegar, Ebbw Vale, Aberdare and the rest coalesced with Merthyr into a network of iron centres, a network that was increasingly self-contained and decreasingly reliant on older iron centres. Merthyr was eventually to stand alone and acquire some modicum of civic dignity in the later ninteenth century. But by then it stood on the edge of the South Wales coalfield, rather than the heart of the British iron industry.

Note

1 J. G. Wood, *The Principal Rivers of Wales Illustrated*, London, 1813, Vol. 1, p. 57.

2

MERTHYR TYDFIL IN THE MID-NINETEENTH CENTURY: THE STRUGGLE FOR PUBLIC HEALTH

R. K. Grant

Source: *Welsh History Review*, 14 (1988–9), pp. 571–86, 588–94

[. . .] T. W. Rammell, an inspector of the General Board [of Health], reported in 1850 that

> Merthyr, having sprung up rapidly from a village to a town, without any precautions being taken for the removal of the increased masses of filth necessarily produced by an increased population, nor even for the escape of the ordinary surface water; without regulation for the laying out of streets, or the building of tenements; . . . from all these circumstances combined, a rural spot of considerable beauty, and with more than the average natural facilities for drainage and water-supply, has become transformed into a crowded and filthy manufacturing town, with an amount of mortality higher than any other commercial or manufacturing town in the kingdom.[1] [. . .]

As Rammell pointed out, urgent problems had emerged from the spectacular increase in the population of Merthyr Tydfil during the previous decades. These problems were, of course, shared by many of the great industrial towns of England, which have been described as being at that time 'burial pits of the human species'.[2] [. . .]

Edwin Chadwick's *Report on the Sanitary Conditions of the Labouring Population* (1842) had shown the connection between overcrowding, bad housing and dirt in towns and the high incidence of disease there; he urged the consequent necessity of government action to provide piped water and sewers. As a result, the Home Secretary reluctantly appointed a Royal Commission in 1844; Sir Henry de la Beche, one of the commissioners, brought out his *Report* in the following year. He drew attention to the lack of a public water supply, to accumulations of filth in the streets, and to the resulting 'offensive smells'. 'The rarities of privies' was said to be 'one of the marked characteristics of the town'.[3] [. . .]

It might have been thought that the river Taff would have provided ample water for these rapidly growing communities of industrial workers. But in fact the ironmasters had established a monopoly of its waters. Richard Crawshay told Thomas Rammell:

> The principal power employed at the Cyfarthfa works, and at the Ynysfach works, is water-power. There are in all 7 water-wheels [. . .] Over the largest wheels about 40 tons of water per minute are passed, when the water is plentiful. In the dry season, which is chiefly in June, July and August, we are at times short of water, and are obliged to pump back the water which has already passed over the wheels, by two additional (steam) engines [. . .]

Rammell further relates that 'Upon one occasion near Merthyr, so scarce and so valuable was water, as compared with human labour, that a number of workpeople were employed to ascend the furnace by ladders, and to perform the office of the water by descending in the balance-bucket, as a counterpoise to the ascending materials'.

The ironmasters, therefore, jealously husbanded the water upon which their industry depended [. . .] Industrial pollution was very great. T. J. Dyke, a local surgeon, described to him in graphic detail how the 'Dowlais brook', having 'its present origin in numerous ponds which have been made by the Dowlais Iron Company, for the use of the ironworks . . . passes through the works . . . and joins the Morlais, black and filthy. The Morlais now runs in a deep dingle to the Pen-y-darren ironworks, through which it also passes, and is made use of in the works. Charged with an additional quantity of dirt, it reaches Pontmorlais' and finally 'rolls a heavy, black and filthy stream into the Taff.'[4]

Rammell concluded that 'the poorer classes are put to great labour and loss of time in collecting very scanty quantities [of water] for domestic purposes'. Drinking water was obtained 'precariously' from 'springs or spouts' providing a meagre and wholly inadequate supply, 'in summer especially, . . . the water dripping from the spouts, or slowly accumulating in the small hollows around, having to be carefully collected in the pitchers or kettles'. Many of these 'springs or spouts', he continued, 'which are mostly used by the poorer classes . . . lie at a considerable distance from the houses, and much time is occupied in going to and returning from them . . . To the female members of the community usually falls the task of procuring water from the springs, and the enormous amount of labour, inconvenience and loss of time to which they are daily and nightly subjected in doing so, would hardly be credited in any civilized community.'

The rector of Dowlais, the Rev. Evan Jenkins, supplied grim and shocking details:

> During winter, there are from 6 to 8 spouts, some half a mile, some a mile distant from the houses, but in summer they are reduced to three, the remainder being dried up. At these water-spouts [. . .] I have seen 50, 80 and as many as 100 people waiting for their turn, and some then obliged to go away without any water at all. They have been known to wait up the whole of the night. In the case of women having a young family, they are left at home at these times to take care of themselves. [. . .] They have no other supply of water whatever fit to drink in summer time, and have no alternative but to wait.

The water used for 'brewing, washing and other domestic purposes' was derived from 'rain-water collected in casks and butts', and from ponds and the Glamorganshire Canal. T. J. Dyke testified that water from the last two sources was 'eminently impure, and wholly unfitted, even after boiling, for human use, inasmuch as it contains a quantity of animal and vegetable matter dissolved in it'. Such was the desperate shortage of water that people took it from stagnant ponds and pools, even though they were polluted by industrial waste such as oil and tar, by horse-dung, and even dead animals. Some better-off people had access to private wells, but even they were frequently contaminated by seepage from nearby cess-pits, privies, pigsties and other nuisances.[5]

The second *Report* of the Royal Commission on public health in towns, published in 1845, had revealed lamentable deficiencies in the water-supplies of fifty of them; but Merthyr Tydfil appears to have been among the worst. Dr. William Kay, the town's first Medical Officer of Health, reported in 1854 that, despite the natural opportunities offered by the Taff, it had the reputation of being 'one of the dirtiest towns in Her Majesty's dominions' and declared that a public water supply was 'essential to the health, cleanliness and domestic comfort' of the 'labouring classes of Merthyr'[6] [. . .]

Turning to another aspect of this dark picture of barbarous neglect of the basic principles of municipal public health, Merthyr Tydfil may at this time 'almost be said to be wholly without provision for drainage'.[7] A local surgeon told Rammell:

> There is but one main sewer, that in Victoria-street; I believe it receives the sewage-water from about 50 houses; it is deep, and the bottom is washed out by the rain-water. It discharges itself into the Taff river. The orifice of its outlet is above the surface of the water in the Taff at ordinary times, but when the river is flooded it is covered. There is consequently a draught up the sewer, which passes (charged with the gases given out by the sewage-water) through the lateral drains into the houses, and through the gully-holes at the head of the sewer into the air in the High-street. Until lately, the noxious effluvia from these gully-holes . . . were insupportably offensive, and most perceptively tainted the atmosphere in the neighbouring houses. These effluvia so evidently caused the illnesses in a family who resided close to the street in this situation, that I considered it my duty to advise them to leave the house, which they did. Their health has materially improved since their removal.[8]

T. J. Dyke was following the miasmatic theory, that disease is caused by noxious exhalations from rotting refuse and waste matter; the great social reformer, Edwin Chadwick, summed it up by saying, 'All smell is disease'.[9]

Many workers' cottages were built, as we have seen, without any sanitary provision at all. According to T. J. Dyke, 'the better class of houses . . . are generally provided with privies and cess-pools; the remainder, with very few exceptions, are totally unprovided with either, the inhabitants making use of chamber utensils, which they empty into the streets before the doors, sometimes into the river, or, in the case of males, relieving themselves upon any of the numerous cinder-heaps which abound in all parts of the town, or by the sides of walls, or backs of houses, all without the commonest regard to decency. Children are placed out in open chairs in the street to perform their necessary operations.'[10] So the Taff and its tributaries, heavily polluted by industrial waste, were also open sewers. [. . .]

There was no provision for street cleaning and refuse collection [. . .]

The result of all these conditions was a high mortality rate, especially among babies and young children. In 1849, the year of the cholera, 34 per cent of the total number of deaths in Merthyr Tydfil parish were of children under five – that is, 998 out of 2,925. The average general rate in Merthyr Tydfil between 1848 and 1853 was 34.7 per thousand inhabitants, a frightful rate which brought the town well within the provisions of the Public Health Act of 1848. Its causes have been identified as the occupations and relative poverty of the iron-workers and colliers and their families who constituted the great majority of the population of the town. Sir Henry de la Beche calculated in 1845 that the average expectation of life of tradesmen in the town was 32 years, but that of an artisan, including puddlers and colliers, was only 17. Merthyr Tydfil, of course, was only one of the many industrial towns in Britain in the nineteenth century where high birth rates were more than counterbalanced by terrifyingly high mortality rates. [. . .]

The most terrifying of the epidemic diseases which raged in these towns was cholera. It was 'swift, dramatic, highly lethal while it lasted, and extremely contagious', and it 'struck terror into the minds of the middle and upper classes who ruled the cities and the country'.[11] Cholera had made its appearance in Merthyr Tydfil in 1832; it struck again in the summer of 1849, which had been very dry. Drought obviously facilitated the outbreak of this dreaded disease because, as sources of water dwindled, people were driven to take it from contaminated sources. [. . .]

Merthyr physicians and surgeons were, of course, much concerned to ascertain the causes of this terrible epidemic; it was not then generally realised that drinking contaminated water spread the disease. The 'contagionists' attached greatest importance to isolating sufferers; the 'miasmatists' believed that disease was caused by 'miasma', or noxious exhalations from putrid matter. Hence the emphasis on removal from towns of the more obvious causes of bad smells, upon street cleansing, sewage removal and a piped supply of clean water. [. . .]

Cholera provided the motivating force, the Public Health Act the means of remedy. Seventeen towns in south Wales, as a direct result of the 1849 epidemic, adopted, or petitioned to adopt, the Act. Of these towns, Merthyr Tydfil was one. The necessary number of Merthyr ratepayers petitioned the General Board, which sent down Thomas Webster Rammell to make a 'Preliminary Inquiry into the Sewerage, Drainage and Supply of Water, and the Sanitary Condition of the Inhabitants of the Town'. He called a public meeting on 15 May 1849, that is to say, ten days before the first case of cholera in Merthyr Tydfil was reported. The meeting opened in the parish vestry-room, which was the only public room in the town at that time. A large crowd assembled [. . .]

The rector of Dowlais spoke strongly in favour of measures to provide a public water-supply and sewerage system. But there was also strong opposition. Lewis Lewis, vice-chairman of the Board of Guardians, representing the viewpoint of the more conservative members of the propertied rate-paying class, argued that implementation of the Act would necessitate the appointment of salaried officials and so increase the rates. Iron-workers and coal-miners present were afraid that landlords would increase rents, to recoup the cost of compulsory improvements to their properties. The workmen's spokesman said, 'The high rate of mortality which had been alluded to was caused by the mode of life of the people, working underground, and by want of sufficiency of food, and not by want of sanitary laws. What they wanted was more meat'.[12] His viewpoint was endorsed by 'the pioneer dietary surveys conducted by Edward Smith in the eighteen-sixties', which 'pointed to a clear association between ill-health and poor nutrition'.[13]

Rammell submitted his *Report* to the General Board of Health on 26 March 1850; it was comprehensive and gave clear directions for public health in Merthyr Tydfil. The Tâf Fechan was to be dammed near Pontsticill, to provide every house in Merthyr and Dowlais with piped water; plugs were to be provided in the streets for cleansing and for fire-fighting. A sewerage system for these communities was projected; the principal main sewer was to discharge itself into the Taff at a point below the Plymouth works. The cost of these works was to be met initially by a loan to the Board from the Commissioners of Public Works; interest, capital repayment and maintenance charges would be paid by water and sewerage rates levied upon the property-owners of the parish.

Accordingly, the rate-payers of Merthyr Tydfil in October 1850 elected their local Board of Health, following the provisions of the 1848 Public Health Act. The Board at the outset was dominated by 'the four iron masters who contributed above half the district rates'[14] [. . .] the town had no corporate institutions such as a town council through which they could have developed political consciousness. Furthermore, until 1866 the Public Health Acts were permissive only; central government could not compel local authorities to proceed with essential public health works. [. . .]

The waterworks were now being pushed forward, largely through the leadership of G. T. Clark, resident Trustee of the Dowlais Iron Company, who had restored its fortunes after J. J. Guest's death; he became chairman of the Board in 1862. Hawkesley, who was appointed Waterworks Engineer by the Board in January 1859, was instructed to draw up new plans along the lines demanded by the ironmasters. His scheme was carried out, under his directions, by Mr. Samuel Harpur, the contractor. Water was drawn directly from the Callan brook at Pentrewernen, and conveyed in iron pipes to filter-beds and a service reservoir at Penybryn. Merthyr Tydfil and Lower Dowlais were supplied thence by gravity, but the water supply for Upper Dowlais had to be pumped up to a service reservoir at Dowlais Top. This scheme was completed by September 1861.

To compensate the ironmasters and their canal for water extracted from the Taff for public use, the Pentwyn reservoir, of 63 million cubic feet capacity, was to be constructed by damming the Tâf Fechan near Dôlygaer. The ironmasters were to have first call upon the stored waters, but the reservoir would be 'so connected with the main pipe as to be capable, in times of very unusual drought, of yielding a portion of its contents for the use of the Town'.[15] The construction of the dam was begun in May with the sinking of trial shafts, but it presented serious and recurrent problems. It was unfortunately built over a great geological fault crossing the valley; consequently there was from the beginning a very serious leakage of water from the reservoir. It was completed in February 1863, but Hawkesley found it necessary to submit a final report in April of the following year, assuring the Board that the Pentwyn reservoir was safe, that the embankment was solid, and that there was no leakage. In fact, repairs to the dam involving very considerable expenditure had to be undertaken at intervals until 1923. The problem was finally resolved when in 1927 the river was dammed further south at Ponsticill, to form the Tâf Fechan Reservoir.

Once completed, the Pentwyn reservoir constituted a great store of water, vital for industrial as well as domestic purposes. The Board's chairman, G. T. Clark, was a doughty champion of public health, but neither was he neglectful of the interests of the Dowlais ironworks. So, in September 1864 William Crawshay, Richard Fothergill, proprietor of the Plymouth ironworks, and the Glamorganshire Canal Company obtained an interim injunction in the court of Chancery restraining the Board from supplying water to the Dowlais Iron Company for industrial purposes. The Board offered a compromise. They would not supply water to ironworks when the level in the reservoir fell below a stated point, with the proviso that 'an ample supply of water for domestic purposes shall take precedence of any Supply for Machinery or Motive Power'. But the Crawshays stood on their rights under the Merthyr Tydfil Waterworks Act of 1858. By a majority vote, the Board then accepted Crawshay's amended terms, whereby the supply of water from the reservoir for domestic purposes was to have precedence 'except in so far as . . . the Cyfarthfa and Plymouth Ironworks and the Glamorganshire Canal Company . . . are at present under the said Act entitled to such precedent Supply'. G. T. Clark, having been outvoted, resigned the chair and his seat on the Board. But at the next annual meeting of the Board in the following March, he was back on the Board and in the chair once again.

Notwithstanding the delays caused by problems at the dam, piped water was being supplied to the people of Merthyr Tydfil as early as March 1861, by taking it directly from the Callan brook; this required the consent of the Crawshays and Anthony Hill, since the Pentwyn reservoir had not then been completed. In July 1861 Henry Wrenn, the superintendent of police, reported that the 'fire-plugs, hose and fire-fighting apparatus' supplied to the police in Merthyr Tydfil were 'effectual . . . The water was carried with great force over all the highest buildings and the inhabitants appreciated this increased security against fire.'[16]

Good progress was made with laying cast-iron water-mains; 40 miles had been laid by January 1865. As early as April 1862 the Board was accepting tenders for lead piping and taps, and by November 2,698 houses in Dowlais and Penydarren and 2,133 in Merthyr Tydfil had been connected to the mains, after application to the Board by the householders. Many property owners were reluctant to pay the new water charges. In April 1864 the Board's surveyor was reporting a number of houses still unsupplied with water, 'whilst many of the tenants are obtaining a surreptitious supply from your works'.[17] The Board's 'Inspector of the Water Supply' began to serve notices on householders not connected with the water mains that 'under section 76 of the Public Health Act . . . a proper water supply can be furnished . . . at a rate not exceeding 2d. for each house'. If property owners defaulted the Board undertook the work themselves and recovered the cost from them. It could not proceed in this way, however, unless the surveyors could show that the houses in question were without a supply of water. He reported in February 1867 that 4,193 houses were supplied by 790 standpipes, and so escaped the compulsory procedure. [. . .]

The difficult and protracted task of providing the town with a sewerage system still faced the Board. It received a letter from its chairman, G. T. Clark, on 2 June 1864, in which he referred to the continuing high mortality rate in the town, especially amongst infants. It was the highest in Wales, and higher than many of the great English towns, such as Sheffield, Leeds and Birmingham, which were 'least remarkable for their health'. 'Much of this excess of mortality', he wrote, 'may be traced distinctly to the want of proper house drains and street sewers'.[18] [. . .]

Progress had at first been slow, but under the chairmanship of G. T. Clark substantial and systematic advances were made. In September 1864 the Board's surveyor, Samuel Harpur, was instructed to prepare plans and estimates for the sewerage of Merthyr Tydfil, Dowlais, Abercanaid, Pentrebach and Troedyrhiw. Fifty thousand yards of main sewers were to be constructed, with the ancillary provision of 'gully shoots and grates, ventilation and inspection shafts etc.' Surface water drains and sewers were to be kept separate. In August of the following year, the Board borrowed £27,000 from the Atlas Assurance Society to finance the scheme. The first contractor encountered unexpected difficulties, such as having to cut trenches through hard rock; nevertheless, by August 1866 ten miles of brick sewers had been constructed, and nineteen miles of sewage pipes laid. A new contractor was appointed, and the work went steadily forward. By October 1868 the surveyor reported that 'the Sewerage Works were now completed'.[19]

The system discharged itself into the Taff. The outfall was near the railway station in the town itself, and this naturally gave great offence. In June T. J. Dyke, Medical Officer of Health, asked in his Report,

> that the discharge opening of the main Sewer shall at the earliest possible time, be carried to a somewhat longer distance from the Town. Bearing in mind the number of persons who daily pass into and out of the Southern Entrance of the Taff Vale Railway Station by the various trains, and that these trains pass into the Station at a place which is immediately over the Sewer mouth, it is most unjust to the Passengers to subject them to the possibility of inhaling the poisonous Gases given off by the Sewage.[20]

The outfall was subsequently moved downriver to Troedyrhiw. But opposition to using the Taff as an open sewer continued, [. . .]

An alternative method of disposing of Merthyr Tydfil's sewage had, therefore, to be found. In October 1864 the Board's surveyor, Samuel Harpur, produced a scheme for spreading the liquified sewage by irrigation over the surface of fields on the Penydarren estate, for Dowlais, and over 'flat meadows' below Troedyrhiw, for Merthyr Tydfil. The opinion of Mr. William Lee, 'an eminent engineer', was sought; he considered that 'applying fresh sewage to land will not be more offensive than spreading farmyard manure'. There was powerful support for this strange contention. Edwin Chadwick himself was an enthusiastic supporter of the use of untreated sewage, in liquid form, as a field manure; he thought that public health measures should be self-financing. But local landowners like Mr. Wyndham Lewis objected strongly that the scheme would create a public nuisance, and it had to be abandoned. So in February 1869 the surveyor produced yet another scheme - Dr. Frankland's 'Intermittent Downward Filtration System' - for the treatment of sewage. A sewage farm was constructed about half a mile south of Troedyrhiw. The sewage was to be filtered and chemically treated in covered tanks; when dried, the final product was to be used as fertilizer. Later in the same year, plans were adopted for the establishment of a sewage farm lower down the valley, outside the borough, on 'barren mountain common' [. . .] 'The principal volume of the Sewage' was to be conveyed there by a twenty-four inch conduit. [. . .]

The exertions of the Board were intensified by the re-appearance of cholera in 1854 and again in 1866, which earned the town unfavourable comments from London. G. T. Clark, in a letter of 2 June 1867 to the Medical Officer of the Privy Council, sought to defend Merthyr's record in public health. 'That our exertions have not been without fruits is evident', he wrote. The cholera epidemic in 1849 had carried off 1,452 victims out of a population of 46,378; but in 1854 only 424 patients had died, and in 1866 there were only 136 fatal cases out of an increased population of 55,000. [. . .]

Notes

1 T. W. Rammell, *Report to the General Board of Health . . . on the Sanitary Condition of the Inhabitants of . . . Merthyr Tydfil*, London, 1850, p. 12.

2 H. J. Dyos, *Exploring the Urban Past*, Cambridge, 1982, pp. 72–3.

3 *Report on the State of Bristol . . . Merthyr Tydfil and Brecon*, London, 1845, pp. 77–9.

4 Rammell, *Report*, pp. 34–6.

5 *Ibid.*, pp. 36–7.

6 Minutes, Merthyr Local Board of Health, I, pp. 353, 356.

7 Rammell, *Report*, pp. 22–4.

8 *Ibid.*, p. 28.

9 N. R. Longmate, *King Cholera*, London, 1966, pp. 67 and 160.

10 Rammell, *Report*, p. 30.

11 M. W. Flinn (ed.), *The Sanitary Condition of the Labouring Population of Great Britain, by Edwin Chadwick, 1842*, Edinburgh, 1965, Introduction, p. 10.

12 Rammell, *Report*, pp. 3–5.

13 G. Rosen, 'Disease, Debility and Death', in H. J. Dyos and M. Wolff (eds), *The Victorian City*, London, 1973, vol. 2, p. 625.

14 Minutes, Merthyr Local Board of Health, IV, p. 12.

15 *Ibid.*, III, pp. 212, 216; IV, pp. 11–12.

16 *Ibid.*, IV, pp. 103, 107, 143–4.

17 *Ibid.*, V, p. 154.

18 *Ibid.*, VI, pp. 18–19.

19 *Ibid.*, VII, p. 248.

20 *Ibid.*, VII, p. 87.

3

INCREASING FIRE-RISKS IN CITIES IN AN AGE OF INDUSTRIALIZATION

C. F. Young

Source: C. F. Young, *Fires, Fire Engines and Fire Brigades,* London, 1866, pp. 57–63

Fireproof structures

The construction of buildings entirely fireproof and uninflammable has long been a favourite idea, and one on which no small amount of thought and money have been expended, but unfortunately with anything but satisfactory results. So far as practice and experience have hitherto taught us, no safety from the destructive ravages of fire has been or can be obtained by the use of the most uninflammable materials for building purposes, unless some other modes of employing them be adopted than those hitherto used. Most of the tremendous and disastrous conflagrations of late years have either originated in or extended to these so-called 'fireproof' structures; and as in these the most valuable goods and materials have been stored, on account of the assumed security afforded thereby, the losses have been correspondingly heavy.

In most of the large London fires which have occurred of late years in these so-called 'fireproof' buildings, it has generally been found that, when once the fire has got hold, it is a waste of power to attempt to save the premises in which the fire broke out, and the exertions of the men have consequently been found to be better employed in preventing the fire from spreading to the adjoining premises; for we read that at the great fire in Gresham Street, 'finding it a forlorn hope to save the principal buildings in which the fire was raging, the efforts of the firemen were at length directed to protect the surrounding property.'

When these buildings, even those the most scientifically and expensively constructed, are, when on fire, once allowed to 'get out of hand,' as it may be termed, their destruction and that of their contents is inevitable. The stones crack and crumble with the intense heat; the iron beams push or pull everything down; the brick arches fall to pieces; concrete is nowhere: thus, in practice, it is found that in a very short period, unless the most lavish supply of water and powerful and efficient means of throwing it on the burning materials are quickly and readily available at the earliest moment the fire is discovered, the whole mass becomes a huge undistinguishable heap of burning and smoking ruins.

These 'fireproof' buildings, as at present constructed, are invariably found, in case of catching fire, to 'hold' the fire, and by this means render their total destruction, and that of their contents, a dead certainty; and they are to all intents and purposes most effective 'blast furnaces,' whenever they, or rather their contents, unfortunately become ignited, the result being, as before described, the complete demolition of the structure and the entire destruction of the contents.

It is worthy of remark that floorcloth factories – the inflammable character of whose contents is well known – are required to be built of wood, as it is considered far better for the entire building and contents originally ignited to be destroyed, than by making them 'fireproof,' and of a nature calculated to hold the fire, to run thereby the risk of burning a whole neighbourhood.

The desirability and importance of fireproof construction, especially from a commercial point of view, cannot for a moment be denied; but a great desideratum, also in the usual parlance, 'from a commercial point of view,' is that it should be 'cheap,' whilst, practically, it should, in addition, be possessed of those of being thoroughly efficient and easily applicable to buildings of all kinds.

It may be laid down as an axiom, that no one substance in nature is absolutely incombustible. The variation or difference will be found in the degree, greater or less, belonging to each substance; and also in the circumstances under which it is placed, or the position in which it is situated in a building on fire. A material which under one condition will be little injured by fire, will under another be totally destroyed. [. . .]

Iron, as at present applied, is found to bend and twist in a most extraordinary manner when exposed to the action of fire; and it matters but little whether it be wrought or cast, both being subject to the same operation; but cast iron has the additional disadvantage of being liable to crack and break, when exposed in a heated state to the influence of a stream of cold water. These peculiarities will show that their position in a building should be such as to render the mischief caused by them as small as possible, and prevent the additional damage from their expansion, contraction, and twisting, increasing the loss caused by the fire itself. [. . .]

The more combustible the character of the goods to be stowed in a building, the more incombustible should be its construction, and the more homogeneous the character of the material employed. It is advisable to avoid the use of iron columns, whether cast or wrought, as supports in such a building; and if brick is used, then it should be of the best quality, and put together in the most substantial manner. Stone is objectionable, unless coated with some nonconducting material, which should be thick enough to withstand a great degree of heat, and adhesive enough not to scale off when exposed to it. Experience has shown that stone and cast iron, unless protected by some incombustible substance, are not fit for fireproof buildings – the stone splitting, cracking, and crumbling under the combined action of heat and water, and the iron expanding, twisting, melting, or cracking under the same circumstances.

Brickwork, unless of very substantial character, stands but little chance when exposed to the intense heat met with in large fires. The mortar crumbles out, the walls bulge and crack, and the material at present called brick, and used as such in London, crumbles to dust, and, as it were, melts away, infallibly bringing down the best part of the structure in which this stuff is employed, and exposed to the action of the fire.

Iron, whether wrought or cast, constantly melts when forming parts of buildings destroyed by large conflagrations. In cases of sugar-houses and warehouses this frequently happens, and in all instances where iron has not been thus destroyed it has suffered materially.

In the great fire at Gateshead in 1854, the bonded warehouse which was so totally destroyed was considered to be a double fireproof structure, being lined throughout with iron sheeting, and supported on metal pillars and floors. The brickwork parted from the sheeting by the heat and crumbled away, and nothing remained but the red-hot skeleton or shell of the building. [. . .]

A more desirable and important point to be properly carried out, and one deserving of no small amount of serious consideration, is the providing of separate and proper storehouses for the dangerous and inflammable materials now daily introduced in large quantities by the requirements of manufactures or trade. It is a question which should be speedily ventilated, and some reliable means adopted by which the necessity of storing in the thickly inhabited parts of the metropolis of saltpetre, sulphur, petroleum, jute, and other equally dangerous substances, should be done away with, and thus remove the fearful risk we hourly run of the repetition of another Tooley Street conflagration, and that too of a far more serious and destructive character.

Hogsheads upon hogsheads of the petroleum, now become such an extensive article of trade, may be found stored in some of the most thickly populated parts of the metropolis, in positions where the ignition of but a small portion must involve the burning of a whole neighbourhood, and the destruction or endangering of human life to an extent frightful to contemplate. The great petroleum fire at Philadelphia is a sample of what may be expected here from this cause if things remain as they are. There can be no reason why such materials, possessing an inflammable character, should not be required to be kept to themselves, in positions and in buildings solely laid out for the purpose; so that in case of their igniting, the mischief might be confined to the building and locality where the fire originated.

Oils, spirits, and liquid materials of an inflammable nature would be best stored in tanks below the surface of the ground, by which means they would be prevented from running about the neighbourhood in a stream of fire, as has so often happened in the great conflagrations in the metropolis during the past few years. A light building erected over these tanks would be all that is needed, and in case of catching fire they would be down in a few moments, there being no body of material to hold the fire, and conduce thereby to lighting the whole neighbourhood. Some position should be chosen where there are no buildings in close proximity, and it should be forbidden by law to erect any within a certain distance; so that if it be impossible to prevent them taking fire, it should be at least possible to prevent the mischief extending to other buildings. [. . .]

In constructing warehouses, the floors might be divided by strong partition walls into numerous rooms, closed by double iron doors. [. . .]

It has been proposed to leave openings in the roofs of such buildings, which could be opened to let out the heated air and smoke in the event of the building or any part of its contents catching fire; but however pretty and reasonable such a plan may sound in theory, there can be little doubt but that in practice it would ensure and quickly carry out the complete destruction of the building and its contents. Such suggestions as these are entirely opposed to the principles which have been found so frequently to diminish the destructive effects of fires, [. . .] namely, the prevention of the access of air or the creation of a current or draught in a building or room which has become ignited, and a little reflection on the subject can hardly fail to show their absurdity.

If it were possible so to construct a building that, on the breaking out of a fire, it could be so hermetically closed to the atmospheric air that its entrance could be prevented, there is no doubt that the fire would go out of itself, inasmuch as the air required to support the combustion would be quickly expended, and, from the product not allowing fire to exist in it, the extinction would be a positive certainty. This, however, being an impossibility under the present system, it is no reason why as close an approximation as possible to the principle above named should not be tried for; for even if it did not act so perfectly as to completely extinguish the fire, still the absence of draughts and of the rapid access of fresh air would keep the fire in hand sufficiently long to enable sufficient means to be taken for extinguishing it. [. . .]

One great cause of the rapidity with which a fire spreads in modern built houses is the match-box style in which they are nearly all constructed - lath-and-plaster partitions, slight floors and joists, in fact a mass of kindling, which, on the application - it may almost be said the sight - of fire, is in a blaze in a moment, and often burnt out before any means can be taken to extinguish it.

The present system of construction adds in no small degree to their liability to ignite, and ensures their total destruction when ignited. The use of hollow wood partitions, lightly plastered, dividing one room from another; hollow wooden floors; wood or lath and plaster everywhere, cannot do otherwise than ensure a good bonfire whenever it happens that such a building catches fire. By means of these open hollow spaces the fire rapidly spreads from one point to another, finding a hidden convenient passage, up or through which it quickly passes, the current of air acting like the draught of a chimney in a stove; and the result is a rapid and most effectual mode of thoroughly igniting a building.

A writer on fires [. . .] has truly said: 'We want a new Building Act, a classification of goods according to their combustive specialities, and a Fire Brigade educated in the philosophy of their art.'

In Paris, the floors, ceilings, partitions, staircases, etc., have all the hollow places filled in with a concrete formed of rubble, stone, and plaster of Paris, by which means each portion of the building is made soundproof, and the spreading of the fire prevented. The importance of making all the parts of a building usually left hollow, air-tight, so as to prevent the formation of a current of air to feed the fire, if unfortunately a house should become ignited, is too self-evident to need lengthened comment; and the constant instances of its efficiency, wherever it has been tried, must be taken as a satisfactory proof of its importance.

Mr. William Hosking, in his excellent work, 'A Guide to the Proper Regulation of Buildings in Towns,' gives a full description of the plan followed in Paris in forming partitions, floors, etc., recommends it for adoption here, and remarks that floors are constructed in Nottingham in a similar manner, where the houses are said never to be burnt, and are free from damp and vermin. He also makes some excellent remarks on the use of iron in building-construction, and the dangers resulting from its injudicious application. [. . .]

4

VICTORIAN CITIES AND THE RAILWAYS: APPROACHES TO A HISTORICAL INTERPRETATION

John R. Kellett

Source: John R. Kellett, *The Impact of Railways on Victorian Cities*, London, Routledge & Kegan Paul, 1969, pp. 1–20

1. Introduction

One of the main problems confronting urban historians in Britain at the present moment is that of discovering ways in which valid comparisons can be made between the major British cities of the nineteenth century without denying or reducing their highly individual character. Their economic bases are similar only by coincidence and in limited measure; often they are totally dissimilar. Their forms of enterprise, the response of their business communities and the policies of their civic administrations vary widely. Comparisons which do not take account of these divergences can easily be misleading. Yet though the local histories of Liverpool, Manchester, Birmingham and Glasgow each have their unique flavour the urban historian cannot be content to treat them individually, in isolation. They, and other British cities, did not live in separate worlds in the nineteenth century, but were all subjected to certain common economic events.

One such major incident in the life of all British cities between 1830 and 1900 was the impact of railways upon the urban fabric and economy. By tracing the varying responses of British cities to this event comparisons can be made which often illuminate the subsequent form and direction of urban growth in individual cases, and yet which bring out the common, overriding influences to which all major cities were subjected. After all, a distinguishing feature of the revolutionary new mode of transport was that, unlike the profitable Victorian omnibus and carting businesses, it did not perform its service within the existing framework of social overhead capital. It had to provide a completely new transport network within each town and new generating points for traffic.

In the course of this transformation of urban traffic the railways made a massive and tangible impact upon the fabric of each major British city. By 1890 the principal railway companies had expended over £100,000,000, more than one-eighth of all railway capital, on the provision of terminals, had bought thousands of acres of central land, and undertaken the direct work of urban demolition and reconstruction on a large scale. In most cities they had become the owners of up to eight or ten *per cent* of central land, and indirectly influenced the functions of up to twenty *per cent.* The plans of British towns no matter how individual and diverse before 1830, are uniformly superinscribed within a generation by the gigantic geometrical brush-strokes of the engineers' curving approach lines and cut-offs, and franked with the same bulky and intrusive termini, sidings and marshalling-yards. Even in the most cursory map analysis these block-like specialised areas, and the rivers of steel flowing between them, stand out by their scale and artificiality, and by the durable, inconvertible nature of their function. In an environment where so much development was small-scale and left to piece-meal speculation, the railway builders and the great estate developers were by far the most important individual figures. Fowler, the Stephensons, Vignoles, Hawkshaw, Sacré, were the conscious moulders of Victorian cities.

2. Delimitation of subject

Before proceeding further with a study of the effects of railway building upon Victorian cities it is necessary to delimit the practical scope such an enquiry can encompass. It does not seem wise or practicable, for example, to attempt to calculate the indirect effects of railways upon cities. Yet this is not to deny their importance. In each town, large or small, the economic base was affected by the fuller pursuit of specialisation which the cheap transport linkages made possible: and since the logistic problems of supplying towns with food, building materials, fuel, raw materials and labour were solved, a rapid rate of urban growth could be maintained, in spite of the new dimensions of mid-century towns. There were accentuated problems of administration and public health in the 1840s and 1850s, but no signs of the slackening rate of physical growth some had predicted.

This profound indirect influence upon the internal functions and the overall size of towns must be conceded: but insoluble difficulties arise when we try to demonstrate it specifically. It is one thing to follow a narrative of the opening of a certain terminus by a story of rapid growth of population and output, illustrated statistically, and to assume tacitly that the one partially results from the other. It is quite another task to distinguish the components of growth which are due to the railways from those which are due to independent factors.

One way this difficulty may be tackled is to select special examples of towns like Crewe, Swindon, Wolverton or Redhill, all of them railway creations, with railway companies providing 25–30 per cent of total employment, dominating both manufacturing and service industry, closely engaged in housing, and even taking over the 'management' of the town. These railway towns are interesting and impressive examples if one wishes to make out an extreme case for the impact of railways upon urban development. In each town the influence of the railway is so magnified that the need to isolate it from other factors hardly arises, unless one seeks to explain the later progress of economic diversification. But such company towns are by no means as common in Britain as their counterparts in the United States; and though of great analytical interest, they are of questionable value as a general guide.

Another approach would be to select for study the second flight railway towns – those more representative towns which had a firm and independent economic base of their own but gained abnormally, for some reason or other, from their railway linkage. Examples of this type are Middlesbrough and Barrow, where raw materials were unlocked by unusually early and effective railway enterprise, or Carlisle and Derby, where certain route advantages, which already gave a marked nodality, were further confirmed by the railway's arrival. Such towns would be arguably more representative than Crewe or Swindon, and it would be possible with them to put forward a strong case for the railways as the predominant growth factor.

It is far more difficult, however, to isolate and assign a measurable component for growth if we select our cities with reference simply to their size and importance and not with a view to illustrating the effects of railways. How much did the five principal British cities owe to the railways? London, Birmingham, Manchester, Glasgow and Liverpool could hardly be described as creations of the railway. Indeed, one manifest characteristic of the early railway age in this country was that, at the time the railway arrived, the major British towns were already further committed to machine production than most of their European or Transatlantic counterparts.

Yet if we examine, not the incalculable indirect effects of the nineteenth-century railways but, more modestly, their specific impact on the major British cities there remains much that is amenable to study. What decided the location of railway facilities and the timing of their provision in a town? Who owned the land on which the railways were built, what did it cost and why did those costs rise steeply? Is any pattern discernible in the process of land acquisition on which urban railway building depended? How critical was the particular stage of rivalry or monopoly prevailing between railway companies serving a town? What direct influence did railway building have upon the old central core of a city? To what extent did the railways on the one hand, demolish, on the other, preserve but dilapidate the existing urban fabric? Was any attempt made to understand the indirect social costs and benefits which urban railway building incurred? Was the railway's role in stimulating suburban extension as important as is usually assumed; and how, precisely, did the provision of services by profit-making companies link up with the promotion of suburban building? Evidence exists to suggest answers to all these questions.

3. Early problems of railway access

The initial phase of railway development in most towns in the 1830s was characterised by good opportunities for approach and terminus building not fully utilised because of unduly low capitalisation for the task in hand, this lack of means being accompanied by a corresponding lack of influence in dealing with proprietors. The chief commercial need at this time was, above all, to get a portion of the line into operation as quickly as possible and to see some return for the heavy initial expenses on Parliamentary charges, rolling stock and permanent way.

The typical stations of this period were all [. . .] on the outskirts of the then built-up areas. The main consideration in their siting was to achieve the cheapest and simplest approach and terminus, with the minimum disturbance of property, even if [. . .] this involved a final stretch of line served by a tunnel and stationary cable engine. Usually the last part of the approach line ran through open land leased out in small market-gardens or through slum terrace and cellar dwellings leased in batches of eight or ten to middlemen, but, in each case, conveniently owned by a few larger landowners whose goodwill had been secured. The termini themselves were mere departure sheds with clumsy roofing covering only the track and leaving the passenger platform exposed. Even at the finest and most spectacular of the London termini, Euston's splendid arch led only to the ramshackle collection of one-storey brick ticket offices which so much engaged Pugin's scorn.

Yet despite their anxiety to avoid adding to the total cost of their projects by disproportionate expenses entering the city, the railway companies of the mid-1830s immediately encountered the shock of urban land prices. [. . .] The extension of the London and Birmingham from Camden Town to Euston alone, little more than a mile, cost £380,000 and had to be supported by a special toll above the statutory fares. [. . .]

Yet the obvious solution to these difficulties, the sharing of terminal approaches and facilities, could not be made workable without a measure of goodwill which the competitive nature of company promotion precluded. In London the Great Western and the London and Birmingham companies wisely decided not to share Euston as had originally been planned: the long and bitter rivalry between the companies and their technical disagreements over gauges would have led to scenes of unimaginable friction and confusion. But the four south-eastern railway companies which did share London Bridge station found it quite impossible to agree on the terms of joint-ownership. [. . .]

By the end of the thirties there were already indications that sharing of termini by genuinely independent companies would raise endless difficulties. 'Joint Stations', a witness said to the 1852 Parliamentary Committee, 'have been a more fruitful source of quarrels, litigation, Chancery proceedings and disagreements than almost anything else connected with the railway system'.[1] Appreciation of this point is of critical importance in understanding the network of urban railways and termini. [. . .]

The scuffle for railway access to Britain's major cities in the 1830s had already disclosed three features of considerable significance for future urban development. In the first place all railway promoters had grossly underestimated the expenses involved and so their schemes often fell short, or remained modest and peripheral. The opportunity was lost for a direct penetration to the central areas at a time when the operation could have been carried out painlessly. Within a decade the original stations on the outskirts were encrusted, like flies in amber, by the rapid extension of building. The approach routes became longer, more expensive and more limited in choice. At the same time difficulties in station sharing had arisen which ensured that future terminal projects would be put forward which made sense in the railway board-rooms but not in terms of urban transport economics. Finally, the intrusive land acquisition and change of use the approach lines and termini required had already encountered a resistance compounded of inertia and financial opportunism. [. . .]

Already, by the 1830s, the railway companies had begun to seek a way to avoid this resistance and delay by laying out routes through the areas where they had managed to find groups of substantial proprietors. These formed a species of fissure in the complex pattern of ownership in Britain's great cities along which the railways could chisel their way to the fringe of the town centre. These 'fissures' frequently provide, both in the 1830s and later, the specific explanation for route and site choices within the very broad limits suggested by geographical and engineering factors.

Yet in spite of the unexpected costs of the 1830s the returns to early establishment or improvement of terminals were very great both in terms of traffic and profits. The one mile Hunt's Bank extension and Victoria Station of 1839 [. . .] was constructed for £475,000, equal to $2\frac{1}{2}$ per cent off the Manchester and Leeds company's 8.8 per cent annual distributed profit for one year; a small price to pay for an investment which would assure the company's long-term prospects and establish a firm and permanent foothold in Manchester. [. . .]

4. Terminals in the 1840s

By the 1840s the modest nature of the railways' installations was becoming apparent, as the profit-making and traffic-generating potential of steam transport unfolded in each great city. 'The possession of good railway communications', the Board of Trade's 1845 report on railway schemes in Lancashire concluded, 'has now become almost as much a matter of necessity as the adoption of the most improved machinery to enable a manufacturing community to contend on equal terms with its rivals and to maintain its footing.'[2] In Manchester, where fluctuations of taste had always made time most important in the staple industry, 'the reduction in the period for the return of orders and the shortening of the process of textile manufacture itself' which early railways brought, led to rapid expansion of employment, saving of capital and even, it was claimed, improvements in machinery.[3]

In Birmingham it seemed equally urgent to improve facilities for different reasons. 'Economy of transit should be carried to the greatest possible extent, *more particularly in a district so far inland*', the city magistrates stated. Similar evidence is available for Glasgow where the need to compete on equal terms with Manchester and Liverpool is stressed. The competitive inter-city advantages brought by good terminal facilities were sufficiently clear by the mid-1840s to quieten criticism and unlock funds for the assault upon the central areas of each city.

In London, in 1846, no fewer than 19 projects for urban railways were put forward, involving so great a change in land use that the Select Committees considering each bill on its individual merits were reinforced by a Royal Commission to consider the general effects of railway schemes upon 'the thoroughfares of the metropolis and the property and comfort of its inhabitants.' In Liverpool, likewise, the urgent need for 'extending station accommodation and giving increased facilities for ingress and egress', and the immense expense of achieving it, was made the main argument for justifying the 'union of capital' of the three companies terminating there, and the formation in 1846 of the L. & N.W., the largest railway in Britain, with a capital of over £17 millions.

With the huge combined capitals amalgamation provided, and the flood of new investment during 'railway mania', the railway companies had ample means to command space even in the bidding against well-established commercial land users. The central business districts, where high rents and land values had always kept a species of monopoly, found themselves unable to exclude the railways by the customary sanction of offering central business district land prices. Nearly 1,000 acres of land had been authorised 'to be taken for Warehouses, Stations etc.,' by various acts in the 1830s. Most of this had been taken up by the mid-1840s and stations built which cost, on the average, three times as much per route mile as their European equivalents in 1844. The railways' land hunger added considerably to this in the next five years. By 1849 the 26 largest railway companies had laid out £19,240,000 on land purchase, the relative costs of this land varying in relation to constructional costs from 64 per cent for the mostly urban Manchester South Junction railway, to a mere 15 per cent for a line which was rural or had running powers into urban termini like the Eastern Union company. In general, land costs ran at about a quarter of the amounts expended upon the actual construction of railways, and were half as large again as the cost of rolling stock, engines and plant. [. . .]

The particular need experienced by the major railway companies in the mid-1840s was for improved and more central stations for passengers, who still provided 61 per cent of overall revenues. The rapid growth in volume of local, short-distance traffic, also placed the 'annexe' stations, a mile or more from the centre, at a further discount. An irresistible solution to these traffic problems in the forties was to convert the old terminals to goods depots and to extend the lines further into the core of the city to new passenger stations. As the cities extended outwards (and for many of them the 1840s was the fastest decade of their growth), so the railways penetrated further into their central business districts. The practice was widespread [. . .] and quite deliberate, as Robert Stephenson testified: the extension from Curzon Street to New Street in Birmingham was exactly modelled on the advance from Camden Town to Euston at the southern end of the London–Birmingham line.

Of the five major cities, Birmingham came out of this phase of central terminus building most transformed, with two new central stations forming landmarks within which the shopping and business area tended to be confined. Robert Stephenson's New Street station [. . .] had the advantage of being an extension by a well-established company and of 'removing a certain class of the inhabitants living just behind the principal and best streets', by its chosen route through the slums and brothels behind Navigation Street. The contribution to social costs, which railway demolition in this area constituted, secured the approval of the Birmingham Commissioners, who were already considering building a commodious new road out to the old Curzon Street station. 'I need scarcely ask you', counsel for the Bill remarked, rhetorically 'whether having been prepared to make that sacrifice of the Town's funds to achieve that object you would think it a very desirable thing to have the Station brought into the Town?' The properties on the terminal site were mostly owned by small, non-resident proprietors in batches of two to eight, but municipal goodwill, compulsory purchase orders and tempting cash offers broke up this maze of titles.

Brunel's rival central station at Snow Hill was also carried through, if not with the entire favour of the Birmingham Commissioners, at least with the militant partisanship of a faction of them, together with the MP for Birmingham, Richard Spooner. They pointed to the limitations of New Street, a passenger station only, very convenient for the carriage traffic from the middle-class Edgbaston area, but not able to cope with the small masters, bringing in their manufactures 'on their backs and in carts to the neighbourhood of Snow Hill.' Above all, a second station, of equal dimensions, and linked with the G.W.R., would 'free Birmingham from the thraldom of the network which the London and Birmingham railway propose to throw over it.'

Brunel and MacLean had the slight advantage of dealing with only one owner, William Inge, whose workshops, houses and yards, and five or six warehouses, covered the terminal site: but their mile-long approach routes by tunnel and viaduct affected nearly 1800 properties.

In the other major cities in the 1840s the transformation was less complete. In Glasgow, though there were several schemes for river crossings and ambitious central stations, none matured. In Manchester the only new station was a minor one, though the existing termini were enlarged and a unique new link railway was set in hand, Joseph Locke's Manchester South Junction Railway, entirely urban, crossing 30 streets on its $1\frac{1}{2}$ mile viaduct. In London [. . .] three ambitious schemes for central termini fell under the veto on central building, but at Waterloo and King's Cross [. . .] two important fringe additions were sanctioned and an inner ring railway commenced. [. . .]

It is clear that in all of these central schemes a critical new factor in the 1840s was the approbation, or, at any rate, the friendly neutrality of the new local authorities. Schemes which were insufficiently canvassed fell through, whatever their merits and whatever the resources of the company. The choice of routes through cities and the dimensions of the terminal sites were subjected to a new influence; the corporate ownership of land (as in Liverpool), or corporate views on desirable areas for demolition (as in Manchester, Birmingham and Glasgow).

The costs of effecting these inroads into the central business district, although they had not grown as rapidly as the railway companies' resources, nevertheless increased beyond all reasonable anticipation. Instead of the £135,000 (Lime Street, Liverpool, 1832) or £475,000 (Victoria, Manchester, 1839) of a decade earlier, £1,000,000 was needed for the new Snow Hill Station at Birmingham (1846); and increases in other towns were of a similar order. Some small part of this increase came from the more substantial nature of the termini. Dobson, Wyatt, Hardwick and Cubitt began to lay out permanent station buildings in the late 1840s in a style and with materials more fitting to those wealthy and massive additions to the Victorian city centre. But most of the expense arose from the rapid inflation of site prices which the railways had themselves precipitated.

5. Railway contractors and property owners in the 1860s

The railway companies' purchase of a sizeable percentage of the central business districts left a mark not merely on the Victorian townscape, but also on the companies themselves, and affected the timing and character of their last major assault on the fabric of British cities, in the 1860s. The profits, even of the largest companies, tended to sag for a time where an ambitious programme of terminus building had been undertaken. [. . .]

A result of the increased cost of sites in the larger cities in the 1840s was that the projecting of such schemes tended to become a separate speculation, almost the monopoly preserve of a few great contractors and engineers. John Hawkshaw and John Fowler were two ex-engineers who, by dint of their expertise and their ability to act as intermediaries between the railway companies and the large contractors, were able to turn entrepreneur and amass great personal fortunes during this phase of city railway building. Hawkshaw and Fowler between them accounted for the building of Charing Cross, Cannon Street and Victoria stations in London, [. . .] Central station, Liverpool and St. Enoch's and Central stations, Glasgow [. . .]. In each case a separate company was floated to build the terminus and few miles of necessary approaches. Work was carried out on credit, and share capital for these specialised terminus companies was sometimes advanced (in return for monopoly privileges) by Waring Brothers, Betts and Peto, the great contractors, who had by now developed a vested interest in demolishing and remaking cities. In these ways risk capital was concentrated by those most directly concerned with urban railway building in the 1860s, and sufficient leverage obtained for the increasingly difficult task of breaking into the cities' inner districts.

Even so a great deal of ingenuity was needed to find the fissures in the ownership pattern along which approach routes could be made. Fortunately, for the projectors of the 1860s, a few large individual titles to land still existed. The Marquis of Westminster owned the greater part of the property along the route of the Victoria Station and Pimlico Railway Company. The decision to cross the Thames by striking westward from London Bridge [. . .] and setting up the Charing Cross and Cannon Street termini was also influenced by the location of large block holdings. The Archbishop of Canterbury's and the Right Rev. Lord Bishop of Winchester's vast slum holdings in Southwark and Lambeth alone made it possible to contemplate driving a surface railway two miles across south London. In Manchester and in Glasgow the last large central holdings in personal ownership were tempted onto the market by railway prices in the 1860s.

Increasingly, however, the railways came to rely for urban land upon corporate bodies, the Hungerford Market Co. and the Drapers' Company, St. Thomas's Hospital and Emanuel College, which owned the terminal sites of Charing Cross, Cannon Street and Holborn Viaduct stations respectively in London; the municipal corporations in Liverpool and Glasgow for the new central sites there; the Grosvenor and the Rochdale canal companies, on whose basins Victoria Station, London, and the projected M.S. & L. Central Station in Manchester [. . .] were laid out. The 1860s, however, was the last decade in which such large units of ownership could readily be found: afterwards the most important developments in urban railways were underground.

6. *Effects on the central and inner districts*

The opening of stations in the central business districts of the provincial cities, and on the immediate periphery of the central area in London, produced both a redistribution of land uses and a re-alignment and stimulation of internal traffic routes. The Select Committee on Railways had already spoken, in 1844, of 'the control which Railway Companies possess by means of their Station-yards over the traffic of the neighbouring districts.' The professional carrying companies redistributing railway loads by horse waggon, Pickfords, Kenworthy's and Baches found themselves, like many passenger cab drivers, engaged in the 1840s in endless squabbles concerning access and facilities; and litigation continued even after it had been recommended that 'reasonable preference may be given to favoured agents, but vehicles tendering themselves to take up or set down Passengers, Luggage and Parcels ought not to be excluded.' Inter-station traffic came to form up to half the livelihood of some cab owners and influenced the framework of central horse omnibus routes. Where river bridges interposed between stations and the business areas, as at Jamaica Street, Glasgow, or at London Bridge, the traffic-generating power of the new termini was vividly underlined.

More slowly, a realignment of land uses around the central termini and along their access routes took place. The early misgivings of insurance companies, three-quarters of which increased their rates for sites adjacent to the railways in the 1840s, were soon overcome, and the new transport medium's capacity to attract some expanding land uses and to repel others was given free rein. Attracted to the termini were retail shopkeepers and transit warehouses; repelled were residential and, on the whole, business users. Business users and specialised warehouses built, like those in Manchester, for sale and exhibition rather than mere storage, preferred to keep a reasonable distance between themselves and the railway. Those heavy industries whose bulk requirements called for special rail facilities tended to falter at the central land prices and seek riverside or suburban accommodation.

This gradual redistribution of uses was sometimes given specific lines of demarcation by a massive new feature of the urban scene in the 1840s, the railway viaduct, striding past working class houses at rooftop level, 'pinning down' areas socially and intersecting them physically. Occasionally these great stone viaducts were constructed to secure the right elevation for crossing a river or for entering a terminal site, but usually their main purpose was to avoid street closures, and it was this feature which gave them their extremely widespread use. Each major town has its lengthy stretches of elevated track, but perhaps the most striking example to take would be that of the Manchester South Junction.

The MSJ had been conceived of by Joseph Locke as 'a species of bridge' to provide a cross-town communication, and there could be no doubt of the need it satisfied in the transport network. By the 1860s it was so crowded with east/west traffic that an hour's wait on its arches was not unusual. But the effects on the area it crossed were dilapidating in the extreme. Those arches which were let were used for 'smithies, marine stores, stables, mortar mills, the storage of old tubs, casks and lumber, and other low class trades.' The condition of the unlet arches was even more notorious. According to the city surveyor, speaking in 1866, 'No improvement has taken place in that district in the proximity of the viaduct. We have still the low class of property which was there years ago when the viaduct was built.' Other parts of Manchester had increased in value by 75 per cent in the 20 years after the MSJ had been formed, he estimated, but values had remained static around the viaduct.

In other cities where the line was not elevated, but passed at ground level through the property enclaves where streets had not been laid out, the effect was hardly less marked. Since the Parliamentary 'limits of deviation' made a generous allowance for railway needs, the companies often found themselves with a belt of superfluous land for disposal, usually to non-residential users. This helped them to recoup some of their expenditure, but increased the barrier effects of such routes. Segregation 'on the wrong side of the tracks' could have a disastrous effect upon the development value of vacant land. By mid-century smaller Parliamentary bills already show 'shadow areas' being created by the criss-crossing of supplementary lines driven relatively cheaply through a wilderness of so-called 'gardens', middens, claypits, 'scavengers yards with old boilers thereon.' Within two decades the outward growth of building had made these disused areas potentially valuable central land, but the maze of tracks now prevented the land's development. The completion of urban link railways and cut-offs, which continued during the 'Great Depression' of the 1870s, came to cover scores of acres of east and north Liverpool, east Birmingham, south Manchester, Glasgow and London with a tangle of interconnecting lines and sidings, crystallising these areas' dereliction.

7. *Inter-city comparisons*

A lengthy sequel to the work of the first 30 years of urban railway building could be written. Continued inter-company rivalry and the endless possibilities for friction which shared stations involved led to a few more large 'secessionist' stations later in the century, of which the most important were Central, Glasgow [. . .] built for tactical reasons by the Caledonian Railway Company, and Exchange, Manchester [. . .] built end-on to Victoria. Some stations were substantially enlarged and their approaches broadened between 1870 and 1900 to cater for the doubling of traffic over those thirty years. The territorial expansion of goods accommodation, moreover, grew more quickly and in a more direct relationship to the increase of traffic than did the passenger stations in the last decades of the century, and this was reflected in the expanding yards and sidings along Thames, Mersey and Clydeside. Again it was only in the last two decades that, with the slow growth of cheap ticket commuter traffic, the railway companies began to make a marked impression, in London, at any rate, upon the Victorian suburbs, influencing their social composition and their direction and rates of growth in a way which is only now receiving attention from economic historians.

Yet, on the whole, the main physical impact of railways upon the heart of the Victorian city and upon the urban property market was over by the end of the 1860s. The claims to land staked during the previous thirty years were sufficient, and the facilities established sufficiently flexible, with improved traffic control, to cater for the transport needs of the rest of the century without further re-shaping of the central areas.

Comparison between the emergent railway systems of each of the five cities brings out marked local differences in provisions. Liverpool was the nearest to a monopoly town, held in the grip of the L. & N.W. almost completely until the mid-1860s, when the M. S. & L. and the Cheshire Lines Committee introduced a competitive element. [. . .]

In Manchester, by contrast, the picture is one of the most extreme competition, with the L. & N.W., Midland, G.N., Lancashire and Yorkshire and M. S. & L. railways forming shifting temporary alliances, building stations of tactical necessity, introducing projects whose main purpose was to block attempts to improve services by rivals, making no attempt to co-ordinate timetables, and even (some alleged) arranging them to clash deliberately. [. . .]

In Birmingham, although the early days of shared stations produced similar scenes, including a classic pushing match between the locomotives of rival companies at the opening of New Street Station, these early excesses soon abated to leave Birmingham the best provided town. The provision of the principal stations and through routes was carried through early, on paper by 1847, in iron and masonry by the early 1850s, and despite minor criticisms, its central stations catered for half a century's traffic growth. Certainly there are no signs that inadequacies in her railway linkages held back Birmingham's development in the later nineteenth century, quite the reverse. A balance of genuine competition between the L. & N.W. and G.W., perhaps arising partly from the incompatibility of gauges, provided all the advantages claimed universally for the *laissez-faire* system. Even Birmingham was not without its follies: the famous Duddeston viaduct [. . .] setting off on towering arches across the city to nowhere, never completed but still standing, without a rail laid, a monument to the tactical needs of a passing phase of railway competition.

In Glasgow a more substantial monument was left in the form of St. Enoch's Station, built by the G. & S.W. and its satellite Union company in a vain attempt to exclude the Caledonian Railway Company. In Glasgow the disposition of competitive forces was so equal and the deadlock between them so bitter as to lead to the frustration of all comprehensive projects in the 1840s and to needless reduplication later.

In London the formation of main line termini and urban link railways fell into a special category for two reasons: it was by far the largest city and tended to magnify external diseconomies intolerably; it was also the seat of the national government and came more particularly under the oversight of central authority. Since it was also the focus for nearly all the great railway companies it was subjected even more directly than Manchester to the disruptive force of *laissez-faire* competition. Schemes put forward in the one year 1863 proposed to raise £33 million, lay out 174 miles of track in the Metropolitan area, build four new bridges across the Thames and schedule one quarter of all lands and buildings in the City of London for compulsory purchase and demolition. Parliament could not stand back idly whilst London and Westminster were torn to pieces by competing companies and so in the 1860s, as in the 1840s, fresh parameters for future urban railways were imposed.

The individual and distinctive nature of the impact of railways upon each of the five major British cities cannot be glossed over, but at the same time certain overriding similarities can be discerned in the problems faced and in the general effects upon the fabric of each city. It would be equally foolish either to underestimate the former or to overlook the latter.

Differences in the competitive postures of the companies exploiting each city and in the amount of surveillance exercised by local and central authorities left their mark, quite literally, upon the development of each city. But this diversity should not obscure the many common features: the similar effects upon the central business districts and their traffic, the demolition of certain areas of low quality housing in each city and the perpetuation of others, the similar problems of securing land title along the access routes and the similar effects upon the urban land market. Here are topics which draw the study of British cities closer together and afford an extensive field for urban comparisons.

Notes

1 House of Commons, 1852–3, XXXVIII, p. 137.

2 House of Lords, 1845, XXXIX, p. 238.

3 *Ibid.*, p. 237.

5

THE CASE AGAINST THE HISTORICAL INEVITABILITY OF SUBURBS

F. M. L. Thompson

Source: F. M. L. Thompson, 'Introduction', in F. M. L. Thompson (ed.), *The Rise of Suburbia*, Leicester University Press, 1982, pp. 5–7, 10–12, 15–16, 19–20

There is an underlying assumption, sometimes made explicit, that the vast increase in the numbers of town-dwellers, or at any rate a high proportion of that increase in any town of considerable size, could in the nature of things – given that more people must occupy more space, and that the space within existing town limits whether defined administratively or geographically was already full – only be housed in suburbs established on the outskirts. Having provided some explanation of the growth in overall urban population, whether in demographic terms or in terms of the economy of the particular town and the sources of expansion in its employment and incomes, the questions for the suburban historian become those of which sections of the net increase in total population moved to the suburbs and why, and which parts of the surrounding space were turned into suburban residential areas, when, by what stages, by whom, and why in some particular shape rather than any other.

Indeed, it makes very good sense in the British context to consider population growth as fundamental, and to look for some threshold in town size beyond which a traditional urban structure of a unified and physically, socially, and economically integrated townscape gave way to a modern arrangement of central town and dependent suburbs. For the nineteenth century it is valid to see suburbanization as a facet of urbanization, a necessary part of and condition for the wider process. It would be hazardous to make any firm pronouncement without local knowledge of the history and topography of every large town, but by mid-century it is likely that every place with more than 50,000 inhabitants thought of itself as possessing some suburbs; Brighton had its Kemp Town and Hove, Newcastle its Jesmond and Gosforth, Hull its Cottingham when they reached about that size. Nevertheless, before accepting that suburban development was a necessary consequence of the scale of urbanization it is worth asking whether the connexion is as strong as the nineteenth-century experience suggests.

There were certainly theoretical alternatives to the suburban mode for accommodating expansion in urban populations, and other countries provide working examples of them in practice. The development of residential suburbs of distinctive appearance and distinctive class was one form of lateral expansion. Another form of lateral expansion was by simple accretion at the town edges of buildings and street patterns that reproduced and continued the character of the established town, in new quarters with mixed residential, commercial, and industrial functions, and with intermixed residents from different social classes. This, apparently, is what happened in most

American cities until the 1870s; they remained compact 'walking' cities, despite considerable growth in size, with no pronounced social – as distinct from ethnic – segregation. It was only with the appearance of the new technology of mass transit by tramways that socially segregated districts began to develop, the vehicle of segregation being the suburb. Towns could also expand upwards rather than sideways, throwing up vertical streets in tenement and apartment blocks. This was the way with most continental, and the large Scottish towns; by this means Paris and Vienna remained classically compact cities, in contrast to London, the scattered city, and one stepped outside Rome, for example, in the later nineteenth century literally into an empty surrounding desert. That English towns did not follow these models was an act of choice, not of necessity.

The largest English towns – Liverpool, Manchester, Birmingham, Leeds, and Sheffield, as well as London – were suburbanized, and socially segregated, half a century or so before the arrival of cheap mass transit, which was not developed effectively in the English setting before the 1890s. It can be argued that horse omnibuses and railways, if they chose to offer suburban train services, provided transport that was neither for the masses nor cheap but was nonetheless sufficient to carry the better-off minority out to suburbs. There is a great deal in this argument, but since the Americans possessed the same pre-tramway transport technology without using it to support the same style of living, omnibus and suburban train should be regarded as permitting, rather than creating, the suburb. In any case it was entirely possible for a place like Camberwell, $1\frac{1}{2}$ to 2 miles from the City, to remain quite substantially a 'walking suburb' at least until the 1870s, and yet to be unmistakably suburban in character and function regardless of the degree of reliance on walking to work. The high density, high rise development of continental and Scottish cities, on the other hand, has been explained by high building-land values and, in some cases, stringent building regulations designed to prevent building outside municipal limits for reasons of internal security and social control. The building regulations were important, and were not paralleled in Britain; Scottish municipal regulations were generally tougher than English ones, which outside London tended not to exist at all until mid-century, but they acted to increase building costs through control of materials and structures and did not regulate the availability of sites. Where building was allowed, building land values were markedly higher in continental towns than in England, and the ready availability of supplies of relatively cheap building land with a gradient of values generally descending from city centres was a prerequisite for suburban development. On the other hand, recent experience suggests that it requires a very sharp rise indeed in plot value in relation to building cost, raising the site element from a traditional 5 to 10 per cent of overall cost to one-third or even one-half, before high density tall buildings become the only economic use of building land. In Scotland, moreover, the generally higher building-land values did not inhibit the development of Kelvinside with its suburban densities, suburban profiles, and suburban qualities, once a demand for that kind of thing appeared among the wealthier Glaswegian middle class. High land prices encourage high buildings, and low land prices are favourable to suburban development, but neither can be regarded as decisive influences [. . .]

Plentiful supplies of comparatively cheap building land and a building industry capable of producing speculatively for a range of incomes permitted the growth of suburbs, but there do not seem to have been any changes in these factors sufficiently sharp or pronounced to have caused that growth to get under way. It is possible, however, that the effective supply of building land was increased by transport improvements opening up fresh and more remote areas for settlement, without any change in the inclinations or policies of landowners being required. In a general sense modern suburbia could not have happened without the omnibus, commuter train, and tram, powerfully reinforced in the twentieth century by the motor; without such means, the geographical spread and the growing distances between home and work could not have been supported. While transport improvements and extensions clearly sustained the general process of suburban expansion and intensification, this does not necessarily imply a causal connexion or indicate which way it operated. New transport ventures are rather more likely to be designed to cater for an established traffic than to create an entirely new one, even though once operating they have great potential for stimulating large increases. In any event when new railways did produce a clear suburban response, as they did in Bromley in the 1850s and in west London in the 1860s and 1870s, this happened long after the attractions of the suburban way of living had been successfully demonstrated, and the new transport promoted the development of an already tested product in a fresh locality. It is less easy to imagine new forms of transport poking their fingers out into the country in the hope of initiating a way of life that had not yet been tried and shown to be socially and commercially successful. On the other hand, flexible and adaptable forms of transport were at hand in the crucial period at the beginning of the nineteenth century that were capable of playing their part in enlarging the residential area through extension of their services, so soon as a demand for them arose.

By the early years of the century there was already a well-developed network of short-stage coach services in the London area, running between such peripheral villages as Paddington, Camberwell, Clapham, Islington, or Edmonton and the City and West End. Fares were high, coaches small, and journeys slow with long stops at the public houses which were the picking up points, so that this kind of passenger transport was not suited to the regular commuter who would have needed to use a private carriage or his own feet for a truly suburban daily journey to work. Nevertheless the short-stage provided the essential link with the centre for the more occasional traveller, for shopping, visiting, or entertainment, which permitted Paddington or Clapham families to evolve a suburban pattern of life. The horse-drawn omnibus, introduced in 1829 on the Paddington–City route, was a development of the short-stage concept, adding a larger capacity vehicle with rear entry for easy access, boarding and alighting at the passenger's wish, faster journeys, and lower fares. Bus services expanded rapidly in the 1830s, freed from the previous prohibition of picking up or setting down in central London, and quickly became established as one of the most important determinants of the character of early suburban development. Buses made possible daily journeys to work at times of day, journey speeds, and fares which were convenient to the affluent to middling middle-class of professional men, civil servants, merchants, bankers, larger shopkeepers, and perhaps some senior clerks. Buses, therefore, allowed families in such groups to live at a distance from work and dispense with private carriages, and permitted those who could never have aspired to own private carriages to do likewise. They permitted middle-class neighbourhoods to function without coach-houses or mews, households to be run with only female servants, and houses to be smaller and less expensive because quarters for male horse servants were not needed. These all became standard features of early Victorian suburbs, and they were all dependent on the horse bus.

The horse may rightly be credited with much influence over the form which the suburban environment took, and with permitting its colonization of growing territories in the 30 years or so before the 1850s, when suburban train services first began to have any significant effect in allowing or stimulating further suburban growth. It would be going too far, however, to suggest that the availability of horse-drawn public passenger transport was decisive in triggering the birth of the suburb. The sequence of events was that the suburb came first and the short-stage coach or omnibus followed once the potential passengers were established. It is true that this came to be such an automatic development that people came to live in districts with very poor, or no transport services in the confident expectation that a new bus route would be opened as soon as there was a sufficient concentration of residents to furnish profitable customers, so that settlement in advance of transport was simply a matter of timing and the mechanics of business and no proof that settlement went ahead without regard to transport services. This reasoning could, however, be applied with little modification to the pre-suburban world, in which the organization and equipment of the horse and coach trades were fully capable by the later eighteenth century of responding to the emergence of new bodies of customers by providing new services. While suburban development needed some form of transport services in order to take root and flourish, it is thus likely that in the initial phases at least it was the development which called forth appropriate kinds of transport, and unlikely that there had been any powerful but latent suburbanizing force held in check by any absence of enterprise or innovation in transport.

If, therefore, the building and occupation of districts of single family homes, and hence the origins of suburbia, cannot be satisfactorily explained by independent changes in the supply conditions, it is natural to suggest that the initiating impulse came from the side of changes in housing demand [. . .]

[. . .] it would be sensible to look to the suburban garden for the roots of the demand for suburban living, something which brought the possibilities of privacy and seclusion with it but which was desired in a straightforward way for its own sake because it was a piece of tangible evidence, however minute, that the dream of being a townsman living in the country was something more than just an illusion. The essential quality of the new suburbs was that they were on the edge of the country with open views beyond, even if subsequent development leapfrogged past them and hemmed them in as inner suburbs. An essential attribute of the single family house was the garden, preferably one in front to impress the outside world with a display of neatly-tended possession of some land, and one at the back for the family to enjoy. It can be suggested that a desire for individual gardens was surfacing among the potential suburbanites at just about the time when building suppliers chose to put the article on the market. It was in the 1790s that the countryside was ceasing to be feared or despised as boorish, backward, or hostile and was coming to be admired by cultivated opinion as the home of all that was natural and virtuous. In 1810 John Nash outlined his scheme for the development of Regent's Park, using the well-tried and conventional building forms of Georgian London, terraces and crescents, but placing them in a new rural setting instead of in a gridiron of streets and squares, so that each resident should have the illusion of looking out on his own country park. Even more important, in 1824 Nash tacked Park Villages East and West on to the imposing formality of the elegant Park terraces, villages which were indeed an aristocratic garden suburb in miniature, rusticated cottages, each one different, each one in its garden, planted in an urban context. Even here *rus in urbe* was realized on the ground for the fashionable aristocracy and very wealthy, a model not only for developers and builders to imitate but also a pattern for the ambitious middle classes to seek to emulate. An aristocratic fashionplate – and in its early, contemporary, stage St John's Wood villadom was almost equally wealthy - it can be argued, transmuted middle-class yearnings for a whiff of the country, which had hitherto seemed unacceptable and inappropriate in town dwellers, into a positive and respectable demand. There were, after all, vast numbers of town-dwellers who had come from rural backgrounds; some of them, presumably, hankered after the country they had left, and some of these had the means to indulge such nostalgia. In late eighteenth-century London from one half to two-thirds of the adults had been born in the country; and it is probable that the mid-nineteenth-century pattern of internal migration which shows that at least two-thirds of the residents of south London and Liverpool suburbs, and presumably other suburbs likewise, had moved to the suburbs from neighbouring and largely rural areas and had not moved out from the central zone, also applied earlier.

Some portion of the new suburbanites undoubtedly did desert the old town centres, escaping from increasing dirt, noise, stench, and disease, dissatisfied with the social confusion of mixed residential areas and with the inconvenience of traditional town houses for the style of life they wanted to pursue. But if they were heavily outnumbered by those coming direct to the suburbs from rural and small town surroundings, it is to these that we should look for the mainspring of suburban housing demand, and there is little difficulty in supposing that they were more interested in clutching at some small reminder of country life than in seeking an environment suited to the practice of an ideology of which they were most likely not yet aware [. . .]

[. . .] It is arguable that suburban growth and the suburban life was set in successful motion by the more imitative and self-effacing sections of the middle class in pursuit of the illusion of bringing the country and gentrification into the urban setting, more intent on appearing to merge themselves unobtrusively into a superior class than on seeking means to express their own class identity. [. . .]

It has long been recognized that transport services played an important part among the general influences on suburban growth, but the exact nature of that part, and whether improved transport was an essential, causal, or permissive element in suburbanization, have been matters of dispute. Improved transport in the nineteenth century meant, above all, railways, and the prevailing view is that only in a few exceptional cases can railways be regarded as an important cause of suburban growth, and that generally 'the development of suburbs . . . preceded the provision of railway services, by periods of at least a decade or two for each of the larger cities'.[1] In part this view relates to the suburbs of the large provincial cities, which, at least until 1900, were little more than two to three miles from city centres, too short a distance for suburban trains to be practicable; the unimportance of railways in provincial suburban growth is confirmed in the study of north Leeds, where the intrusion of the Leeds–Thirsk railway was an impediment rather than an assistance to development. In part, however, the view concerns London's suburbs, and rests on the stringent assumption that railways only caused suburban growth when railway policy on fares and services took an actively promotional line, an assumption that virtually reduces the railway suburbs to Edmonton and Walthamstow, where the Great Eastern's workmen's fares and workmen's trains reluctantly but decisively promoted the rapid development of working-class suburbs after 1864. The outer suburbs at more than five or six miles from the centre could not have developed as dormitories without commuter rail services, however skeletal, and even if railway companies did little or nothing to encourage such growth through special fares or convenient services the presence of a railway should be regarded as a necessary, although not a sufficient, condition for outer suburban growth. The building of railways preceded the development of suburbs in Bromley, Acton and Ealing, and Bexley, but the interval between opening a station and substantial suburban expansion varied between a few years and many decades, and the absence of any immediate railway-triggered expansion in Bexley confirms the proposition that railways made the outer suburban dormitories possible but did not create them.

Where other conditions were favourable - an attractive location, an established nucleus of village or small market town, landowners keen to act as developers, a handful of existing residents with city connexions, and a propitious moment in the trade cycle - the promotion of a railway could be the catalyst of expansion, producing a genuine railway suburb. This happened in Bromley, where local landowners were the chief promoters of the railway which was opened in 1858, ushering in a decade of rapid growth spearheaded by a small number of affluent middle-class households able to afford the high cost of commuting. Very soon, however, it became apparent that residential expansion was proceeding in spite of the inconvenience, inefficiency, and inadequacy of train services, and Bromley passed into the same situation as Ealing until the 1860s, where the Great Western completely neglected its suburban services, or Bexley in the 1930s, where headlong housing expansion ran well in advance of train services or access to stations, a situation in which transport services followed after development was under way. A much more complicated statement about the relationship between transport and development emerges from the close analysis of railway promotions and train services in outer west London, and here it is possible to see the interdependence of the two, with the promotion of new lines in advance of suburban housing both by speculative landowners and by a railway company reliant on suburban traffic, the District, and the improvement of services by main line companies at hours and frequencies to suit the commuters' needs, acting as stimuli to revive building activity from a preceding depression. Here railway services acted as a chief instrument for communicating upswings in the building cycle to the district in the middle phases of its development, even if they had been of little importance in starting the initial development.

In the inner suburbs where distances were too short for railway operation, and in the provinces, horse buses were of great importance to middle-class commuters and these were invariably introduced only after suburban development had taken hold. From 1870 onwards horse trams were being rapidly introduced, and as a means of cheap mass transit they had a much more widespread effect than workmen's trains and fares in enabling the lower middle class and the artisans to push out into suburbia and to threaten the exclusiveness of middle-class suburbs. The encroachment of trams was resisted in Ealing, as in Edgbaston, Hampstead, Kemp Town, Kelvinside, and Victoria Park, because it threatened to bring in a lower class of people and bring on social deterioration, and as in other places the resistance was in the end unsuccessful. By the early 1900s electric tramways were being projected into virgin territory, well ahead of suburban settlement. While it is salutary to be reminded that in Leeds the horse trams did not run early enough in the morning for working-class commuters, and were largely used by the middle classes, in general they were a working-class form of transport and a reminder that suburbs did not remain exclusively middle-class. Further research may well show that tramway suburbs were more significant than railway suburbs, not perhaps as entirely new settlements called forth out of green fields, for no transport service seems to have been capable of doing that save in exceptional circumstances, but as places which experienced a dramatic transformation in social character and physical scale as a direct result of tramway penetration.

Note

1 John R. Kellett, *The Impact of Railways on Victorian Cities*, London, Routledge & Kegan Paul, 1969, p. 354.

6

THE DEVELOPMENT OF MANCHESTER'S INDUSTRIAL REGION

F. Vigier

Source: F. Vigier, *Change and Apathy: Liverpool and Manchester during the Industrial Revolution*, Cambridge, Mass., M.I.T. Press, 1970, pp. 87–8, 90–9, 127–33, 137, 139

The changes brought about by the introduction of machinery in the scattered and decentralized cotton industry were at first far from dramatic. The first major improvement, John Kay's 'flying shuttle,' was adapted to the weaving of cotton cloth around 1760. It increased the production of weavers and allowed almost total flexibility in the width of cloth that could be woven by one man, since the shuttle was activated by a central lever rather than being thrown by hand. [. . .] It had two far-reaching consequences. First, the relative complexity of the new machinery required a larger capital outlay, which increased the weavers' dependence on the 'manufacturers' who could provide them with a cash advance to purchase the equipment or with the equipment itself. Second, the higher output of the fly-shuttle increased the demand for cotton yarn, causing a chronic shortage that had to be met by effecting improvements in cleaning and spinning the raw staple in order to increase production.

Hargreaves's spinning jenny, patented in 1770 but in use for several years before, mechanically reproduced the hand operations of a spinner. It allowed the simultaneous spinning of several yarns, each on its own spindle, and was simple enough to be operated by children. Although it multiplied the output of the human hand by a factor of eight to twelve, it was not able to produce thread of sufficient strength and fineness for the warp. Arkwright's introduction of spinning by rollers was a different approach to the problem insofar as the technique did not depend on the human hand as a source of power. While the jenny, like Kay's improved loom, was still a handicraft type of machine, suitable for cottage work, Arkwright's water frame (1769 and 1775) was a factory-type machine that was powered most efficiently from a relatively large, central source. The introduction of the water frame marked a decisive change in pace; from then on, all technological improvements – such as Arkwright's carding machine and Samuel Crompton's 'mule,' which combined the principles of the jenny and the water frame – depended for their successful utilization on the economies of scale that only factories could provide. Moreover, for the first time, a surplus of yarn was produced, and England became an exporter not only of cotton cloth but of yarn as well.

The availability of an external source of power sufficient to activate the machinery was the principal locational requirement of the cotton industry. The presence of a waterfall became indispensable. The numerous streams of Lancashire offered ideal conditions, but with few exceptions, the existing towns and villages did not possess an adequate waterfall, and as a result, factories were built in the countryside, drawing their labor force from nearby villages and hamlets. When it was more convenient, the factory owners constructed new workers' housing near their plants. [. . .]

The impact of the new machines was felt by the closing decades of the century. In 1750, there were only six urban places within a twenty-mile radius of Manchester, that is, towns of fairly dense settlement with a population greater than 2000; in 1775, there were twelve such towns, and in 1821 there were sixteen. [. . .]

The urbanization of the Manchester cluster occurred in phases, each reflecting the technological and economic restructuring of the area's textile trade. Prior to about 1775, Manchester's dominance was not due to manufacturing, which continued to be carried out in the countryside or in adjacent towns.[1] The growth potential of any single town within the cluster was directly related to the presence of existing manufacturing activities and hence to a combination of ecological and historical factors. This is exemplified by the high growth rate of such traditional industrial towns as Bolton, Bury, Stockport, and Oldham. The development of Manchester itself was tied to the overall level of economic activity within its cluster, since it remained the principal marketing center, largely because of its accessibility to Liverpool after the navigational improvement of the Mersey and Irwell rivers in the 1720s.

The first technological innovations introduced between 1775 and 1811 led to the rapid growth of the cluster's older manufacturing towns. Simultaneously, the implantation of mills in such rural places as Tyldesley, Altrincham, and Middleton precipitated their urbanization. [. . .]

The repercussions of industrialization can be found in the close relation between urbanization and the geographical distribution of textile mills and in the spreading network of canals and roads that were being constructed to facilitate the shipment of goods between Manchester and the surrounding region (see Figures 6.1 and 6.2). [. . .] After 1800, when the use of the steam engine allowed greater flexibility in the location of factories, Manchester industrialized quickly and started to grow faster than the combined towns of its cluster. By 1821, at which time no fewer than 66 textile factories were located within its boundaries, Manchester contained almost 50 per cent of the urban population of its cluster (as opposed to 32 per cent in 1775), and its economic dominance as a manufacturing and marketing center had been firmly established.

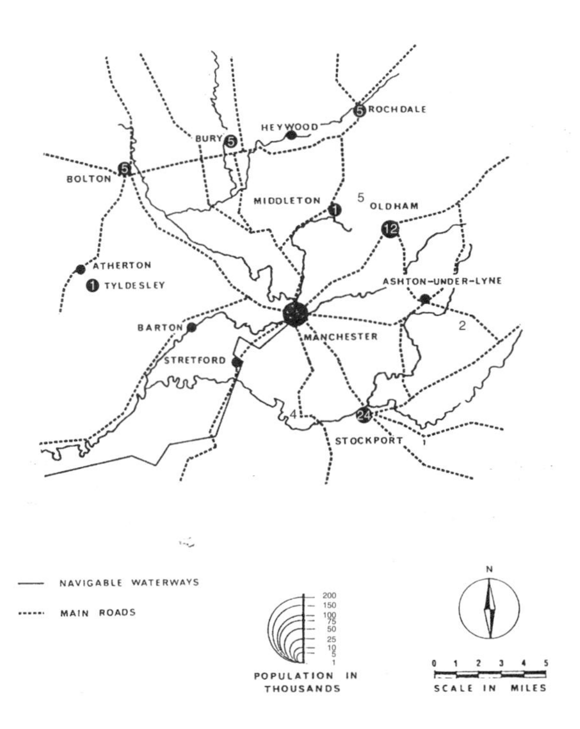

Figure 6.1 Manchester cluster, 1775
Note: Numbers refer to textile mills

Preindustrial Manchester

The rise of Manchester from the 'mere village' described by Daniel Defoe in 1727 to a thriving commercial town during the second half of the eighteenth century, and to one of England's

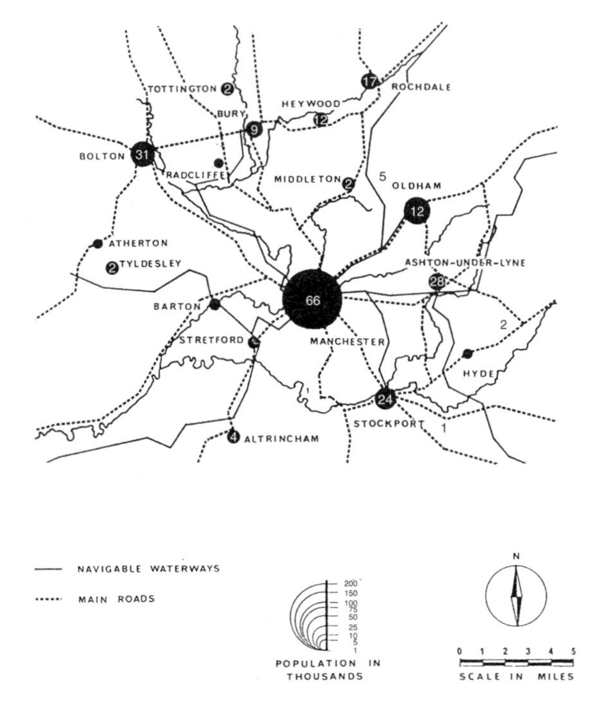

Figure 6.2 Manchester cluster, 1821
Note: Numbers refer to textile mills

most important industrial centers and its largest city after London in the nineteenth century, is attributable as much to the ability of its inhabitants to adapt to changing conditions as to the ecology of the region, which was favorable to early industrialization. [. . .]

In the last decade of the eighteenth century [. . .] the introduction of manufacturing within Manchester itself, made possible by the steam engine, which was a more than adequate substitute for the natural power whose absence had retarded its industrialization, caused a sudden explosion of the population, which quadrupled in the next twenty-five years and then doubled every twenty years. Simultaneously, Manchester's rising economic dominance within its cluster is shown by the steady increase in its percentage of the urbanized population: 44.9 per cent in 1801, 49.2 per cent in 1821, and 47.5 per cent in 1851.[2]

The developed area of mid-eighteenth-century Manchester consisted of less than a quarter of a square mile (including Salford), a rough circle with a radius of approximately twelve hundred feet. The most highly developed part of the town was between Dean's Gate, Brown Street, and the bottom of Shude Hill. The town was expanding rapidly to the south because of the docking facilities at the bottom of Kay Street, which were used by ships of up to fifty tons linking Manchester and Liverpool along the newly improved Irwell River. St. Anne's Square and the surrounding streets formed one of a number of typical residential districts for middle-class merchants, with 'pleasant . . . new streets with houses of red-brick and stone.' [. . .] A new exchange was being constructed on Market Place by Sir Oswald Mosley, the lord of the manor, 'for the accommodation of the merchants and manufacturers of the town.' In fact, the whole town seems to have had the appearance of a vast construction yard. [. . .]

In spite of all this activity, or perhaps because of it, eighteenth-century Manchester was lacking in most public amenities. [. . .]

When the first wave of industrialization occurred in their neighborhood, increasing the local production of cotton goods, the Manchester merchants defended their town's economic dominance as a distribution center by expanding their market area to cover the whole of England and even Europe. Their interest in transportation (the Irwell and Mersey Navigation Act, for example, was sponsored jointly by the merchants of Liverpool and Manchester) reflected their awareness of the struggle necessary to survive economically in a period of change. When industrialization at home became possible, thanks to James Watt's stationary steam engine, both the necessary capital and a keen awareness of the international cotton trade allowed a rapid transformation of the town from a regional market to 'the Metropolis of Manufacturers.' Liverpool's ascendance as a major port resulted from its oligarchical corporation's substantial public investments, particularly in the construction of docking and warehousing facilities; Manchester, devoid of any but the most rudimentary forms of local government, showed no corresponding pattern of governmental interest and effort in its economic development, but rather a collection of individual endeavors.

Surprising as it may seem to us today, the absence of local controls was thought to have definite advantages. Foremost was the lack of regulations restricting trade to freemen of the town. While Liverpool merchants could, and did, control potential competitors by refusing them the freedom of their borough or exacting a fee for the temporary privilege of engaging in trade, Manchester was ready to welcome all comers attracted by the economic opportunities it offered, be they merchants or workmen, Englishmen or foreigners. There was a dramatic expression of this hospitable attitude in 1799, when a German merchant [and cotton manufacturer], Carl Brandt, was elected 'boroughreeve' (mayor) of the town [. . .]

Although Manchester's population explosion did not take place until the closing decades of the century, by the 1770s its growth was already accompanied by many of the difficulties that became characteristic of the industrial towns and prompted nineteenth-century reformers to take action. The mercantile town was transforming itself into the finishing center for the region's cotton goods, and the manufacture of laces and hats was also extensively carried out. These early manufacturing activities had started to attract workers who were crowding the existing housing and beginning to spill over into new quarters, built hastily with little or no care for the comfort or health of their inhabitants. [. . .]

The chronology of these migrations during the early years of the industrial revolution is somewhat nebulous, but it seems likely that Manchester was one of the first Lancashire towns to experience the successive waves of migrants from the rural areas of Lancashire, Cheshire, Wales, and Ireland that were to swell urban population in the ensuing decades. The appearance of a substantial number of paupers and the problems of feeding a growing population were therefore of great concern [. . .]

The rise of the metropolis

Once steam power became available, it was only to be expected that the cotton middlemen would expand their activities from commercial to manufacturing pursuits. They could increase their profits by manufacturing their own yarn on a large scale, and they had the capital available to finance the relatively high cost of the new machinery. Finally, the development of the power loom in the early 1800s led to large industrial concentrations, 'mixed firms' combining weaving sheds and spinning mills, consumers as well as producers of yarn. This growth was extraordinarily rapid: in 1794 there were only 3 cotton mills in Manchester; in 1802 there were 26. Moreover, mechanization created a demand for subsidiary activities, particularly the manufacture and servicing of machinery. Numerous firms specializing in engineering and foundry work settled in Manchester during this period, serving not only the town's industries but the cluster's other urbanized areas as well. By 1795 there were at least 6 major foundries of regional importance in Manchester and Salford in addition to numerous smaller establishments still at a handicraft level.

The growth of industry in Manchester, combined with the town's function as a regional market, prompted an increasing interest in transportation [. . .] the Duke of Bridgewater's canal [. . .] halved the cost of coal through the construction of a first link between the Worsley pits and Manchester in 1764, and which connected the town to its port of Liverpool in 1776. Industrialization brought an upsurge of canal-building activity. Between 1791 and 1800, 8 acts of Parliament were passed for canal companies in the Manchester region and 6 more between 1800 and 1819. Construction was rapid: in 1801 Salford was connected to Bolton and Bury, and in 1804 the canal between Rochdale and Manchester was completed. The towns of the Manchester cluster were fully connected not only to their overseas markets but also to the rest of England via the Grand Trunk Canal. Improvement of the road network was slower but gained momentum in the first two decades of the century. [. . .]

The lack of continuous statistical series on industrial employment, even in as important a sector of the English economy as the cotton industry, is an obstacle difficult to surmount. [. . .]

I have attempted to trace the growth in the number of factory workers in Manchester. These results, summarized in Table 6.1, show the remarkable increase in factory employment that occurred at the turn of the century. This increase is particularly remarkable in that it is a 'deflated' figure, including neither employment in the finishing processes of the cotton trade nor the growing importance of such subsidiary industries as metal and machine fabrication.

A further indication of the development of industry is the more frequent presentment to the Court Leet, starting in 1801, of nuisances directly attributable to factories. In 1801 alone, ten cotton-spinning factories and four other industrial concerns were presented for being

> possessed of a certain ffurnace [*sic*] which they used to . . . burn and cause and procure to be burned . . . large quantities of Coal and did thereby then and there [make] great quantities of Smoke and soot which then and there issued from the said ffurnace into and upon the Dwellinghouses of divers of his Majesty's liege Subjects . . . to the great damage and Common Nuisance of the Inhabitants.[3]

I have already mentioned the extraordinarily rapid increase in population that accompanied Manchester's industrialization, an increase whose full impact was felt only after 1780. The direct result of this population explosion was the extension of the urbanized area beyond the densely settled boundaries of Manchester and Salford. Figure 6.3 shows the density of settlement in Manchester parish between 1775 and 1831, together with the distribution of textile mills in the area. It should be noted that even after the introduction of steam power, the mills still had to be close to a plentiful source of water, which was used for the boilers and to clean the cotton fiber. The town's piped water supply being hopelessly deficient, proximity to the Irwell, the Medlock, and the Irk was in most cases a necessity for the mills. Since these rivers formed the town's boundaries, the adjacent townships started to urbanize, first Ardwick, Chorlton-upon-Medlock, and Hulme, and later Cheetham, Harpurhey, and Newton, as shown in Table 6.2. While there were few opportunities for employment in these towns (with the exception of Chorlton-upon-Medlock, which had twelve mills in 1821) the industry on the Manchester side made them attractively located residential areas

Table 6.1 Growth of Cotton-factory Employment in Manchester

	1790	*1801*	*1811*	*1821*
Number of mills	1–2	23	34	66
Number of hands[a]	1240	48,300	28,300	51,800
Percentage of labor force	3.5	90.0	48.0	30.0

Note: [a] Spinners, weavers, and printers.

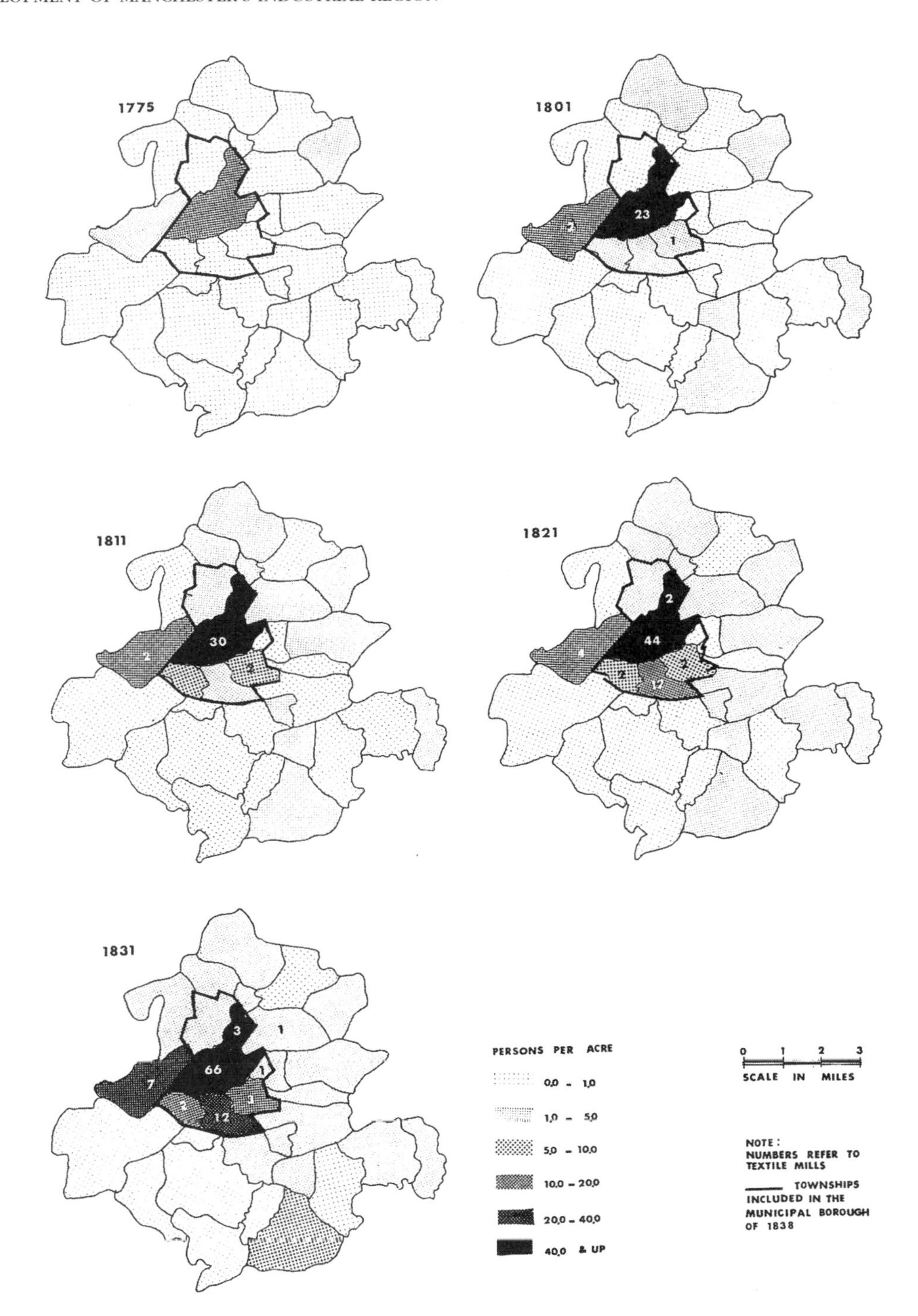

Figure 6.3 Manchester parish, 1775–1831
Note: Numbers refer to textile mills

Table 6.2 Gross Population Density – Manchester and Suburbs

	1774	*1801* *(persons per acre)*	*1811*	*1821*
Core				
Manchester	14.3	44.6	50.4	68.5
Salford	3.5	10.1	14.2	19.1
Industrializing suburbs				
Ardwick[a]	0.5	3.6	5.6	7.1
Chorlton-upon-Medlock[a]	0.4	1.1	4.1	11.0
Hulme[a]	0.3	3.3	6.4	8.8
Other suburbs				
Beswick[a] -Bradford	0.6	0.6	0.7	0.6
Cheetham[a]	0.6	0.8	1.3	2.3
Harpurhey	0.5	0.7	1.0	1.8
Newton	0.4	0.9	1.2	1.7

Note: [a] Included in the municipal borough formed in 1838.

for low-income groups as the density of Manchester itself increased.

By the turn of the century, crowded conditions were already causing upper-income groups to migrate to suburban locations. [. . .] These pleasant surroundings were in sharp contrast to the conditions that prevailed in Manchester itself and in its urbanizing suburbs. The influx of population, attracted by the new factories, had not only caused hitherto unknown levels of overcrowding in the older parts of the town but also an unprecedented building boom of cheap speculative housing next to the industrial districts. Twenty-five hundred new houses were constructed between 1774 and 1788; about 4000 more between 1788 and 1801; and 6000 more between 1801 and 1821. Even then, the workingmen and their families were being crowded into every available cellar and room, partly because of the disparity between the wages they were able to bring home and the rents charged by speculators, and partly because of the necessity to live close to the factories in the absence of any means of public transport. The results were appalling. [. . .]

[. . .] In the eastern portion of the town, the location of industry along existing streams created a centripetal development pattern in which speculative low-income housing tended to nestle close to the mills. By the 1820s the built-up area of Manchester covered slightly over three-quarters of a circle with a one-mile radius, approximately a threefold areal increase over the preindustrial town of the 1770s [. . .]

Notes

1 Manchester did not acquire manufacturing importance until the last decade of the eighteenth century and the introduction of the steam engine, which offset its lack of adequate streams for water power. 'In 1789, the first steam engine for spinning cotton was erected in Manchester, and from that year the manufacturing prosperity of the town may date its rise.' John Reilly, *The History of Manchester*, London, J. G. Bell, 1861, p. 257.

2 The drop in 1851 reflects the suburbanization of adjacent areas with the introduction of commuter railroads.

3 John P. Earwaker, *The Court Leet Records of the Manor of Manchester*, H. Blacklock, Manchester, 1888, vol. 9, p. 174.

7

WATER FOR MANCHESTER

J. A. Hassan and E. R. Wilson

Source: J. A. Hassan and E. R. Wilson, 'The Longdendale Water Scheme 1848–1884', *Industrial Archaeology* 14 (1979), pp. 102–4, 106–9, 111–13, 115–17, 119

The reservoirs constructed by Manchester Corporation in the Longdendale valley between 1848 and 1877 represent a pioneering achievement in the history of hydraulic engineering. The six mile chain of large artificial lakes, each backing to the dam of the preceding one, are an impressive sight to travellers crossing the Pennines between Hyde and Barnsley. In fact the scale of the scheme is more extensive than the Longdendale works suggest as storage and service reservoirs throughout the region are supplied from the valley. By 1884 the project involved sixteen reservoirs extending to over 854 acres, the five in Longdendale covering 462 acres. The works stand comparison with some of the greatest constructions of the Victorian railway engineers.

The scheme was the brainchild of John Frederic La Trobe Bateman (1810–1889). Bateman was involved in many great hydraulic and land reclamation projects, but Longdendale was his most significant achievement. The scheme's importance lay partly in its unprecedented size. In the third quarter of the nineteenth century there were already single reservoirs, such as at the Croton Waterworks in New York and elsewhere in the British Isles, which were bigger than those which served Manchester.

Yet taken as a whole the Longdendale chain constituted, it was alleged, the most extensive system of artificial lakes certainly in Europe if not for a time in the world. In the 1860s the Manchester waterworks, wholly or partially supplying an area with a population exceeding one million, were the largest in this country. Nevertheless, as has been pointed out, the Longdendale waterworks may not appear especially impressive by modern standards. More important than scale, however, were the innovatory concepts underlying the project. Before 1848 there were one or two minor gravitation schemes in existence. But this was the first *large-scale* water supply undertaking on the gravitation principle, collecting and impounding water in a relatively remote hilly district and dispatching it – in this case over a distance of 18 miles – to a large centre of consumption. Moreover despite many unforeseen difficulties the undertaking was for consumers an economical and for the Corporation a financially successful one. [. . .]

Manchester Corporation had been induced to take over control of the town's water supply in the 1840s owing to mounting dissatisfaction with the service offered by the local water company. In 1846 water was provided internally (and intermittently) to only 10,918 houses out of 46,577 in the borough, with a further 12,937 houses relying on street taps.

When another private company proposed to compete for the supply the Corporation decided to seek Parliamentary authority to take over the old Manchester and Salford

Waterworks Company and obtain a monopoly of the town's water supply on the basis of an ambitious and costly scheme to exploit the resources of Longdendale. Such a scheme had already been recommended by Bateman to the Company, who however lacked the vision and resources to pursue such a project. The Corporation obtained its Acts of Parliament in 1847 and 1848. Bateman was to be consulting engineer. Thus Manchester became one of the first local authorities in the country to develop a major, large-scale municipal water undertaking.

Initially only three artificial lakes were planned in Longdendale, Woodhead, Torside and Rhodes Wood. In November 1848 the Waterworks and Gaol Construction (later Waterworks) Committee took over management of the works. Already construction was proceeding energetically at Woodhead Reservoir and at Mottram Tunnel (which took the main pipes under high ground out of the valley on the way to Manchester). About 1,100 men were employed. Heavy rains and floods caused serious damage at Woodhead in 1849. Yet despite numerous interruptions and delays Bateman had pressed forward the work sufficiently vigorously so as to fulfil his promise that Longdendale water would be available in 1850. In fact it began to flow into Manchester in December of that year. In February 1851 the Waterworks Committee reported with satisfaction that the various works were progressing rapidly and satisfactorily and that several important portions were completed. Thus the project was partially operational by the same year that the borough formally took over the water supply from the old waterworks company.

Despite this achievement, however, the challenges and pitfalls of the great scheme were only just beginning to reveal their extent. As early as February 1852 Bateman admitted that 'many of the nicer arrangements of the works will require designing afresh'. [. . .] Longdendale valley at the outset had seemed ideal, topographically and geologically, as a water catchment basin, drawing on a drainage surface of 30 square miles of Pennine uplands. But superficial appearance concealed fundamental problems. Originally it was not appreciated that the valley constituted the main 'fissure' in a geologically troubled district. Another feature was the many landslips which had occurred in the past from the gradual perishing of soft shale beds which occurred in the millstone grit series. This led subsequently to harder material falling down and resting upon landslip debris. [. . .]

The first failings occurred at Woodhead. The site was selected on account of the narrowness of the valley at that point, and the slight expense which a high embankment would involve. Bores had indicated that the ground beneath consisted of solid rock, but the specimens were from large, loose boulders. The ground was, in fact, composed of debris from an old landslip, and its shifting, shaley and dangerous character became evident when an attempt was made to construct a puddle trench for the embankment across the bottom of the valley. At one stage the whole south side of the valley, along with the main line railway, threatened to slip into the river basin. Nevertheless the work was continued [. . .] and the embankment completed. When the reservoir was filled with water in 1851 there were considerable leakages in the embankment, causing discolouration of the water which passed below and threatening to render it useless for consumption. Over the following years there were repeated, but vain, attempts to make the dam watertight [. . .] only in 1858 was it formally recognised that existing arrangements were irrevocably deficient. Then Bateman proposed the construction of a new and improved embankment some way below the existing one and to have the two united as one. The result is the strange plateau across the valley which seems to hold back the Woodhead lake. [. . .] The search, through trial pits and bores, for

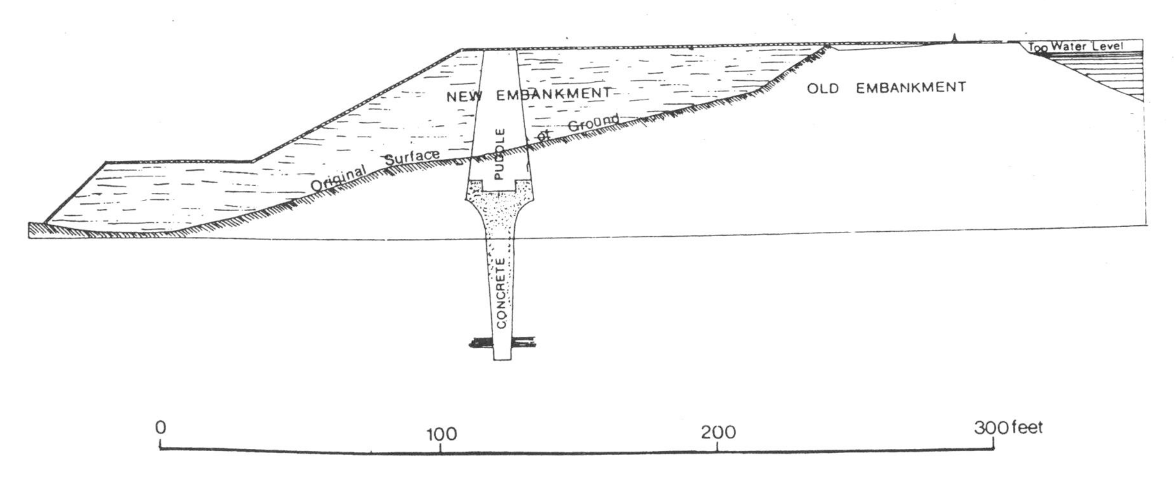

Figure 7.1 A transverse cross-section of the Woodhead Embankment showing the old embankment completed 1851 and the new embankment completed 1877

ground that would provide a secure foundation for the new embankment turned out to be an arduous business. Not until 1863 was a satisfactory site determined upon and preliminary work commenced. [. . .]

Much higher standards of construction were adopted than for the first Longdendale dams. Dislocated and loose strata made the formation of a puddle trench for the new embankment difficult. The trench had to be excavated to fifty or sixty feet below the old river course before a satisfactory (shale) foundation was reached. The trench was filled with 'hydraulic' concrete so that the embankment would be able to withstand the 'wasting or penetrating power of 160 feet pressure of water.' Concrete was introduced in this manner to a depth of 110 feet in the embankment. [. . .] Everything was done to make the work as 'sound and as watertight as possible.'

This major reconstruction was ultimately successful and when the reservoir was filled on completion in 1877 no leakage occurred – to the immense relief of everyone concerned.

At this point it would be helpful to provide further observations on the general principles of the Longdendale scheme. In theory its purpose was to provide for the most direct dispatch as possible of pure spring water from the Pennine uplands to Manchester. One of the innovations which Bateman devised to expedite this result was the introduction into the larger reservoirs at varying heights of syphons, with valves inside, for the purpose of drawing off water from near the surface which is the most pure and least turbid. As the water level drops lower syphons become operational. Because of these arrangements and because all the reservoirs, including the service reservoirs near Manchester, were made sufficiently deep as to prevent the growth of vegetation in the water, very pure water was in fact delivered in Manchester. Consequently Longdendale water was sent into the city unfiltered. The saving in cost was regarded as a benefit. The only further precaution taken was the straining of water through fine copper wire gauze (which could be removed and cleansed as required) before being admitted into inhabitants' homes. [. . .]

Originally it was planned that water impounded in the reservoirs would be mainly drawn upon for the supply of owners of water rights. Millowners on the River Etherow below the dams had been guaranteed a stipulated flow under the Corporation's Water

Acts to 'compensate' them for Manchester having acquired the privilege of abstracting water from the river. At one very early stage it had been envisaged that only Woodhead Reservoir would be built and this solely for the supply of such compensation water. In reality water stored in the reservoirs was also drawn upon for Manchester's daily supplies, especially during summer when the flow from the springs diminished. Nevertheless the role of the Longdendale reservoirs in guaranteeing compensation supplies was crucial. It was not until 1870 that for the first time the amount delivered to the city actually exceeded the amount released to millowners as compensation. Torside and Rhodes Wood were used exclusively for the supply of compensation water until 1873. [. . .]

One way of regaining water so 'lost' to Manchester was to buy back from the millowners some of the compensation water [. . .] The millowners accepted this on payment of the total sum of £50,000. The Waterworks Committee were delighted with the outcome of the negotiations which meant that 4.6 million gallons per day were thereby saved for the inhabitants of Manchester. [. . .]

The second method of mitigating the compensation problem was to create additional storage capacity so that relatively less of the total capacity of the project would be devoted to ensuring the delivery of compensation supplies. Bateman strongly advocated this course.

Vale House and Bottoms Reservoirs were built below Rhodes Wood Reservoir for the sole purpose of supplying compensation water. They were in fact too low to supply Manchester by gravitation, but they released the higher reservoirs for this purpose.

Parliamentary authority was granted and construction was started in 1865. Vale House was finished in 1870 and Bottoms was finally completed in 1876. [. . .] About 600 men were employed on the works. It was reported:

> The introduction of steam cranes and other arrangements for economising labour have been very successful[1] [. . .]

In the construction of the two reservoirs lessons learnt in the earlier work were now applied. Concrete foundations for the embankments were used from the outset. In the current state of hydraulic engineering in the conditions of Longdendale it was very difficult to keep reservoirs watertight which exceeded the depth of sixty feet. Hence two smaller reservoirs at slightly greater cost instead of one large reservoir were built. The entire operation proceeded in a much smoother fashion than in the creation of the first three lakes, and the standards of workmanship and finish were apparently of a high order. [. . .]

Besides being path-breaking in its general conception and scale the Longdendale scheme also involved numerous specific innovations in water technology. The unprecedented scale and associated safety problems demanded that many milestones in waterworks engineering would have to be passed. One major and novel difficulty was that of containing the water in aqueducts, which operated under a considerable pressure, should a burst occur at a higher point. No suitable automatic device was available. The firm of Sir W. G. Armstrong was commissioned to investigate the problem, and they came up with the invention of the self-acting closing valve which answered the task. These were attached to the great mains entering Manchester at Hyde.

Other pioneering arrangements employed by Bateman in Longdendale were the large three-part valves, also 'beautifully executed' by Armstrongs of Newcastle, and fixed at Woodhead in 1851. They enabled one man, rather than a team, to operate the valves which controlled the discharge of water under very heavy pressure from the reservoirs. Anonymous 'reflux valves' were used to prevent water running backwards through a burst. [. . .]

Another innovation first employed in the scheme was improved pipe-jointing methods.

The impact of the Longdendale project on the Manchester region was considerable. Average annual output of water under the old company in 1840–7 was equivalent to approximately 1.97 million [gallons per day]. By 1875–82 it was 17.3 million [gallons per day]. In 1846 an intermittent internal supply of water had been laid on to 23.5 per cent of all Manchester houses. By 1881 a constant internal supply was provided for at least 80 per cent of all Manchester homes.

The arrival of relatively inexpensive and abundant pure water had important consequences. The improved supply led, as had been anticipated, to an increased demand. The laying of mains to the suburbs of the city and the supply of water to 'houses of better description fitted with water-closets and other domestic conveniences' lead to an increased consumption per head. As Bateman reported in 1874:

> The quantity of water used per head . . . is constantly upon the increase. The domestic habits of the people are undergoing great changes by the freer use of water for the purposes of ablution and by the introduction of water closet and baths, while at the same time great use of water is made by the residents in the suburbs, who not only use water for ordinary domestic purposes, but for stable, cattle, and gardens, all of which may be considered legitimate purposes and for which the supply of water to a town should be employed.[2]

However, the effect on the urban environment should not be exaggerated. Improved water supply was a necessary but not sufficient condition for improved public health. As long as sanitary reform made poor progress in Manchester in other areas, notably in refuse disposal, sewage treatment and river pollution, then the increased water supply made little environmental impact. The sanitarians of the 1840s doubtless exaggerated the role of water supply. Contrary to expectations mortality did not decline with the onset of an improved supply. In fact the crude death rate remained above 30 per thousand between 1840 and 1868.

The transformation in the water supply, however, did bring clear gains to property owners and manufacturers. The increased water supply was directly associated with an improved service from a reformed fire brigade. According to Bateman the proportion of property destroyed by fire was 21.3 per cent of the value of the property attacked in 1846–50, while by 1877–8 it had fallen to less than 2.0 per cent. Manufacturers and traders benefited through the provision at relatively low cost of a vital raw material. In 1849 the Manchester and Salford Waterworks Company laid on a supply to only 1,742 'public institutions,' including factories and warehouses, out of a total of 8,456 public institutions. By 1878 the number of commercial and industrial customers had grown to 16,557. The nineteen largest customers then were all firms of either calico printers, dyers, or bleachers and dyers.

The success of the Longdendale project had wider implications. Firstly, it lent force to the collectivist tide in favour of municipal control of water supply, and other local authorities followed Manchester's lead in taking over this utility. While in 1851 the water industry remained unconvinced regarding the desirability of public control, by the time the Royal Commission of 1867–9 reported, expert opinion had come down firmly in its favour. Moreover the financial arrangements adopted at Manchester (and Liverpool and Glasgow) – the levying of a public water rate on all property owners and a domestic rate on consumers – were put forward as a model by the commissioners.

Secondly, although the Royal Commission was less than enthusiastic regarding long-distance gravitation schemes (certainly as a solution to London's needs) the proven viability of the Longdendale project did pave the way for the launching of even more ambitious projects elsewhere. [. . .] By the late 1860s even longer distance projects were being discussed, and Bateman and others were drawing attention to the potential of the Welsh and Cumbrian mountains. The first local authority to seriously consider Wales as a source of water was Liverpool, and Bala Lake had been advocated as early as 1864. In 1874 Liverpool commissioned Bateman to advise on the various schemes which had been proposed to the Corporation. Bateman's response was to strongly recommend a joint Liverpool and Manchester scheme to tap Ullswater. Liverpool instead committed itself to Wales starting the work of damming the Vyrnwy valley in 1881.

By 1877 it was said of Longdendale:

> . . . almost every available site in the whole valley above the manufacturing population has been converted into a reservoir, and the utmost quantity which the district will yield has been obtained for Manchester.[3]

The potential of the valley for the large-scale organisation of water supplies had been exploited to the full. But demand was accelerating. Thus almost as the scheme received its finishing touches it was becoming evident that it would not be able to meet Manchester's growing requirements for much longer.

Bateman urgently warned that demand would soon outrun supply: he gave eight to nine years in 1868 and six to seven years in 1875. [. . .] Additional drainage area was acquired by Act of Parliament in 1865. Raising the embankments by a few feet added at little cost valuable expansions in storage capability. Additional storage and service reservoirs were built, notably at Audenshaw. But the more the impounded and turbid flood waters of the storage reservoirs were relied upon, the greater was the deterioration in the quality of the supply, with a relative decline in the use of pure spring water. [. . .] By 1875 Bateman was unofficially warning influential parties of the 'catastrophe' which would shortly occur if his warnings were not heeded and his proposals – to promote a scheme drawing upon the Lake District – were not taken up.

The Waterworks Committee were by now convinced of Bateman's arguments. Due to the fact that Liverpool could not be brought into a joint project it was proposed that a scheme to exploit the resources of Thirlmere should be adopted, the decision being made by the Committee in 1876.

The Bill authorising Manchester to dam and draw upon Thirlmere received the Royal Assent in 1879, and Cumbrian water was flowing into Manchester by 1895. The Thirlmere undertaking, with its 90 mile-long aqueduct, remains an outstanding example of an ambitious waterworks project. Long-distance gravitation water supply, however, was first established as a viable proposition in Longdendale. Thirlmere eclipses Longdendale in scale and capacity, but not in principle. [. . .]

Notes

1 Manchester Central Reference Library, Records of the Manchester Corporation Waterworks Department: Minutes of the Waterworks Committee, vol. 14, Bateman's report, 18 March 1868.

2 *Ibid.*, vol. 20, Bateman's report, 6 October 1874.

3 Bateman, Report to the Waterworks Committee, 7 February 1877, pp. 3–4.

8

URBAN GROWTH IN INDUSTRIALISING SCOTLAND, 1750–1840

T. M. Devine

Source: T. M. Devine, 'Urbanisation', in T. M. Devine and R. Mitchison (eds), *People and Society in Scotland: I, 1760–1830*, Edinburgh, Economic and Social History Society of Scotland and John Donald, 1988, pp. 32–7, 39–41

The bases of urban growth

Why Scotland should experience such a precocious rate of urban growth is a question which requires detailed consideration since its consequences for the long-run development of Scottish society were so profound. The essential foundation, though not the principal direct cause, was the revolution in agriculture which occurred in parallel with town and city expansion. Urbanisation could not have taken place without a substantial increase in food production to sustain the needs of those who did not cultivate their own food supplies. At the same time, agrarian productivity had to improve in order to release a growing proportion of the population for non-agricultural tasks in towns and cities. [. . .] However, for much of the period of this analysis, the urban masses mainly relied on grain, milk, potatoes and meat supplied from Scottish farms. They were fed through a rise in both the production and productivity of agriculture achieved by a reorganisation in farm structure, a more effective deployment of labour and higher yields derived from improved fallowing, the sowing of root crops and the adoption of new rotation systems. No authoritative measures exist of the precise rate of increase in food production but it must have been very substantial. [. . .]

Agrarian change was a necessary precondition for urbanisation but the process of agricultural reform also contributed directly to town growth at two other levels. First, the increasing orientation of agriculture towards the market further stimulated the function of urban areas as centres of exchange. There was a greater need than before for the commercial, legal and financial facilities which concentrated in towns. Perth, Ayr, Haddington, Dumfries, Stirling and several other towns owed much of their expansion in this period to the increasing requirements for their services from the commercialised agricultural systems of their hinterlands. Regional specialisation in agrarian production also enhanced the need for growing centres of exchange. Inverness, for example, expanded on the basis of its crucial role as the sheep and wool mart of the Highlands as that area became a great specialist centre of pastoral husbandry in the first half of the nineteenth century. Secondly, the prosperity of Scottish agriculture during the Napoleonic Wars boosted the incomes of tenant farmers and inflated the rent rolls of many landowners. The increase in the purchasing power of these classes had major implications for urban growth because it resulted in rising demand for the products of town consumer and luxury industries, and for more and better urban

Figure 8.1 Glasgow c. 1835. The city's population had climbed to over 274,000 and Glasgow was the fastest-growing town of its size in western Europe
Source: Mitchell Library, Glasgow

services in education, in leisure and in the provision of fashionable accommodation.

Yet agrarian improvement was the necessary condition for Scottish urbanisation rather than its principal determinant. Towns which acted mainly as exchange and service centres for rural hinterlands expanded only relatively modestly, at a rate which was only slightly more than the national rate of natural increase. Moreover, the rise in population which occurred in all western European societies from the later eighteenth century encouraged food producers throughout the continent to increase their output to cope with enhanced demand. The nature of the Scottish Agricultural Revolution may have been distinctive but agrarian improvement was too common in Europe at this time to provide the basic explanation for Scotland's exceptional pace of urban development. It is more likely that Scottish town expansion was a direct consequence of Scotland's equally remarkable rate of general economic growth between 1760 and 1830. The Industrial Revolution before 1830 was mainly confined to mainland Britain and it is hardly a coincidence that in this same period urbanisation occurred more vigorously in England and Scotland than in any other European country. Scottish industrialisation and Scottish urban growth were both results of the same economic forces. [. . .]

This process had two interlinked aspects. The first was commercial in origin. In the eighteenth century, Scotland was in a superb geographical position to take advantage of the changing direction of international trade towards the Atlantic world. This momentous alteration in transcontinental commerce was a highly dynamic factor in port development along the whole western coast of Europe from Cork to Cadiz. Scotland was virtually at the crossroads of the new system and the Clyde ports grew rapidly to become the great tobacco emporia of the United Kingdom until diversifying later into the importation of sugar and cotton. It was no coincidence that in the later eighteenth century four of the five fastest-growing towns in Scotland were in the Clyde basin. Commercial success was bound to foster urban expansion. [. . .]

But, in the long run, the expansion of manufacturing industry was even more critical for urbanisation than the stimulus derived from international and interregional commerce. Of the thirteen largest towns in early nineteenth-century Scotland, five at least trebled their population size between c. 1750 and 1821. In addition to Greenock these were Glasgow (from 31,700 to 147,000), Paisley (6,800 to 47,000), Kilmarnock (4,400 to 12,700) and Falkirk (3,900 to 11,500). Greenock apart, the inhabitants of all these towns mainly depended either directly or indirectly on manufacturing industry. It was the larger industrial towns and the constellation of smaller urban areas with which they were associated which set the pace of Scottish urbanisation. [. . .] The water-powered cotton-spinning factories of the last quarter of the eighteenth century were more often to be found in rural settlements [. . .] The continued presence of industry in a variety of forms in the countryside helps to explain why a majority of the Scottish people still lived outside large urban areas by 1830.

Yet, in the long run there were obvious advantages in industrial concentration in towns. [. . .] firms saved the costs of providing accommodation and other facilities for their workers from their own resources; they were guaranteed access to a huge pool of labour and transport costs between sources of supply, finishing trades and repair shops could be markedly reduced or virtually eliminated [. . .] These advantages built up a dynamic for urban expansion even before 1800. Thereafter the new technology of steam propulsion and conspicuous progress in transport developments through the construction of canals and roads steadily intensified the forces making for urban concentration. In cotton-spinning, and eventually in other textile industries, steam power encouraged industrial settlements on the coalfields and removed the one major obstacle which had previous[ly] constricted the expansion of manufacturing in the larger towns. Glasgow provides the most dramatic case of the pattern of change. In 1795 the city had eleven cotton-spinning complexes, but rural Renfrewshire had twelve. The fundamental need to have secure access to water power obviously diluted Glasgow's other attractions as a centre of textile industrial production. However, steam-based technology was rapidly adopted after 1800 and concentration accelerated on an enormous scale in the city and its immediate environs. By 1839 there were 192 cotton mills in Scotland employing 31,000 workers. All but seventeen were located in Renfrew and Lanark and ninety-eight were in or near Glasgow. In Paisley, or its vicinity, there was a further great network of forty factories employing almost 5000 workers. [. . .] Before 1830 textile manufacturing was the principal motor of this process of agglomeration. [. . .]

The urban structure

Despite fast urban growth there remained considerable continuity between the old world and the new. The four major cities of the early nineteenth century, Edinburgh, Glasgow, Aberdeen and Dundee, were also the biggest Scottish burghs of the seventeenth century [. . .] The biggest urban areas, therefore, were all ancient places and the traditional county and regional capitals also continued to play a role whether as centres of administration, local government or as markets for prosperous agricultural hinterlands. But by 1830, the Scottish urban system had also developed some characteristic features typical of the new era.

First, urbanisation was mainly concentrated

in the narrow belt of land in the western and eastern Lowlands. Between 1801 and 1841 never less than 83 per cent of the entire Scottish urban population (defined as those inhabiting towns of 5000 or more) lived in this region. Within the area there was heavy concentration in Glasgow and Edinburgh where, as early as 1800, 60 per cent of Scottish urban dwellers resided. This remarkable pattern had implications for the demographic structure of Scottish society because population concentration on such a scale could not have taken place without considerable redistribution of people over a relatively short period of time. Thus, whereas the percentage of total Scottish population in the central Lowlands rose from 37 per cent to 47 per cent of the whole between c. 1750 and 1821, it fell from 51 per cent to 41 per cent in northern Scotland and remained roughly static at 11 per cent in the southern region of the country over the same period. The modern population profile of Scotland was beginning to take shape. Within the urbanising zone the fastest growth among the largest towns was in the west, with four towns in that area at least trebling in population. Paisley expanded more than six times, Greenock more than five, and Glasgow grew fourfold.

Second, there was wide diversity within the urban structure. Any attempt at neat categorisation of Scottish towns is bound to be arbitrary [. . .] In very broad terms, most Scottish towns fitted into three categories: first, the four major cities; second, industrial towns; third, local capitals in historic sites which performed marketing and service functions for their immediate neighbourhoods. [. . .] Of these groups, the industrial staple towns and some of the cities were most likely to suffer the adverse consequences of expansion which are often associated with urbanisation at this time. [. . .] those who dwelt in the regional centres were better placed. Their typically moderate rate of population growth ensured that the existing organisation of sewage and waste disposal was not so easily overwhelmed as elsewhere, though it must be stressed that sanitary problems were a familiar feature of all Scottish towns, whatever their size, at this time. [. . .]

The main contrasts in the occupational structure of Scotland's four cities are best studied from Table 8.1. One similarity, however, is

Table 8.1 Occupational Structure in Scottish Cities, 1841

Percentage of workforce in:	*Glasgow*	*Edinburgh*	*Dundee*	*Aberdeen*
Printing & Publishing	1.12	3.88	0.56	0.91
Engineering, Toolmaking and Metals	7.17	6.07	5.59	6.32
Shipbuilding	0.35	0.17	1.14	1.24
Coachbuilding	0.40	0.92	0.21	0.34
Building	5.84	5.73	6.05	5.99
Furniture & Woodmaking	1.06	2.73	0.77	0.87
Chemicals	1.22	0.24	0.19	0.37
Food, Drink & Tobacco	5.24	8.31	5.27	4.66
Textiles & Clothing	37.56	13.04	50.54	34.68
Other Manufacturing	2.90	3.02	1.29	3.18
General Labouring	8.40	3.69	3.84	6.87

Note: Occupational classification in the 1841 census is questionable and imprecise. The precise figures presented here are unlikely to be wholly accurate; they provide an impression of overall structures rather than an exact measurement of them.
Source: Census of 1841 (*Parliamentary Papers*, 1844, XXVII) and R. Rodger, 'Employment, Wages and Poverty in the Scottish Cities, 1841–1914', Appendix 1, in George Gordon (ed.), *Perspectives of the Scottish City* (Aberdeen, 1985).

worth emphasising initially. The dominance of textile employment, except in the case of Edinburgh, is very obvious from the tabulation, and underscores [. . .] the essential importance of cotton, linen and, to a lesser extent, woollen manufacture in Scotland's first phase of urbanisation. The relative stability and balance of Edinburgh's economic base is also apparent. The majority of the employed population worked in small-scale consumer industries many of which depended on demand from a middle and upper class clientele. Domestic service was far and away the largest employer of female labour. This occupational pattern reflected Edinburgh's metropolitan status, the significant numbers of salaried professionals in the city who represented a large pool of demand for services and luxury consumer goods and the major functions of the capital in the areas of law, banking and education. Poverty and destitution were endemic in certain parts of the city, notably in the Old Town and elsewhere, but the ordinary people of the capital were less likely to experience the full horrors of cyclical unemployment on the same scale as their counterparts in Dundee and Glasgow. [. . .]

Glasgow and Dundee were alike in their heavy dependence on textile employment, the speed of their growth in the nineteenth century (though Dundee grew most rapidly from the 1830s and 1840s) and the relative weakness of the professional element in their occupational structures. It was these two cities which suffered very severe problems of health, sanitation and poverty. In the short term, and especially before 1840, Glasgow's difficulties attracted most attention.

9

IMPROVING THE CLYDE: THE EIGHTEENTH-CENTURY PHASE

John F. Riddell

Source: John F. Riddell, *Clyde Navigation: A History of the Development and Deepening of the River Clyde*, Edinburgh, John Donald, 1979, pp. 30–4, 36–43

Despite the lack of success so far obtained with the deepening of the Clyde, Glasgow was by no means prepared to forget the river as a navigation. [. . .] Some of the plans put forward were quite fanciful. One ingenious mind suggested that large waterproof bags should be manufactured and kept ready at Port Glasgow for placing along the sides of vessels wanting to go upriver. By the use of these bags, the inventor claimed, ships could be made to float higher out of the water and would thus be able to pass over the numerous shoals in the channel. A calculation submitted with his design demonstrated that a vessel with a draught of 10 to 15 feet could have this reduced by their use to as little as five feet.

Not surprisingly, nothing further came of this and similar suggestions and it was not until 1764 that one which appeared to be at least practical was considered. The idea came from a local man, a Dr. Wark, and involved the construction of a training wall at the Broomielaw Harbour. Wark considered that such a wall might be used to force the stream into a narrower channel and so induce higher velocities in the flow. The greater speed of the current would then produce an increase in the erosive or scouring action of the river, leading eventually to a channel of greater depth. At this time the banks of the Clyde at Glasgow were some two to three hundred feet apart and, except in times of flood, the river tended to meander sluggishly over the whole of this breadth. Any reduction in the width of the channel by a confinement of the banks was quite likely to produce an increase in depth in order to maintain the original cross-section area.

To the magistrates and councillors, by now quite inured to the many and various proposals of the 'experimenters', Wark's scheme appeared to possess the very desirable advantages of both practicality and economy. [. . .]

[But it is not known] whether the training walls proposed by Dr. Wark were in fact constructed. Certainly no reference to them is made in the reports on the river prepared in later years, the credit for the successful introduction of this method of deepening being given instead to a gentleman from Chester named John Golborne.

It was in September 1768 that John Golborne came to Glasgow for the first time. The purpose of his visit was to discuss with the magistrates a proposal for the construction of a ship canal across Central Scotland. James Brindley, the father of the canal age in Britain, was then his companion, although Golborne himself was not inexperienced in canal work. He was adviser to the River Dee Company and had continued the works of channel training and land reclamation begun on that estuary in the 1730s. [. . .] [H]is visit to Glasgow encouraged the magistrates to seek advice on problems other than a cross-country ship canal. The improvement of the River Clyde and of the silted-up harbour at Port Glasgow were of more immediate importance, and [. . .] Golborne was quite pleased to make an examination of the waterway and harbour. On 30 November, 1768 he duly presented a report of his findings and recommendations to the town council.

The report was optimistic. Regarding the harbour at Port Glasgow, Golborne suggested that some improvement might be obtained if one of the small streams which flowed into the Clyde at the town could be diverted through the mud-filled harbour basin. The Renfrewshire hills rise steeply above Port Glasgow and the streams which flow off them have considerable energy by the time they reach the narrow strip of raised beach on which the town is situated. The energy of one of these fast-flowing streams, Golborne proposed, could be used to scour away the accumulated sediment from the harbour and wash it out to sea. Any further deposition of material would also be prevented, thus removing the problem for all time. Such a straightforward solution greatly appealed to the Glasgow magistrates and the required diversion, together with an extension to the quayage at Port Glasgow, was carried out successfully under an Act obtained in 1772.

The major part of Golborne's report was about the Clyde itself. 'The River Clyde,' he reminded the magistrates, [. . .] 'is at present in a state of nature, and for want of due attention the channel has been suffered to expand too much.' Because the sides of the river consisted of much softer material than that forming the bed, the current had always tended to erode and scour these sides, while the bed had been left firm. The river had thus 'gained in breadth what it was wanting in depth'. As to the depths obtaining in the autumn of 1768, Golborne noted that between the deep water at Port Glasgow and the fords at the Dumbuck shoal the channel was in no place less than 12 feet deep at low water and thus required little attention. But at the shoal, described by Golborne as 'the first and grand obstacle to navigation', the division of the river into two separate low water channels resulted in the depth available falling away to 3 and in some parts to only 2 feet. The bed of the estuary at Dumbuck was found to be covered with a hard crust of sand and gravel on which the current, reduced in velocity by the division of the stream, could make no impression.

Upstream of the Dumbuck shoal the two channels re-united into one at Longhaugh Point, a flat bulge on the southern shore, and from there eastwards to the rocky outcrop, on which was situated Dunglass Castle, Golborne recorded depths of 16, 18 and even 20 feet. Off the rock itself, no less than 24 feet of water existed, even at low tide. Another shoal was encountered upstream at the Kilpatrick Sands, however, and again the depth reduced to only a few feet. Above Old Kilpatrick, 8 feet was sounded and this depth continued to the Newshot Isle, a large, flat island, little more than a grassed-over shoal. At this island the river divided itself once more and, in both the north and south channels round the island, Golborne found that the bed of gravel and boulders reached up to within 30 inches of the water surface. Above Newshot Isle, up to Glasgow, the sands and shoals in the river and the depths over them were noted to be little changed from the time of Smeaton's survey some thirteen years before.

As to how this highly variable and extremely natural channel might be brought up to a navigable standard, John Golborne was in no doubt. 'I shall proceed on these principles of assisting nature when she cannot do her own work, by removing the stones and hard gravel from the bottom of the river where it is shallow, and by contracting the channel where it is worn too wide'. [. . .]

[His main aim was to reduce] the effective

Figure 9.1 The Clyde, looking downstream to Bowling and Dumbuck, 1922. The remains of some of Golborne's dykes can be seen beyond the perch

breadth of the channel, so that the resulting increase in current velocity might accelerate the erosion of those parts of the bed at that time most resistant to the unconfined flow. Golborne proposed that the required reduction in breadth could be most easily obtained by the construction of projecting jetties or dykes [. . .] Smeaton's comment that this was a method of deepening 'much practised in England' also serves to indicate where Golborne is most likely to have obtained his idea. Indeed, it is reasonable to suppose that he had used the method himself on the Dee, for evidence of such groynes and dykes still exists on that estuary.

Regarding the shoal at Dumbuck – 'the grand obstacle' – Golborne suggested that an improvement could be obtained if one of the two channels was blocked off completely, thus concentrating the flow in the other:

> . . . if a jetty were extended over the south channel, to confine the current, and the hard crust of gravel removed by dredging, the reflowing current would then act with greater force, and grind down a deep and capacious channel; and this must plainly appear to any intelligent person.

Further dykes were to be built from Dumbuck right up to Glasgow to increase the scouring action. The dykes were to be sited in pairs, one extending out from the north bank and one opposite to it from the south, to gain the maximum reduction in breadth compatible with the flow. The dykes, or jetties, were to be of half-tide height, that is up to the level of low water, but Golborne considered that the spaces between each (that is between ones on the same side of the channel) would soon be infilled with sand and silt, narrowing the river to a 'proper breadth' over its whole length and at the same time producing valuable 'firm land' along its banks. About the improvement in depth which might be obtained, Golborne wrote in his report:

> By these means, easy and simple in themselves, without laying a restraint on nature, I humbly conceive that the River Clyde may be deepened so as to have 4 feet, or perhaps 5 feet depth up to the Broomielaw.

[. . .] the Glasgow magistrates were not long in agreeing to the implementation of the [. . .] proposals for the deepening of the channel. They appeared to be workable, and, while not inexpensive, were not costly in terms of the advantages they would bring to the town. Trade was expanding every day, especially the trade in tobacco, and the shipping owned by Clyde merchants now totalled more than sixty thousand tons. Glasgow was a 'flourishing city'. Its importance and influence were also increasing, and when the magistrates came to seek parliamentary approval for the new proposals, little resistance was offered to their bill. On 9 January, 1770, in almost record time, Glasgow obtained its Second Act for the Improvement of the Clyde Navigation.

The preamble to this Act explained that since the previous one had been obtained in 1759, a more accurate survey of the river had been undertaken and that the magistrates had now been advised that

> by contracting the channel of the said River Clyde, and building and erecting jetties, banks, walls, works and fences in and upon the said River, and dredging the same in proper places between the lower end of Dumbuck ford and the bridge of Glasgow, the said River Clyde may be further deepened, and the navigation thereof more effectively improved than by any lock or dam, and so as there shall be 7 feet in every part of the said River at neap tides.

[. . .] the magistrates were empowered to demand [a river duty of] one shilling and eight pence per ton. An additional rate of one penny per ton on goods loaded or unloaded at the Broomielaw Quay was also allowed. These rates were only to be levied, however, so long as 7 feet of water existed throughout the channel at high water of a neap tide, again making any income from the Clyde conditional upon its improvement. Any three magistrates or users of the river could apply for a survey to be made to ensure that the depth was maintained, the Sheriff Depute of Lanarkshire, Dunbartonshire or Renfrewshire being appointed 'the sole and final judge of the depth'. [. . .]

Prior to the 1770 Act the responsibility for the improvement of the Clyde had rested with the town council of Glasgow on a somewhat unofficial basis. This state of affairs was most unsatisfactory from Glasgow's viewpoint, 'and in consideration of the great labour they must be at and of the risque their common stock must run, in improving the Navigation', the lord provost, magistrates and council of Glasgow, and their successors, were now appointed Statutory Trustees of the Clyde with powers to demand dues from the users of the river, and 'to alter, direct and make' the channel throughout the thirteen miles between the lower end of the Dumbuck Shoal and Glasgow Bridge.

That there might be the specified minimum 7 feet of water at high tide for 'ships, vessels, barges and lighters to come and go', the new Trustees were given the right to construct jetties, walls and dykes to contain the channel, to excavate soil from the banks and bed, to deposit this excavated material on the most convenient part of the banks and to

> plant the banks on each side of the Clyde with willows or other shrubs for the safety or preservation of the banks, and for preventing the same from being hurt or carried away by the river.

Yet another clause allowed for the repair and enlargement of the quay at the Broomielaw and for an extension of the Harbour to the south bank of the Clyde. [. . .]

John Golborne had suggested that a minimum high water depth of 7 feet was possible,

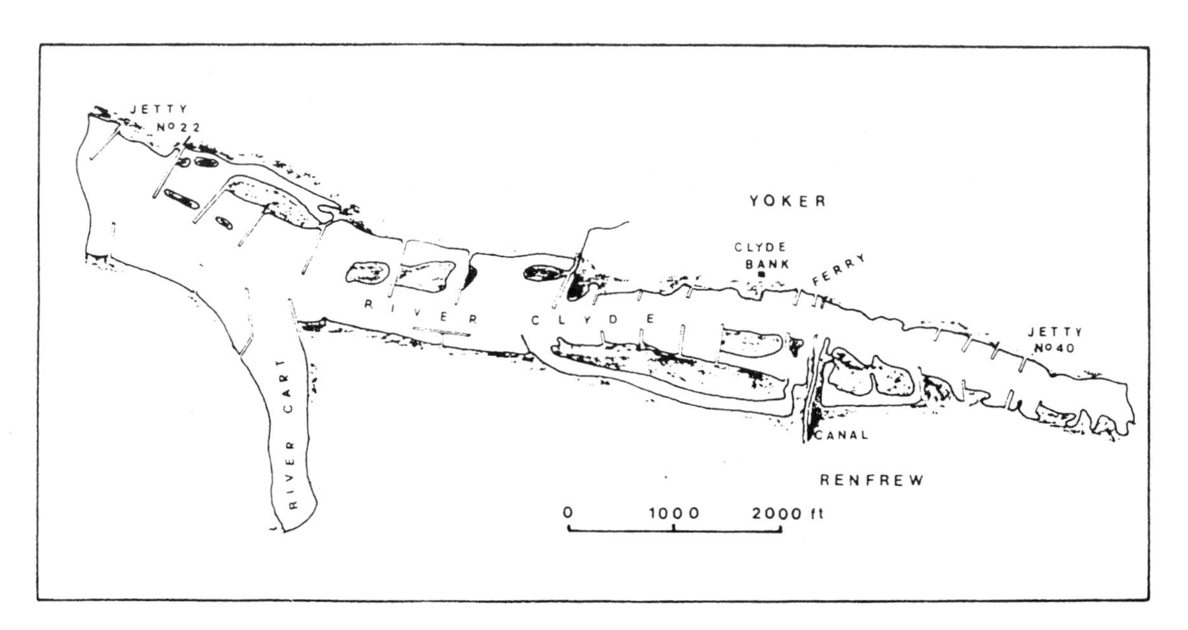

Figure 9.2 A reach of the Clyde showing Golborne's lateral dykes

and it was to gain this overall depth that work was now commenced. As was the usual practice of the times, Golborne was appointed both contractor and consultant, and on 30 November, 1770 an agreement for the improvement was signed by him and the Glasgow magistrates. The value of the contract was £11,000, and under its terms the contractor undertook to create '7 feet of water at least at neap tides in every part of the river' within four years of the date of signing. Golborne was required 'to furnish and provide all proper and necessary machines, instruments, materials and utensils, proper workmen and every other thing requisite for making the river deep'. He was also to give 'his personal and punctual attention' to the contract for at least six months each year. [. . .] To see to the day-to-day management of the contract, Golborne appointed his nephew James, a young man who had already worked with his uncle on similar projects in England.

The Golbornes commenced their operations in the spring of 1771, and within a matter of weeks the whole river assumed an air of great activity. Stone quarries were opened up to provide solid material for covering the jetties, borrow pits were created to supply the earth and gravel for the centres, and trees were felled to provide timber for supports. Labourers and craftsmen were hired and placed in camps near to the works. There they toiled throughout the summer, excavating and infilling, confining, deepening and reshaping the very course of the channel. Towards the end of the season a number of the projecting dykes were complete, and [. . .] the Glasgow magistrates gathered with Golborne for an initial evaluation of what had been achieved.

The results, so far as could be seen, were truly astonishing. Where the dykes had been built, the increased scour of the narrowed channel had eroded away the bed with such effect that reaches which were once only 2 and 3 feet deep now had depths of 5 and 6 feet. Sand and silt were all washed away by the current to leave a channel deeper than ever before. Where harder material existed, Golborne had introduced special ploughs to break

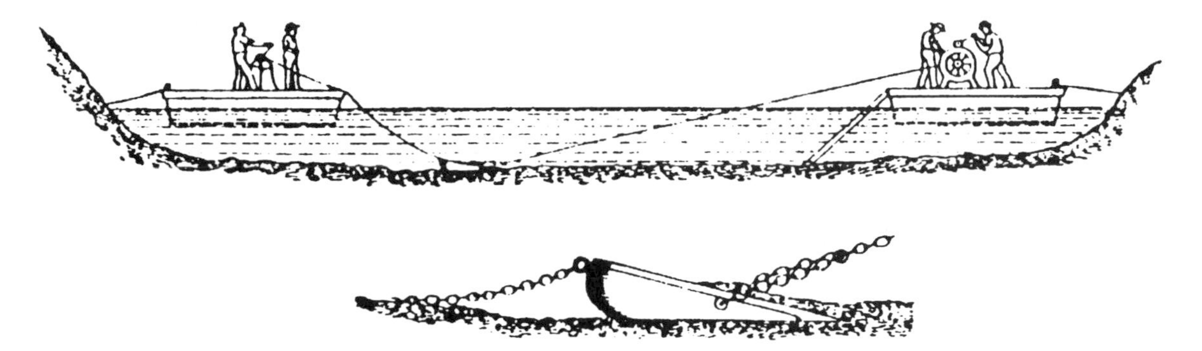

Figure 9.3 A dredging plough, as employed by John Golborne

up the surface and expose it to the current. A primitive form of dredger, these ploughs were observed with fascination by the magistrates, as they had been by Thomas Pennant during his recent tour of Scotland:

> The machines for deepening the river now at work are called ploughs; they are large hollow cases, the back is cast iron, the two ends of wood, the other side open. They are drawn across the river by means of capstans, placed on long wooden frames or planks; and opposite to each other near the banks of the river. They are drawn over empty, returned with the iron side downwards which scrapes the bottom, and brings up at every return half a ton of gravel, depositing it on the bank.

[. . .] Over one hundred separate jetties extended out from the north and south banks of the Clyde by the end of the 1772 season and, as far as the people of Glasgow were concerned, a miraculous 5 feet of water existed throughout the length of the river at high tide. All of the shoals down to Dumbuck had been successfully deepened and at the Broomielaw Harbour the Hirst Shoal had been eroded to such an extent that 4 feet of water existed at low tide where two years before there was only 15 inches. Vessels of up to seventy tons could now reach Glasgow and both lightering and direct shipments were greatly expedited. *The Glasgow Weekly Magazine* in its issue of 20 March, 1773 could therefore inform its readers with pleasure that 'thanks to the operations of Mr. Golborne, three coasting vessels arrived lately at the Broomielaw *directly* from Ireland with oatmeal, without stopping at Greenock as formerly to unload their cargoes'.

The effectiveness of Golborne's system of improvement was without question and the magistrates were soon pressing him to achieve the contracted 7 foot depth throughout the river. Dumbuck Ford was still a serious stumbling block to a deeper Clyde. The damming of the southern channel had not proved to be as effective as was first thought, and despite the attempts to loosen the material with the ploughs, the shoal remained stubbornly in place. Improvements upstream of Dumbuck would be quite wasteful unless the depth over the shoal could also be increased. The problem at Dumbuck was the division of the main stream into two channels. While the construction of a jetty at Longhaugh Point had been of some benefit, great quantities of water continued to spill across the shoal to flow in the southern channel. A significant increase was required in the 3 foot depth which then existed at the Dumbuck Shoal. Not only would large vessels then be able to sail up to the deep water at Dunglass and Bowling but a larger volume of water would be able to pass in and out of the upper estuary in each tide, 'facilitating the operations to bring 7 feet of water to the Broomielaw'.

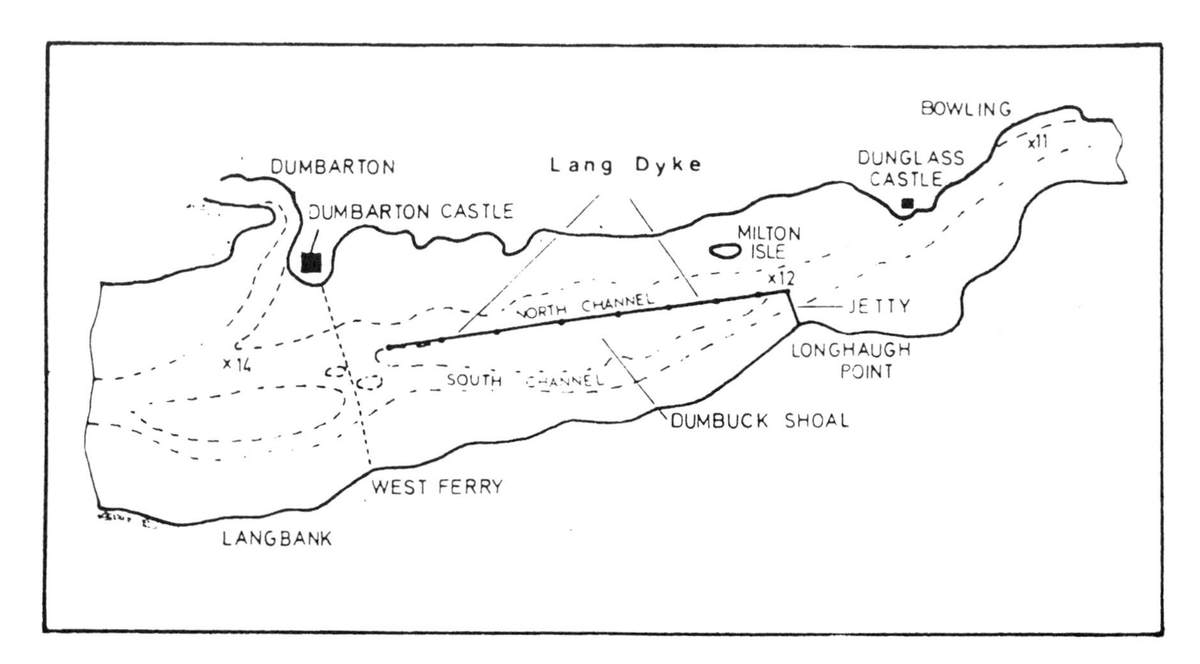

Figure 9.4 The Lang Dyke. A lasting memorial to John Golborne

To achieve this deeper water, Golborne put forward a proposal to construct a longitudinal training wall through the Dumbuck shoal. This wall would extend from the lateral jetty at Longhaugh Point downstream to the beacon marking the lower end of Dumbuck Ford and would provide a solid barrier between the north and south channels. [. . .]

Work on constructing the dyke commenced in June 1773 and such was the speed with which progress was made that all building was completed before the onset of winter. In this short space of time almost 800 yards of half-tide training wall was erected along the south edge of the northern channel to contain the current. Considering that all the stone had to be floated out on barges and then placed underwater, the construction of the training wall – so aptly named the Lang Dyke – was a major achievement. Although added to in later years, the wall remains noticeable to all who sail on the Clyde even now and is in fact the only one of Golborne's dykes to survive the later improvement of the river.

Elsewhere on the channel, the task of training and deepening continued, with some jetties being added to and with some having sections removed as their effect became apparent. The jetty system – still used on rivers to this day – was very much one of trial and error. No hydraulic models were to be built for another hundred years, nor were the principles on which the system depended fully understood. Erosion took place very quickly at some places, with doubling and trebling of the depth being obtained almost overnight. In other parts no movement occurred at all until the dredging ploughs had been brought in to loosen up the bed. In a number of reaches the distance between pairs of jetties was too great and new shoals formed as the material scoured upstream settled out in the slower moving waters. Thus further jetties had to be put in to ensure that the current was always sufficient to achieve and maintain the minimum depths contracted for. As some of these lateral jetties were up to 500 feet long, a vast quantity of stone, earth and timber was used in their construction. It says much for the engineering ability of both James and John Golborne that the work was completed with such speed and economy. [. . .] On 8 December, 1775 a sounding of the river from the lower end of Dumbuck Ford to the Broomielaw was undertaken [. . .] The results of this survey showed that between the stated bounds there was 'more than 7 feet of water at an ordinary neap (high) tide'. [. . .]

10

GLASGOW AND THE CLYDE

John F. Riddell

Source: Peter Reed (ed.), *Glasgow: The Forming of the City*, Edinburgh, Edinburgh University Press, 1993, pp. 41–8, 50, 52–3

With the Union of the parliaments in 1707, tobacco quickly became the main trade of the expanding town of Glasgow. [. . .] This trade in tobacco was of immense importance to the development of Glasgow. The income earned, the trade routes established not only to America but also to Europe where much of the processed leaf was sold, and the setting-up of small businesses to supply the manufactured goods shipped out as return cargo to the States, led to greatly increased prosperity. Those engaged in the trade, the 'tobacco lords', became men of power and influence. They quickly dominated the town's economic and political affairs [. . .]

The tobacco trade was critically dependent upon good transport links. A key part of these links was the River Clyde. It was to the harbour opened at Port Glasgow in 1676 that most of the leaf imported from the States was landed. Situated at the edge of the naturally deep water, Port Glasgow provided a safe harbour for Glasgow's growing fleet of transatlantic sailing ships. The twenty miles of river extending upstream from Port Glasgow to Glasgow's bridge across the Clyde were choked with shoals and shifting sandbanks. Prone to floods and the vagaries of wind and tide, the waterway was not only impassable to the ocean-going vessels but difficult for even the smallest craft to navigate with safety.

With the Clyde blocked by shoals, trade to and from Port Glasgow was dependent upon the pack-horse and waggon. The delays and losses caused by this method of transport were a source of constant difficulty for the tobacco merchants. Yet lying alongside the rough roads was the river. If only this could be just slightly deepened, goods could surely be moved by water transport. As the eighteenth century advanced, the need to improve the link between Glasgow and Port Glasgow became vital to the continuing success of the tobacco trade. With both profits and power, the tobacco lords utilized every resource to secure the required deepening.

Initially, the deepening pressed for was not intended to allow the transatlantic trading vessels access to Glasgow. Rather, it was to improve the river for barges and smaller vessels which could transship cargoes at Port Glasgow and move both tobacco leaf and manufactured goods more quickly, more reliably and more profitably than road transport [. . .]. With the canal age then well under way, it was not surprising that Glasgow first turned to the early canal engineers for advice about improving its river. [. . .] Surveys were undertaken, plans prepared, parliamentary approval obtained and contracts let. Work started in 1760 and, despite setbacks, the channel, or 'navigation' as it became known from its canal-like appearance, was steadily deepened. By 1772, a depth of five feet existed along the whole river at high tide, this increasing to over ten feet by the early years of the nineteenth century. [. . .]

The method of deepening [. . .] created opportunities which played an important role in the economic and physical forming of the city.

By narrowing the channel and placing the excavated material behind the retaining walls, the boundary between the river and the land was for the first time defined and fixed. The intertidal marshes and sandbanks which had existed for so many years had formed a barrier. Subject to winter inundation, the wetlands could be used for no purpose other than hunting and fishing. Fear of flooding kept development well back from the river's variable edge.

Infilling of these flood plains and marshes allowed people and buildings to move nearer the river. Vast new areas of raised, level ground were created by the narrowing and deepening. Estates such as Mavisbank [. . .] along the south shore, and Meadowside [. . .] on the north, had their riparian lands improved and made both useful and valuable. As the availability of suitable land is a major constraint on urban development, Glasgow was indeed fortunate that its river banks were so quickly identified, controlled and made secure from flood.

The shipbuilders took the lead in realizing the opportunities presented by the ready availability of good, level ground along the river. With the advent of steam, marine engineering - the manufacture of engines and boilers - started in Glasgow in the early nineteenth century. The East End, and especially the Gallowgate area, saw the first workshops. The hulls were then still built of wood in the traditional shipbuilding yards of Dumbarton and Greenock. As the Clyde became deeper, and with the coming of iron, both shipbuilding and engineering quickly took advantage of Glasgow's new riverbanks.

John Barclay was the first to commence business, opening a repair slip and later shipyard at Stobcross on the north bank in 1818 [. . .] Three years later, David Napier moved his engine works to a site at Lancefield, about half a mile to the west of the Broomielaw. [. . .]

As the nineteenth century advanced, many more shipyards and engine shops followed. The new sites, unconstrained by steep banks such as bedevilled the yards on the Tyne and Wear, allowed ample space for launchways, workshops and storage. The two Thomson brothers, James and George, laid out their new shipyard at Clyde Villa on the south bank between Glasgow and Govan in 1851. David Tod and John MacGregor moved their yard from Springfield on the south side to the reclaimed ground at Meadowside in 1845. Charles Randolph and John Elder moved from Govan to the flat fields of the Fairfield estate in 1863. Many other yards set up along both sides of the river, concentrating at the most suitable sites for launching the ships [. . .].

The establishment of the shipyards on the new greenfield sites at first meant that the yards were islands of industry in an otherwise rural scene. The shipyard workers travelled daily from the city, first by train or ferry and then by tram and bus. New communities soon arose around the yards. Most of the companies built houses for managers and foremen, and often substantial tenements for their workers. New labour was attracted from the Highlands and from Ireland. Shops, recreation facilities and public parks followed. New burghs formed as the yards' communities merged, until the green fields along both banks of the Clyde had almost totally yielded to industry. Just like its river, Glasgow flowed steadily westwards, carried along by the tide of shipbuilding activity.

The engineers and craftsmen of the upper Clyde from the beginning specialized in the construction of iron screw-propelled steamships. From an output of just over 25,000 tons in 1851, production rose within twenty years to nearly 200,000 tons. The introduction of steel shipbuilding in 1878 took the annual output to some 420,000 tons after just five years of the new material. Between 25,000 and 30,000 men were employed directly in the ten major shipyards in the Glasgow to Renfrew area in 1878, while the number engaged in supporting activities was five times as great. Workshops, providing everything from anchors to the finest furniture, expanded with the yards. Along with the main engine- and boiler-makers, however, these businesses had no need of direct access to the river. Thus, while shipbuilding took housing westwards, the supporting trades were to be found in every part of the city. Even great boilers and engines could be moved by road with relative ease along the level streets which ran parallel to the river.

Shipbuilding made much use of the flat land created along the banks of the Clyde, but the main development was for docks and quays. The deepening and improvement of the Clyde Navigation brought the ships to Glasgow – once there, their cargoes had to be loaded and unloaded easily and quickly. Sufficient and suitable accommodation for the shipping of the time was thus a matter of constant discussion. The facilities provided in the end were considerable, and perhaps became more widely recognized as physical features of Glasgow, both in plan and in elevation, than any other development.

The siting of Glasgow at a relatively narrow point on the River Clyde quite naturally resulted in the first accommodation for shipping being provided by quays erected on and parallel to the river banks. This form of construction was typical of riverine ports as opposed to the box-like harbour basin and breakwater which were commonly built on coastal sites. Unlike the latter, lineal quayage was simple and inexpensive to construct.

It is likely that some form of basic landing stage existed at Glasgow as far back as when the town was formed, although records suggest that it was around 1660 that the first substantial structure was built. This was sited on the north bank in an area known as the Broomielaw, just downstream of where the New (Glasgow) Bridge was built in 1768–72. Part of the Clyde's grassy flood plain, the Broomielaw was used for grazing, recreation and the washing and drying of clothes before its quite logical choice for a quay.

The quay at the Broomielaw was enlarged and strengthened in 1726. The new quay was built from the best available stone, supported by oak and iron beams, and was reputed at the time to be one of the finest to be found in Britain. It was further enlarged in 1792 and, by the turn of the century, was often a scene of intense activity. A crane, weighing machine and shed were erected and improvements made to the road access.

The steady deepening of the river allowed more and larger vessels to reach Glasgow. As the accommodation became constrained, the ships had to berth three and four abreast, with long waits to get alongside to discharge and load. A further extension of the Broomielaw Quay downstream was proposed. This would take the western edge of the shipping berths nearly half a mile distant from Glasgow Bridge. [. . .] [But instead of taking the] shipping accommodation further away, why not [. . .] construct a tidal basin or wet dock off the main channel within the city? Thus began a long debate, with the proponents of further lineal quays emphasizing the economy and ease of construction, and the dock supporters urging that convenience to the users should be the first consideration.

A number of schemes were put forward for docks over the period 1806 to 1809. [. . .] Glasgow was then still very much a town on the north bank of the Clyde. With merchants and traders wanting the ships to be brought as close as possible to their factories and warehouses, dock sites near to the Broomielaw were seen as the best way to provide the required berths. [. . .]

Glasgow Town Council published details of all the proposed schemes and invited comments from interested parties. The debate went on for some time, but in the end it was the economics of construction that decided the issue. Off-river basins were expensive both in land acquisition and in construction; building up the riverbank was cheaper and faster. There then followed a half-century which saw Glasgow Harbour expand steadily westwards by the construction of lineal quays. [. . .]

The arrival of the steamboats put increased pressure on the quay space. [. . .] A plan (Figure 10.1) showing the vessels berthed on 16 March 1840 gives a precise idea of the congestion of the Harbour at this time. [. . .]

Coal shipments, a growing trade for the Clyde, were properly provided for with the opening in 1849 of the south-side General Terminus Quay. This quay was equipped with three coal-loading cranes and was connected via the tracks of the General Terminus & Glasgow Harbour Railway Company to the lines of the main railway. [. . .] The completion of the south-side Mavisbank Quay in 1858 brought to an end the first half-century of quay development. In this period, the available accommodation for ships was increased from 1,200 feet to 13,000 feet, with the harbour stretching westwards from the city centre for over a mile. This was a major achievement, but also one which placed Glasgow in the unique position among major British ports of having all its shipping accommodated in riverside quays rather than docks or tidal basins.

By 1864, some 1.5 million tons of shipping and over a million tons of cargo were being handled at the riverside quays. Except for General Terminus, these quays were equipped with single-storey goods sheds backing in turn onto the main road system. [. . .] The space occupied by the harbour facilities was thus confined to a narrow strip of riverbank with ships loading and discharging on one side and goods arriving and departing by cart and waggon on the other. Storage space was limited. With the roads being used by waiting carts as well as by through traffic, congestion was at times extreme. Goods continued to move predominately east/west, converging on the cross-river bridges.

Some manufacturers did realize the benefits of moving west with the new quays, but in general it was larger premises like timber yards, mills and warehouses that took up occupation of the land beyond the roads. The smaller workshops and factories expanded only slowly to the west, with the most popular sites being in the Anderston and Finnieston areas. New housing was kept well back from the river, and was not obvious to the river-user.

As trade continued to expand and the new berths filled to capacity, the demand for still more accommodation was unabated [. . .] a more realistic solution was again to consider the provision of off-river basins [. . .] the decision was taken to develop the south-side site.

Work started in 1864, and the Windmillcroft Basin, to become more generally known as Kingston Dock, was formally opened three years later. A rectangular basin lying parallel to the Clyde, Kingston Dock was 1,100 feet long and 210 feet wide. The relatively small tidal range on the Clyde – about ten feet at spring tides – enabled the basin to be left open to the river. Neither gates nor a lock were provided, giving the advantage of twenty-four-hour access for shipping.

The quays of Kingston Dock were completed with goods sheds, but, in common with all the riverside berths, no cranes were provided. Ships were expected to use their own gear for loading and discharging. A prominent feature of the [. . .] dock was the hydraulically-operated swing bridge constructed across the entrance to maintain continuity of the road behind the riverside quays. Filled to capacity, Kingston Dock could accommodate some twenty-two 500-ton ships.

Kingston Dock was no sooner brought into use than the Clyde Trust was planning the construction of the dock across the river at

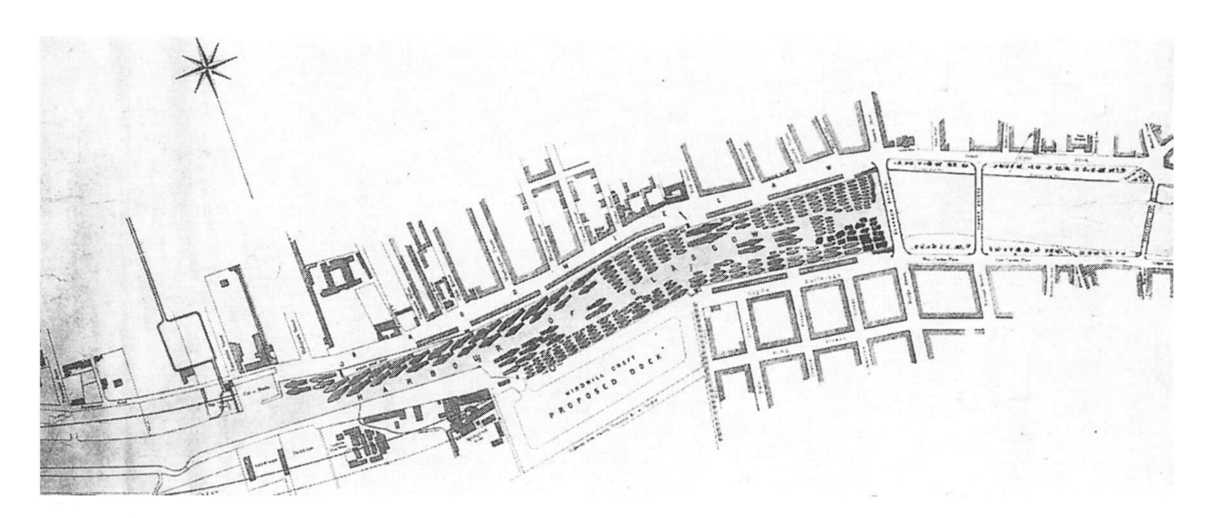

Figure 10.1 Glasgow Harbour on 16 March 1840. The river channel has not yet been widened; note the proposed dock at 'Windmill Croft'
Source: Strathclyde Regional Archives

Figure 10.2 Annan's photograph (c. 1853), one of the first and best known, of Glasgow Harbour. Notice the river steamers at the Broomielaw, in the foreground, and the still predominant sailing ships berthed across the river
Source: Glasgow Museum of Transport

Stobcross. [. . .] This was intended as a much larger project than the south-side basin. It was designed by two eminent engineers, John Bateman and James Deas, the latter being Engineer of the Clyde Trust from 1869 to 1899. [. . .] The Pointhouse Road had [. . .] to be diverted. [. . .]

The construction of the new road involved the building of a massive retaining wall, in parts forty feet high, together with an inclined ramp to give access to the dock from the Dumbarton railway. The wall was completed in 1875.

Excavation of the dock started in 1872, and the first part was formally opened on 18 September 1877 [. . .], the new basin was named Queen's Dock. [. . .]

[It] provided a total of 10,000 lineal feet of quay space with a water depth of twenty feet. At the time of its opening, it was the largest dock in Scotland and the fifth-largest in the United Kingdom. It was fully equipped with goods sheds, and along the north quay a number of powerful cranes were provided for loading coal.

The cranes were powered by high-pressure water, as was the swing bridge sited at the dock entrance. This was a common form of power before the advent of electricity. The high-pressure water was supplied from a hydraulic power station [. . .] the need for still more shipping accommodation soon arose. The site selected for Glasgow's third dock was Cessnock, on the south side to the east of Govan [. . .] the [. . .] proposals produced a storm of opposition. The main objectors were the residents of Govan who realized that the new dock, with its need to divert the Govan Road, would greatly increase their journey time to Glasgow. The arguments continued for some years, until a compromise was reached with the Clyde Trust agreeing to meet all the costs of the new road, to compensate the local tramway operator and to construct a new river steamer pier at Govan.

The dock at Cessnock comprised three parallel basins opening off a canting basin at the west end [. . .] it was opened in stages between 1893 and 1897.

[. . .] the Clyde was a veritable highway [. . .]. All manner of vessels used its waters. Pride of the Clyde were the transatlantic steamers [. . .]. These great passenger ships maintained regular services to the United States and Canada as well as carrying many emigrants. Cargo steamers of all sizes brought in raw materials for the industries of Glasgow and Central Scotland and food for the growing population. They returned downriver with every variety of goods from the mundane coal and pig iron to the most complicated machinery and manufactures, bound for every corner of the world. [. . .]

The impact of the Clyde on the City of Glasgow took many forms. Its role as a highway to the world and as a birthplace for the great ships allowed Glasgow to become the 'Second City of the Empire'. The Clyde brought wealth and prosperity to a wide area. It provided employment and opportunity. The merchants and traders, shipowners and shipbuilders, became household names, men of power and importance [. . .]. Many small businesses flourished, assisting the ships and shipbuilders, providing every variety of service and product. These businesses clustered around the docks and quays, and occupied hundreds of offices throughout the City.

The most widespread effect was in direct employment. Building the ships, handling the cargo, manning the ferries, dredgers and tugs and going to sea was hard work. Being laid off at the first hint of recession, and often dismissed after a launch or completion, and working long hours in the open, in the cold, in the rain and snow, was not pleasant. But it was work. And for the many tens of thousands who earned their living from the Clyde, there was the chance to develop a skill, bring up a family and have a pride in what you did. [. . .]

Nowhere was that success more evident

Figure 10.3 The Clyde Port Authority (formerly Clyde Navigation Trust) headquarters, 1882–1908
Source: Strathclyde Regional Archives

than in the head office building of the Clyde Navigation Trust itself (Figure 10.3). Situated at the corner of Robertson Street and the Broomielaw, this magnificient building [. .] is [. . .] rivalled only by Glasgow's City Chambers [. . .]

Viewed from the river and the Clyde Navigation to the west, the Clyde Trust's offices were a fitting monument to the importance of the river to the City. [. . .]

11

LETTER TO GLASGOW TOWN COUNCIL: PROPOSAL FOR GLASGOW'S WATER-SUPPLY FROM LOCH KATRINE

W. J. Macquorn Rankine and J. Thompson

Source: Rankine and Thompson to Glasgow Town Council, 1 March 1852, Glasgow, in *To Commemorate the Public Services of the late Robert Stewart,* Glasgow, 1868, a privately circulated volume now in the Mitchell Library, Glasgow

For many years it has been generally admitted, that some abundant source of pure water ought to be found, to supersede the Clyde in the supply of Glasgow, and that the mountainous districts which exist in various directions round the City, appear to offer the means, not only of doing this, but of substituting the natural descent of water by gravitation for the costly operation of pumping: a change which would not only effect a considerable annual saving, but would greatly facilitate the extension, throughout the whole of the City, of the system of constant service, now recognized to be essential to economy, comfort, and health. [. . .] Considering that six years have elapsed since the Glasgow Water Company obtained an Act for bringing in water on the principle of gravitation and constant service, and yet that nothing whatsoever has been done by them towards the execution either of that, or of any other project, for improving the water supply of Glasgow [. . .]

[. . .] it is absolutely necessary that it should be undertaken by some new body. [. . .] The proposed scheme would be enabled, partly by the reduction of working expenses, and partly by the facility of providing with little or no increase of outlay, large quantities of water for sale, by bulk, for manufactures and enginepower, to realize a profit out of a much lower water-rent than that levied by the existing Company.

We beg leave, however, to suggest, that the benefit to the public would be still greater, if the management of the supply of water to Glasgow were undertaken by your Corporation [. . .]

This system is well known to have been carried out in Manchester with perfect success, on a scale, at least, as large as that now required for Glasgow [. . .]

We therefore beg leave to lay before you the results of extensive investigations, in which we have for some time been engaged, in order to ascertain the source which best combines the advantages of abundance and purity of the water, facility and security in construction of the works, and economy in their execution, maintenance, and management.

With this object, we have carefully examined those sources which appeared to us the most eligible. Our attention has been directed principally to the ranges of hills to the north of the Clyde, for the following reasons.

First. A great elevation can be obtained at a moderate distance from Glasgow.

Secondly. The prevailing rocks being either igneous, or of the primary formation, the water is exceeding soft, and free from mineral impregnation.

Thirdly. The extent of moss, and of cultivated land is so small, that the quantity of organic matter in the water is imperceptible; and

Fourthly. The entire absence of mines removes all apprehension of danger to the works from subsidence.

Beginning with the ranges nearest to Glasgow, we examined all the available streams of the Campsie and Kilpatrick hills. Here, although the quality of the water is excellent, considerable expense would have to be incurred, and engineering difficulties encountered, in the formation of Store Reservoirs.

We next examined the district at the south-west end of the Grampian chain, containing the sources of the Forth and the Teith, and abounding in large lakes of extreme purity.

The advantages in economy, safety, and durability, of a natural over an artificial store reservoir are so great, that in all cases where a lake can be found, of sufficient purity and abundance, and of a suitable site and elevation, it ought to be preferred to any other source of supply.

Two natural reservoirs of this kind are, from their elevation, available to Glasgow: Loch Lubnaig and Loch Katrine. We have carefully examined both of these lakes, the mountains and streams which supply them, the rivers which they discharge, and the routes by which their waters may be conducted to Glasgow. In every point of comparison, the preference must be given to Loch Katrine, for the following reasons:-

First. Its elevation is higher than that of Loch Lubnaig, although its distance from Glasgow is not greater; consequently, a conduit of less size and cost will suffice to convey a given quantity of water from Loch Katrine than from Loch Lubnaig; and from a conduit of the same size a greater quantity of water will be discharged.

Secondly. Another advantage arising from the higher elevation of Loch Katrine is, the greater facility and cheapness with which the conduit can be carried across the ridges of high ground which lie between Loch Katrine and Glasgow.

In the plans of the Glasgow Water Works Extension, deposited by the Glasgow Water Company in 1845, the comparatively low level of Loch Lubnaig, and the consequent necessity of a moderate slope for the aqueduct, have been so little attended to, that the proposed Town Reservoir is marked at a site so high that it would be impossible to fill it; the aqueduct, as shown on the section, being carried at Craigmaddie (about seven miles from Glasgow) down to a tank lower than the proposed water-level of the reservoir.

Thirdly. Owing to the conformation of the country, the works along the proposed aqueduct will be much lighter on the line from Loch Katrine than on that from Loch Lubnaig, some works on the latter line being so heavy as to amount, practically, almost to a bar to the undertaking. The aqueduct from Loch Katrine has to cross the valleys of the Endrick and the Kelvin only: that from Loch Lubnaig has to cross the valleys, not only of the Kelvin and the Endrick, but of the Forth and the Teith also; the latter two valleys being comparatively of great magnitude, and having a formidable ridge of mountains between them, which would have to be traversed by a tunnel two miles long, through rock of the toughest description. The necessarily low level of the Loch Lubnaig aqueduct is the occasion of another tunnel near Milngavie, of three miles in length, and through whinstone; which tunnel could not be completed in less than five years.

The only tunnel of considerable length required on the Loch Katrine is one of about a mile long.

Fourthly. Advantage can be taken of the form of the natural outlet of Loch Katrine to maintain the water at its winter level, for the purpose of storage, by operations of a simple and inexpensive character; while to produce the same result at Loch Lubnaig would be very expensive and difficult.

Having for these reasons selected Loch Katrine as a natural reservoir for the supply of Glasgow, we shall now describe, generally, the method by which we contemplate to make it available.

As to purity and softness, the water is unparalelled. It is supplied by streams rising in a district of mica slate, almost entirely uncultivated and uninhabited, and contains only *two grains* of solid matter in a gallon.

The hardness, according to Dr. Clarke's scale, is less than one degree, thus excelling the celebrated water of Aberdeen.

It is at all times so clear that filtration might be dispensed with, even after the greatest floods, as we have ascertained by observation.

The extent of the district which drains into Loch Katrine is above 30 square miles, about one sixth of which area is occupied by the lake itself. By the operation already referred to, of storing water up to the winter level of the lake, the present ordinary summer flow in the River Teith might be maintained, for the benefit of the landed proprietors and mill-owners, and a surplus of 40,000,000 gallons per day afforded, the whole or any requisite part of which might be brought into Glasgow.

The aqueduct by which Glasgow might be supplied would be about 36 miles in length; and would consist of a conduit excavated or built for about 24 miles, and of iron pipes for the remaining 12 miles. We propose that the conduit should be made, from the first, large enough to discharge 40,000,000 gallons per day, because very little reduction of expense would arise from making it smaller. Along the 12 miles where pipes are requisite, it will be sufficient, at first, to lay a single line of pipe 4 feet in diameter. The inclination being 5 feet per mile, this pipe will be capable of discharging 20,000,000 gallons per day, or one half of the total available quantity; that is to say, of supplying a population of 600,000, or double the present number, with 25 gallons per individual per day, for domestic and sanitary purposes; amounting, in all, to 15,000,000 gallons per day, with a surplus of 5,000,000 gallons to be sold to large consumers.

This aqueduct should communicate with a Town Reservoir, on one of the small hills immediately to the north of Glasgow, capable of containing one day's supply, and at an elevation of about 250 feet above high water mark, that being sufficient for the extinguishing of fires in any part of Glasgow, without the aid of engines.

The cost of these works, including contingencies, we estimate at £260,000. This is the capital requisite to bring twenty millions of gallons of water per day from Loch Katrine to Glasgow. The cost of laying distributing pipes suitable to the present extent of the city, including contingencies, would be about £200,000.

The capital required will then stand thus:

Aqueduct and Town Reservoirs,	£260,000
Distributing Pipes,	200,000
Total,	£460,000
The interest of this sum, if raised by debentures at 5 per cent., is	£23,000
We estimate the annual Working Expenses, Repairs, and Depreciation, at	6,000
Making in all, to be raised annually	£29,000

The 300,000 inhabitants who now occupy the district north of the Clyde, will consume for domestic and sanitary purposes only 7,500,000 gallons per day. There will therefore remain a surplus (in the present state of the city) of 12,500,000 gallons per day, applicable to manufacturing purposes and to the driving of machinery by water power. For the present, we shall suppose that only *one fifth* of this surplus is sold, or 2,500,000 gallons per day, at the very low charge of £2 10s. 0d. per annum for each thousand gallons per day consumed; or 1d. and two-thirds per thousand gallons.

The want of a supply of water for sanitary purposes, including street-washing, public baths, and the gratuitous supply of public hospitals and paupers, has been much felt in Scotland, and it is believed that scarcely any class of the inhabitants would object to a small rate being levied for this important object. Now, taking the total rental of heritable property in the district upon which this rate might be levied at

£1,000,000, *One Penny per Pound* would produce	£4,000
The sale of water to large consumers, would produce	6,250
They would, therefore, have to be raised by domestic water-rent,	18,750
To make up the requisite income of	£29,000

The amount on which domestic water-rent can be levied, may be estimated at £600,000. [. . .] When the whole of the inhabitants of the district take their supply from the new source, the water-rent may be reduced to [$3\frac{3}{4}$ per cent., or] Ninepence in the Pound.

Supposing the city north of the Clyde to increase to double its present population, or 600,000, the consumption of water for sanitary and domestic purposes would increase to 15,000,000 gallons per day; and the sale of water by bulk to large consumers, may be also estimated at double its original amount, or 5,000,000 gallons per day; thus making up the whole quantity of 20,000,000 gallons per day, which a single pipe of four feet in diameter is capable of delivering with the proposed fall. [. . .]

Supposing the rateable domestic rental, being doubled like the population, to amount to £1,200,000, the above sum would be raised by a water rent of about *two per cent.*

This example shews the progressive reduction of water rent with the increase of population, which would take place under the system proposed by us, and the contrast it bears, in that respect, to the pumping system of the existing Water Company.

On a still larger quantity of water being eventually required, whether for domestic use, or for manufactures and engine power, it can be introduced at a moderate cost, by laying additional lines of pipe for the 12 miles already mentioned, whereby 40,000,000 gallons might be made available.

We may here observe, that the very low price of iron pipes renders the present a peculiarly favourable time for the execution of Water Works.

It thus appears, that the Corporation of Glasgow might provide an ample supply of pure water, sufficient not only for the present inhabitants, but for future generations, at a price much less than that which is now paid for the existing very deficient and impure supply. It may be observed that the funds of the Corporation can incur no risk from the undertaking; for it is proposed that the Lord Provost, Magistrates, and Council should, by themselves, or through a Committee to be named by them, as in the case of the Police, act merely as Parliamentary Trustees for the execution and management of the works. The necessary funds, as already stated, would be raised by debenture, or by loan, on the security of the rates, as was done in the case of the Improvements on the Navigation of the Clyde. [. . .]

As a means of giving a constant, plentiful, and cheap supply of pure soft water for domestic, sanitary, and manufacturing purposes, and at the same time of affording a surplus, capable of being applied to the production of an economical moving power, so as to diminish the number of Steam Engines within the City, and thereby to lessen the impurity of the atmosphere of Glasgow, we respectfully submit this scheme to your consideration.

12

PRODUCTIVE METROPOLIS: LONDON'S INDUSTRIAL CONTRIBUTION

Roy Porter

Source: Roy Porter, *London: a social history*, Penguin, 1994, pp. 140–2, 187–8, 191–8

Thames-side trade fuelled London's industries, not least shipbuilding itself, which required large yards with auxiliary trades - rope-, sail- and mast-makers and repair facilities. [. . .]

Prominent among East End industries were distilling, sugar-refining and brewing. Big distillers, such as Booth's, Gordon's and Nicholson's, settled in Clerkenwell to take advantage of the high-quality well-water. [. . .] Further north in Spitalfields, Joseph Truman established himself around 1680 in Brick Lane, and began building up the Black Eagle, one of the great Georgian breweries. Other major concerns sprang up. [. . .] In 1786 Whitbread's alone produced 150,000 barrels of strong beer [. . .].

London was proud of its quality trades: its cutlery was preferred to Sheffield's; its clocks and watches were renowned. Silk-weaving thrived at Spitalfields under Huguenot management. Though vulnerable to the whirligig of fashion, the industry expanded behind protective tariffs against imported silks, employing at its peak some 12,000 weavers. By 1750 there were 500 master weavers around Spital Square, about 15,000 looms at work, and 50,000 locals dependent on the trade. The industry soon faced severe difficulties, however. Trade down-swings created unemployment, leading to riots. [. . .]

Disturbances continued. In 1769 fourteen ringleaders were apprehended, two being hanged outside a Bethnal Green tavern. The Spitalfields Act of 1772 attempted to rationalize the industry by curtailing recruitment of apprentices and forbidding weavers' unions, but problems continued, and the less skilled branches of the industry collapsed and moved to Macclesfield and other provincial centres where labour was cheap.

From the 1740s London had porcelain factories at Bow (Stratford High Street) and Chelsea, inspired by the Chinese porcelain the East India Company was shipping from the East. In Whitechapel stood the long-established bell foundry, which was sending church bells as far afield as Cologne and Carolina. (It later made 'Big Ben'.) Another superior trade was cabinet-making, boosted by the introduction of mahogany. [. . .] Thomas Chippendale opened his workshops in Long Acre about 1745, later moving to St Martin's Lane. Hepplewhite set up shop in Cripplegate, and Thomas Sheraton arrived in 1790. And there were scores of other skilful cabinet-makers, many around Covent Garden but others in the City [. . .]. 'He employs four hundred apprentices on any work connected with the making of household furniture, joiners, carvers, gilders, mirror-workers, upholsterers, girdlers - who mould the bronze into graceful patterns - and locksmiths,' remarked Sophie von La Roche in

1786, visiting the workshops of the Aldersgate cabinet-maker Thomas Seddon. [. . .] And scores of other crafts bloomed – a *Directory* of 1792 named 492. Weaving might be in decline, but the metropolis was Britain's most diversified craft centre. There were expert furnishers, house-painters and decorators, basket-makers, box-makers, watchmakers, printers and engravers and dozens more. Many precision crafts had connections with shipping: maps, globes, telescopes, barometers, scientific and surgical instruments, clocks, spectacles, plate and jewellery – all were made in London.

Well over 5,000 people were involved in watchmaking alone. The trade was minutely subdivided; the everyday trade located itself in Clerkenwell and St Luke's without Cripplegate, but the top watchmakers were to be found in the main shopping streets – Cornhill, Cheapside and the Strand. There were also London makers who supplied the provinces; Bayley's of Red Lion Street were making around 4,000 watches a year, employing over 100 workmen.

London also had its share of dirty trades. Scattered through the East End and in the poorer Surrey suburbs were bone-boilers, grease-makers, glue-makers, paint-makers and the dyeworks of Southwark ('an odious stinking business'). Trades such as tallow-making consumed the offal of Smithfield, poisoning the atmosphere. Industries formed zones. Sugar-houses crystallized in Ratcliff and St Katharine's, and starch-works in East Smithfield, Whitechapel and Poplar; Ratcliff specialized in glass and gunpowder. The metal trades, engravers and printers converged in Clerkenwell, lime-burning kilns were established in Bethnal Green and Mile End, and brickmaking developed around Brick Lane. The lower reaches of the River Lea were famous for milling. [. . .] The Lea formed an important transport artery, Bromley's mills grinding corn barged down from Hertfordshire. [. . .] Flemish dyeworks and copperas works were set up in Poplar. Across the river in Rotherhithe and Deptford there were ropewalks, coopers' and boat-builders' yards, and oil, colour and soap works.

London trades were labour-intensive. Stepney and Whitechapel housed wharfside workers, dockers and porters. [. . .]

In time, some threats arose to London's manufacturing supremacy. Other trades as well as silk migrated to provincial centres where labour was cheaper. [. . .]

Framework knitting shifted to Nottinghamshire [. . .]. A geographical division of labour was arising: simple processes would be performed in the provinces, half-completed goods then being carted to London for finishing and marketing. [. . .]

[. . .] London avoided the overdependence on any particular industry – like cotton in Lancashire or coal in South Wales – that made prosperity precarious elsewhere.

Yet challenges were looming. Rivals emerged with the Industrial Revolution, which has been called 'a storm that passed over London and broke elsewhere'. In Lancashire, the West Riding and the Midlands, a new economic order was being forged, based upon fossil fuels, iron and steel, textile factories, and heavy industries like engineering and shipbuilding. The new age, it has often been said, belonged to the factory not the workshop, to steam not handicrafts. 'The capital cities,' wrote Fernand Braudel, 'would be present at the forthcoming industrial revolution, but in the role of spectators. Not London but Manchester, Birmingham, Leeds, Glasgow and innumerable small proletarian towns launched the new era.'[1]

This is too crude, however. For one thing, Industrial Revolution or not, the capital hardly lagged in technology and innovation. Late eighteenth-century London had more steam engines than Lancashire – they were used not in textile mills but in flour-mills and for pumping London's water supply. As late as 1850 London's manufacturing output was still unrivalled in Britain, for the obvious reason that the capital's vast population created unsurpassed demand. And the nation's industrial economy was profoundly dependent upon the capital's imports, transport and communications, wholesale and retail networks, finance skills and its service sector more generally. Historians have sometimes written about London's reliance upon 'service' employment as if that meant the capital were somehow less 'productive' than other regions – parasitic in fact. But this is a false assumption. [. . .] Leading historians today are reiterating the prime role of international trade in Britain's nineteenth-century economic miracle: the Port and the City were no less crucial to this than the factories of Lancashire or the coal pits of South Wales.

It is nevertheless true that industries planted elsewhere, on coal and iron deposits, in due course poached manufacturing and engineering business away from London. High costs were becoming a drawback; business began to leave London because outlays were lower elsewhere and labour cheaper. London experienced a slow but ominous decline as a manufacturing centre. Thames shipbuilding, for instance, was unable in the long run to compete with Tyneside and Clydeside. [. . .]

Clouds were thus gathering on the horizon. Yet in terms of production and employment Victorian London thrived, retaining prime place in all major sectors except metals and textiles. A seventh of the total population of England and Wales was living in London in 1871; a sixth of all the manufacturing workers lived there. One out of every three London workers was in manufacturing. According to the 1861 Census, London employed 25,000 in clothing and dressmaking, 95,000 in food-processing, 92,000 as builders, 59,000 in shopkeeping, 35,000 metalworkers, 25,000 furniture-makers, and so forth. The Census revealed large numbers in skilled, specialized trades – 13,000 machine- and tool-makers, 6,000 carriage-makers, and 5,000 musical-instrument-makers [. . .] In all there were 469,000 workers engaged in manufacturing industry – 15 per cent of all those so employed in the whole of England and Wales. [. . .]

Around the docks and the boatyards, many activities developed on the waterfront, making use of coal brought by coaster. [. . .] Meanwhile the railways were trimming costs to undercut the sea trade. Cheap coal and the North Woolwich railway encouraged the siting of industries and utilities in the Silvertown and Canning Town areas.

Among them were gasworks. In 1812, the Gas-Light and Coke Company had become the first gas company chartered to light the City, Westminster and Southwark. Rival companies followed. [. . .] For cheapness, London's eighteen gasworks were mainly sited on the river, or on the Regent's and Grand Surrey Canals. In 1867, the Gas-Light and Coke Company bought a site at East Ham by the Thames, allowing coal to be landed straight from barges. [. . .]

Ready coal, coke and gas supplies stimulated subsidiary industries dependent upon coal by-products. In 1856 Burt, Boulton & Haywood, tar distillers, set up in Silvertown, manufacturing naphtha, pitch, creosote, and disinfectants: they led the world in gas distilling. With the advent of aniline dyes, Burt's bought up Perkin's aniline-dye factory and transferred it to Silvertown; the British Alizarine Works was set up nearby; and in 1879 the Gas-Light and Coke Company built an ammonium-sulphate plant, making nitrogenous fertilizer. Numerous sulphuric-acid works sprouted in West Ham.

Riverside areas thus teemed with distilling, boiling, refining and chemical industries: the Isle of Dogs was 'covered with steam factories'. Chemicals, soap, confectionery, rubber, dye, engineering, rope-making, printing, tin-canister- and sack-making were all prominent. The river, the docks, the North Woolwich railway line and the availability of cheap land together ensured prosperity.

Canals also made their contribution to

London's economy. In 1801 the Grand Junction Canal was extended from Uxbridge to Paddington, thereby providing a direct link between London and the Midlands. [. . .] From 1820 the Regent's Canal joined with the Grand Junction Canal at Paddington, going eastward [. . .] to join the Thames at Limehouse, where another basin was constructed. Much used for coal and timber shipment, it linked all of North London to the national canal network and brought trade to Hackney, Hoxton (where Pickford's planted their depot), St Pancras and Camden Town.

By far the world's biggest food importer and consumer, the metropolis generated huge processing industries to feed its millions. Supplying its daily bread, corn-mills hugged the Thames, Wandle and Lea. Wandsworth's mills were the largest, using both steam and water power, but John Rennie's Albion Mills at Blackfriars were the world's first flour-mills to use steam power (till burnt down in 1791) – a mark of the metropolis's insatiable demand for bread. Sugar-refining grew up in Whitechapel. [. . .] Distilling and brewing were vast business, dominated by a dozen great firms, of which Barclay, Perkins' brewery in Southwark was the largest. [. . .]

In mid-Victorian times London had six major fields of industrial employment – the building trades, clothing and footwear, wood and furniture, metals and engineering, printing and stationery, and precision manufacture (watches, scientific instruments, surgical apparatus, etc.). The accent had long been on high-grade wares. Clerkenwell, [. . .] specialized in tinplate, barometers, thermometers, engraving and printing machinery; its makers of clocks, watches and scientific instruments were renowned. [. . .] Across the river, Bermondsey and Southwark formed the centre of the leather trade, [. . .] especially around Tanner, Morocco and Leathermarket Streets. [. . .] Silk remained significant. In 1851 there were some 5,500 silk-weavers in Bethnal Green and Whitechapel. [. . .]

Cabinet-makers clustered on an axis running north-east from St Paul's, along Cheapside towards Finsbury Square and Shoreditch, and also in the West End. Many high-class establishments employed between 80 and 100 artisans. And a ready-made trade was emerging alongside, aided by steam-powered sawmilling, the opening of the Regent's Canal, and the establishment of new timber yards convenient for Shoreditch and Finsbury. [. . .] heavy engineering and the metallurgical industries continued to expand. [. . .] The biggest works was owned by John Rennie, a Scot who learned his trade with Boulton and Watt in Birmingham before setting up in Holland Street near Blackfriars Road – Southwark, Waterloo and London Bridges were all his work. After his death, his younger son, John, carried on the trade, while his elder son, George, specialized in marine engines for the Admiralty – in 1840, he built the *Dwarf*, the first naval vessel with screw propulsion.

Among London's greatest engineers was Joseph Bramah. A Yorkshireman who trained as a cabinet-maker, Bramah developed an improved water-closet. [. . .] He is primarily remembered, however, for precision locks. His protégé Henry Maudslay began at Woolwich Arsenal. For eight years they worked together, making machine tools and Bramah's patent hydraulic press, which provided engineers with a steady continuous pressure of practically unlimited power. James Nasmyth and Joseph Whitworth also helped ensure London's engineering supremacy. In the 1840s Maudslay's and Rennie's were employing about 1,000 and 400 men respectively.

Down-river, another great engineer was long engaged upon a prodigious feat of engineering. The Rotherhithe-to-Wapping Thames tunnel was built by Marc Isambard Brunel and his son, Isambard Kingdom Brunel, the mastermind of the Great Western Railway. The elder

Brunel ran saw-mills at Battersea and experimented with steamboats. John Penn made marine engines at Greenwich, while Stevens's of Southwark pioneered railway signal equipment. [. . .]

Decentralization became more common after 1850, costs being lower. Dye, glue and chemical works and jute, soap, match and rubber manufactures moved away from Bethnal Green and Whitechapel towards Bow, Old Ford and Hackney Wick. [. . .] By 1850 West Ham, beyond the Lea, was becoming the home of soap, rubber, chemical, bonemeal, paint, glue and tarpaulin manufacture. Soon it could boast a major sugar-refinery, manure factory, creosote factory, vitriol factory, lampblack factory, varnish factory, tar factory, and the Chemical Light Company. West Ham became London's late Victorian industrial success story: it was ideally situated; its water-frontage cut transport costs; it was well provided with docks and railways; and, beyond the supervision of the Metropolitan Board of Works and later the London County Council, its vestry was lax in the enforcement of building, factory and smoke regulations. [. . .]

In 1850 inner London's industries had been second to none [. . .] But thereafter problems multiplied. Space was comparatively costly in the capital. In the 1870s an acre in the City cost about twenty times as much as its suburban equivalent. [. . .] From the 1890s relocations occurred with increasing frequency. In 1895 the printers Ward Lock moved out of Fleet Street to Stoke Newington. Stratford, like West Ham a growth suburb, prospered largely thanks to the railway. In 1847 the Eastern County Railway had moved its locomotive and carriage works there, creating a spaghetti junction of lines and sidings. Stratford's works turned out a carriage a day and a locomotive a week. By 1900 they were employing nearly 7,000 people.

Yet this was the exception. London's industries were typically small. According to the 1851 Census, only seven firms employed over 350 men, and out of 24,323 employers only eighty employed over 100. There were 3,182 masters who employed just one man. Small workshops predominated. Petty capitalism had its advantages, notably affordable start-up costs. [. . .] With technological change, and especially mass production, London's manufactures faced challenges. Under stiffer competition, some collapsed. The leather industry underwent decline, and by the 1880s the Bermondsey tan-yards were depressed. The collapse of traditional heavy engineering was slow but sure after 1850, one firm following another to the industrial north. Innovations, however, still happened in the capital. It was in a laboratory in St Pancras that Henry Bessemer devised his revolutionary process for making steel, and in a little Clerkenwell workshop Hiram Maxim constructed his first machine-gun.

Note

1 F. Braudel, *Capitalism and Material Life 1400–1800*, transl. M. Kochan, Glasgow, Fontana/Collins, 1974, p. 440.

13

THE LONDON MILK TRADE, 1860–1900

E. H. Whetham

Source: *Economic History Review*, 17 (1964-5), pp 369-80

The history of the liquid milk trade in the fifty years before the First World War is an interesting example of the effect of transport costs upon the structure of an industry handling a bulky and highly perishable commodity. The trade grew in these years both in importance and in complexity, as a result, firstly, of the growing demand from urban consumers for liquid milk, and secondly, of the steadily increasing size of London. As bricks and paving stones covered more of the grazing grounds of Middlesex and Surrey, more milk had to be transported over longer distances; problems of deterioration in transit became worse. [. . .]

A further complication in this period was the growing public concern over the quality of the milk supplied. To discourage adulteration was the object of the first legislation dealing with milk; as 'the germ theory' of disease became generally accepted, public health authorities sought, with varying degrees of enthusiasm, to enforce minimum standards of cleanliness and hygiene in cowsheds, dairies and milk-shops.

Transport before 1860

[. . .] In the 1840s and 1850s, milk had been retailed round the streets of London by men and women carrying, on a shoulder yoke, a pair of wooden or tin tubs holding 8 or 10 gallons in all. The milk might have come from an urban back-street; it might have been brought also by yoke from a suburban farm; or the owner of a retail business might send a cart to collect milk from cows grazing at Hendon, Highgate, Dulwich or Norwood. One such concern gathered its milkers in Fleet Street at 1 a.m., brought them and the milk back to the shop there about 6 o'clock for the morning round with yokes, while the second cart started at 10 for the afternoon milk.

The urban dairies, with cows fed in stalls [. . .] were judged to provide the best milk, for their cows were better fed and often better housed than many cows kept out on the suburban grazings. Many customers bought the milk given to children and invalids direct from these cowkeepers, a Select Committee was told in 1874.

A few well-known businesses had established fashionable shops where children could be sent to drink milk, or servants sent to buy milk, from selected shorthorn cows kept either in stalls behind the shop, or on suburban grazings. One such business had been established on the site of the present Albert Hall [. . .] These firms were narrowly balanced between the cost of transporting milk from the increasingly distant fields, and the cost of carrying the greater bulk, firstly of the fodder and litter required by stalled cows, and secondly of the manure they produced which had to be removed from the towns. As long as there was a sufficient number of wealthy or particular customers prepared to pay for 'milk warm from the cow', it was profitable to keep stalled cows which, when decently fed, gave both milk and a saleable carcase for the butcher at the end of their lactation. The poorer parts of the town were supplied with an adulterated product from suburban farms or with the milk taken from cows awaiting sale or slaughter at Smithfield.

The effect of railways

The reduction in transport costs brought about by railways had a complex effect on the better-class trade. In the first place, it became cheaper to move the bulky fodder and manure which was involved in urban dairying. [. . .] Secondly, the suburban farms lost their partial monopoly of the supply both of hay and grazing and also of milk, as supplies could be brought by railway from longer distances. But as long as 'railway milk' began its journey frequently mixed with water, uncooled and contaminated with bacteria, its condition by the time it reached the housewives left the better-class trade in the hands of the urban dairies. Nevertheless, most of the milk bought in towns was used, not for liquid consumption, but for cooking, and the demand for 'railway milk' seems to have steadily increased with the extension of the railway service and the fall in prices.

The development of this trade brought considerable changes firstly in equipment and secondly in organization. Open wooden tubs were clearly unsuitable; the wooden barrels used for beer and other liquids quickly soured. Iron containers were found to be the most durable and most easily scoured. [. . .] These iron hand-made churns were expensive, hard wear made them liable to leak and they quickly became rusty. By 1880, the Dairy Supply Company was offering a churn made from double-tinned steel plate, with a lid designed to keep out dust and rain, as a cheaper and more durable product.

The retail trade

By the mid-1860s the volume of milk supplied by the larger firms in central London had outgrown the capacity of the milk-women with their yokes. Retail rounds were conducted either by a horse-drawn 'float', or from low wheeled carts, pulled or pushed by the roundsman and his boy. In poorer districts, milk was usually ladled by a measured dipper from a wide-mouthed churn into whatever containers the housewives provided, but the better class of trader delivered milk in covered cans [. . .]

This costly equipment for the good-class milk round was provided to supply a commodity sold at 2*d.* or 3*d.* or 4*d.* per quart [. . .]

The wholesale trade

As the centre of London became further removed from grass and fields, the widening gap between retailers and dairy farmers was filled by the new trade of milk wholesaling. Its importance developed greatly in 1865/6, the years of the 'cattle plague' (a combination of rinderpest, pleuro-pneumonia and foot-and-mouth disease) which inflicted heavy losses on the urban dairies. London would have been seriously short of milk in these years if it had not been for the 'railway milk' hurriedly bought from farmers in the home counties and railed to the London stations. [. . .] It was calculated that some 7 million gallons of milk had been brought into London by rail in the course of 1866 from 220 country stations, in order to replace the supplies lost from the urban dairies affected by the cattle plague; by 1880, officials of the Local Government Board estimated that railways carried 20 million gallons of milk annually into London.

Even before 1865 the larger retailers were selling more milk than could be supplied by cows either in their town stalls or out at grass in the suburbs. They necessarily began to buy the output of farmers conveniently placed near railway stations in the home counties [. . .]

The [. . .] sale of liquid milk involved [. . .] a major decision on many farms. The basic price offered per gallon had to be offset firstly against the railway charges to the London terminal. Then there was the delivery, once or twice daily, of the milk from farm to station by horse and cart in charge of a man who was also required to help load the churns on to the train and to bring back the empty churns. There was also a value to be put on the tedious work involved in churning or cheese-making, work usually undertaken by the women of the farm families, but sometimes by hired dairymaids. What was left of the basic price after these charges had then to be balanced against the receipts from butter and cheese and of the use of their by-products. And finally, there was the question of how much winter fodder could be provided to sustain a flow of milk for the winter half-year [. . .]

Cooling and cooling depots

The increase in the distance over which milk travelled into London made acute the problem of souring. [. . .] The decline in the urban preference for 'milk warm from the cow' and the growing transport of milk by railway led many farmers to cool their milk before dispatch, for fear that the buyer might reject it, unpaid, on the London platform. Cooling was usually performed by standing the full churns in water, but about 1870 Lawrence's cooler became widely adopted. This device consisted of a zigzag of metal tubes through which water circulated and over which milk was poured. But many farms had an inadequate water supply for either of these methods; a few used deep wells or had piped local springs, but most farms depended on surface streams and shallow wells which were often contaminated and not much cooler than the atmosphere.

This necessity for cooling before dispatch by rail quickly created another stage in the organization of the London milk trade. Wholesalers began, in the late 1870s, to set up depots at country railway stations where milk delivered by farmers could be properly cooled. The sites chosen had therefore to combine easy access to a dairying region, proximity to a mainline station, and ample water supplies, often from a specially sunk well. By 1879 the North Wiltshire Dairy was opening new buildings at Stratton, Swindon, with its London milk cooled there and not on farms [. . .] in 1882 the Semley Dairy was advertising its twice daily dispatch of cooled milk from 3,000 cows in Dorset for the London market.

Some of these depots, such as that operating at Semley, seem to have been organized by country dairies in touch with the London trade, but others were financed and controlled from London, as wholesalers and the larger retailing firms sought to improve the quality of their supplies. The cost of these depots, with their double handling of milk, was partly offset by a lower rate of wastage, but two other economics were also feasible. Supplies could be better adjusted to the urban demand, for the new telegraph enabled the London firms to send daily instructions on their needs. [. . .] Secondly, the bulking of churns into van loads all directed to one consignee enabled the cooling depots to secure lower rates from the railways than were obtained by farmers each sending two or three churns daily to different salesmen. [. . .]

The dairy companies

The setting up of cooling depots was both a cause and a result of the growing size of firms in the London milk trade. Such buildings required capital; their economies could only be secured by a large and regular flow of milk; there had to be a corresponding increase in the London trade. [. . .] By 1870 there were 8 companies mentioned by the Post Office

Directory in the London dairy trade, among them the Express Country Milk Company controlled by the Barham family who had long operated a dairy at 272 The Strand. [. . .] By [1880], the Amalgamated Dairies Company was also describing itself as a wholesaler, operating from its depot in Derbyshire and the terminals of King's Cross and Finsbury. [. . .] By 1890 companies had become common in the London dairy trade. Thirteen of those listed in the Directory were described as wholesalers, and there were 15 manufacturers of dairy machinery and utensils. [. . .]

Changes in techniques

This development of cooling depots and companies coincided with three technical innovations. The centrifugal separator, first exhibited at the Kilburn Show of the Royal Agricultural Society in 1879, converted the manufacture of butter from a farm craft into a factory trade. Instead of standing milk for 24 hours for the cream to rise, milk could be run through the separator to give a continuous supply of fresh cream; steam-powered churns and mechanical butter-workers were then available to finish the job. With such equipment, dairies and cooling depots could economically convert surplus milk into butter; because the separators worked more efficiently than the human hand. [. . .] The second innovation was the continuous refrigeration plant, which enabled dairy produce and meat to be brought into Britain from America and Australasia as well as from the countries of northern Europe. Imports of dairy produce increased rapidly from the 1870s onwards and prices fell notably after 1883. The farm-house manufacture of dairy products thus became less profitable, and more farmers sought a market in the liquid milk trade. These rising imports and falling prices also discouraged the few co-operative factories recently established by farmers and landowners to make butter and cheese. Some of those set up in Derbyshire and Leicester turned over to the cooling of milk for the urban trade. [. . .]

The third technical innovation was the condensing plant, which, in conjunction with tinplate, provided a form of milk which was indefinitely storable, cheap to transport and quickly became a cheaper alternative to the fresh milk. Three condensing plants were in operation in England by the summer of 1881, of which the one at Aylesbury was reported to be taking 4,000 gallons daily in the flush season from 82 farms. By the turn of the century, butter-making and duck-rearing on the skim milk, the two famous specialities of the Vale of Aylesbury, were almost extinct in that area, and farmers were selling whole milk either to the condensery or to the London wholesalers. The process was also adapted within a few years to produce a variety of other products, such as baby foods and calf-rearing mixtures, which might contain either whole or skim milk. [. . .]

Hygiene and the law

The adulteration of milk first became a legal offence under the Adulteration of Food or Drink Act, 1860, but few local authorities exercised the powers offered them, until compelled to do so by the Adulteration of Food, Drink and Drugs Act, 1872. Chemical analysts were then called upon to discriminate between pure and adulterated milk [. . .] the difficulties of proving where the adulteration had taken place increased with the widening area from which milk was drawn into the London market. [. . .]

The prevention of adulteration was only one step in securing better milk, since bacterial contamination was general before the 'germ theory' of disease was accepted. Cowsheds, dairy plant and milk-shops came under medical inspection when the Public Health Act, the Sale of Food and Drugs Act (both passed in 1875) and regulations issued in 1885 under the Contagious Diseases (Animals) Act, 1878 required elementary cleanliness in those selling food [. . .]

While the cooling of milk was the generally accepted method of prolonging its commercial life, all housewives knew that the souring of milk could also be delayed by boiling it; this process, however, usually separated out the cream in the form of skin and changed the flavour and consistency. Nevertheless, heating milk was developed as an alternative method to cooling it. In the 1890s, a few dairymen began to offer 'sterilized milk' as a speciality for children. This process involved filling glass bottles which were then secured with wired porcelain stoppers (similar to those used for ginger beer bottles) and placed in water gradually raised to boiling point and kept at that temperature for a considerable period. Alternatively, some milk was 'pasteurized' in bulk and then bottled for distribution on the retail rounds. Because rail transport was not economic for bottles, both these processes had to be performed at the urban depots, and therefore involved a considerable centralization of the milk, with a consequent increase in the distance travelled from the depots to the ultimate purchasers. [. . .]

But bottled or specially treated milk provided a small fraction of the trade before 1900; only the medical profession and a few educated families were aware of the gross bacterial infection of milk which contributed to such widespread diseases as infant diarrhoea and tuberculosis. Yet this growing interest in hygiene encouraged consumers to be suspicious of 'cheap' milk and to favour the large firms with cooling depots in country districts, with steam-powered plant for washing churns and delivery cans.

By the last decade of the nineteenth century, the London milk trade seems to have acquired a settled pattern. The suburbs round the metropolis were served partly by wholesalers from the centre but mainly by producer-retailers from the outer ring. The centre of London drew its milk from outside this zone, either direct from farms 30–50 miles away or from the cooling depots established along the railways 50 miles or more from London. [. . .]

Over these forty years, there had been a steady expansion in the demand for liquid milk for the London market, and in the area from which it was drawn. But that area closely followed the pattern of the railway lines along which the milk travelled daily. [. . .]

14

LONDON'S MILK SUPPLY, 1850–1900: A REINTERPRETATION

David Taylor

Source: *Agricultural History* (University of California, Davis), 45 (1971), pp. 33-7

The ever-increasing expansion of London had necessitated, since the late sixteenth century, a large and regular supply of foodstuffs to be brought from the provinces into the capital. One exception to this trend was milk. Unlike other commodities it could not be readily transported over long distances. Consequently, large numbers of cows were kept within the metropolis and its immediate environs. By the mid-nineteenth century contemporaries estimated the number of cows in and around London supplying the inhabitants with milk to be in the order of 24,000. [. . .] At the one extreme were the holdings, like that of Mr. Rhodes of Islington, with an average of 400 cows in milk; this was exceptional. More common was the cowkeeper with 50–80 beasts and a small farm near town. The most common type of holding, however, had no more than 10–12 cows; moreover, it was probably to be found in a dingy backyard in a poor quarter of town. [. . .] In 1865 J. C. Morton calculated that there were some 12,000 cows in the metropolis, scattered throughout the area in parlors ranging from large to small, and from model cleanliness to appalling squalor. [. . .]

For later observers, the outbreak of rinderpest in 1865 has become a turning point in the development of English dairying, marking the change from an internal to an external supply of milk for London, which in turn stimulated change, particularly in the southwest of England. [. . .] This interpretation has been followed by later writers, notwithstanding the unresolved problems and contradictions it contains. [. . .]

[. . .] E. H. Whetham talks of the increase in railborne milk after the outbreak of cattle plague in 1865, but offers no explanation of the fact that she herself mentions - that even before 1865 demand in London was greater than the available supply from urban parlors and suburban farmers.

Any attempt to calculate the total yield from the cows supplying London is fraught with danger: figures for average yields are few and of limited application. Among the highly fed cattle of the London cowhouses an average yield of 750 gallons per annum would probably err on the generous side; for the remaining 12,000 cows in the surrounding districts the figure is unlikely to have been more than 500 gallons per annum. The total gallonage would, therefore, be approximately 15 million gallons in any year. At the same time, the population of London was about 3 million, while per capita consumption of milk, according to Morton, was a mere one-fifth of a pint per day, or 9 gallons a year. Thus total demand at 27 million gallons a year was far in excess of internal supply. In other words, some 36,500 cows would have been needed to meet the demand of the capital. Even if one makes allowance for the watering down of milk supplies (a figure of 20 percent was frequently quoted by contemporary observers) the number of cows required would still be about 30,000. Therefore, unless these figures for yields and consumption are wildly inaccurate, indeed almost inverted (which is unlikely in view of the consensus of both contemporaries and later observers on these points), external supplies of railborne milk must already have been quite considerable - in the order of 5 to 10 million gallons per annum.

The general view, however, has minimized the importance of railborne milk in supplying the inhabitants of London during the years before the large-scale outbreak of rinderpest in 1865-66. By concentrating on the last decades of the century - a period, incidentally, for which there are more readily available data - a false impression of the changing source of London's milk supply is given. [. . .]

In the first place, the dramatic view of an overwhelming upsurge in railborne milk is unconvincing. If railborne milk was so limited in quantity in the early 1860s it is unlikely that the huge expansion implicit in this view could have taken place. The logistical problems of bringing in something like ten times as much milk, albeit for a short time only, would have been so great that one doubts that an increase of such magnitude really took place.

Moreover, even before 1865 demand exceeded internal supply. It seems reasonable to argue that milk was brought into London by rail, though from a limited area, within thirty miles of the capital, *before* the effects of the 1865 rinderpest were felt. Scattered evidence exists to support this view. [. . .] Evidence comes from the reports on the state of agriculture in the counties near the capital, published in the *Journal of the Royal [Agricultural] Society of England.* For example, H. Evershed, in 1853, noted that 'a new trade has been opened in this district [Surrey] since the completion of the South-Western Railway, from the convenience thus afforded of sending milk and vegetables to London . . . Several dairies of 20 to 30 cows are kept, and the milk is sent to the various stations of the South-Western Railway, and conveyed to the Waterloo terminus for the supply of the London Market.'[1] Even when one looks farther afield than London the picture of pre-plague developments remains clear. In 1849 G. Beesley observed how 'large quantities of milk are forwarded daily by the railways, to Liverpool, Manchester, and Bolton - in some instances a distance of forty miles, but ordinarily fifteen to twenty.' Similarly, dairying in Warwickshire revived as a result of the growing demand for liquid milk from Birmingham.

Other examples can be culled from all parts of the country. Wherever there was a large and growing center of population, the surrounding district would almost invariably contain specialist milk-producing areas. As the towns grew in number and size the number of cows housed therein became increasingly inadequate, all the more so as the amount of land available for them often decreased, [. . .] However, one cannot dismiss the cattle plague of 1865-1866 as unimportant. On the contrary, it was an extremely important stimulus, but to trends that were already under way. The extent of the outbreak must not be forgotten. By 30 December 1865 outbreaks of the disease were reported on 9,954 farms in Great Britain. In all, some 76,000 cows were attacked, of whom 14,000 were killed, 40,000 died, and only 7,000 recovered (the rest are unaccounted for). By March 1867, of 12,460 cattle in the metropolitan area, 7,474 had been attacked by the disease and only 315 had recovered, with a further 602 unaccounted for. The need to find alternative supplies led to demands being made on areas as far away as Macclesfield, almost 200 miles from London. At the same time new sources of supply were sought in Derbyshire, Staffordshire, Nottinghamshire, Wiltshire, Gloucestershire, and Somerset. The amount of

Table 14.1 Milk carried by rail, 1865–7

Great Western Railway		*London and North-Western Railway*	
	gallons		gallons
January 1865	8,954	1865	490,320
January 1866	143,600	1866	1,209,284
May 1866	285,918	1867	648,000
June 1866	221,851		
January 1867	143,588		

milk carried by the Great Western Railway increased markedly in a short time. Fussell, in *The English Dairy Farmer*, quotes the [figures in Table 14.1].

However, without figures for years before 1865 and for some years after the outbreak of the cattle plague it is not possible to quantify the undoubted expansion which took place. The increase in railborne milk to London in the following decades seems equally dramatic. By any estimate the figure more than doubled, for by 1891 some 40 million gallons of milk, or more, were being carried by rail each year to the capital. Per capita consumption rose slightly, but a drop in the number of town-housed cows, plus an increase of 1.25 million in the population of London was sufficient to account for this rise.

Against this background, the events of 1865–1866 take on a new perspective. [. . .] The plague, by wiping out half the cows in London and by causing a widespread revulsion against the squalor of many cowhouses, sounded their death knell. At the same time it gave a tremendous boost to the country suppliers, especially those in the more distant parts of England. The very fact that it was possible to increase the supply to London so extensively and so rapidly suggests that an embryonic supply and transportation structure existed before the emergency conditions of 1865–1866. Yet, if one must not minimize the extent of railborne milk in the pre-plague years, equally one must not suppose that the boost received in those years solved all the problems facing the dairy farmer. Even where the railway line was accessible there was still the multiple problem of transport to be overcome. Milk souring was a serious obstacle to the expansion of production for the urban market. To echo Sheldon: 'Some people think that they are indebted to the railways for the supply of country milk. This is a mistake. The milk must still have been produced in the immediate neighbourhood of London had not the method of reducing the temperature of the milk by means of refrigeration been discovered.'[2] [. . .]

Notes

1 H. Evershed, 'The Farming of Surrey', *Journal of the Royal Agricultural Society of England,* 14 (1853), pp. 402–3.

2 J. P. Sheldon, *The Farm and the Dairy*, London, 1889, p. 128.

15

THE GROWTH OF LONDON'S RAILWAY MILK TRADE, c. 1845–1914

P. J. Atkins

Source: *Journal of Transport History*, 4 (1978), pp. 208–13, 215–19, 222–4

As a city of over two million inhabitants, mid-nineteenth-century London had to cast its net far and wide to procure the quantity, quality and variety of food it required. Local self-sufficiency had long since withered in response to the expansion and sheer scale of the built-up area, the easing of communication problems with other parts of the country, and the development of efficient supply organization. One foodstuff, however, still came largely from the immediate environs of the city: sufficient liquid milk to meet the requirements of London was produced within a radius of eight or nine miles of the Royal Exchange, and in the early 1850s there were probably over 20,000 cows kept specifically for this supply.

One might have thought that a potentially promising new form of transport such as the railway would have been exploited with enthusiasm and alacrity by traders in London as a means of importing milk from areas better suited to its production than the cramped and costly urban cowsheds. In fact, the volume of 'railway milk' consumed in the capital grew only slowly for the first 20 or 30 years of its potential availability, and one aim of this paper is to explain the nature of the trade in these early decades. A second aim is to examine the very rapid growth of importation by rail from the 1870s [. . .].

A key factor throughout the period was the highly perishable nature of the milk itself. This was the main reason why milk had been produced in and around the city since time immemorial [. . .] In hot weather milk was often sour and therefore unsaleable within a few hours of leaving the cowshed, and this remained a restriction on the location of its production until a way could be found either to reduce the deleterious effect of heat and therefore inhibit the souring process, or to provide a very rapid and suitable means of transport to the place of consumption.

It seems that milk was first brought by rail to the capital in 1845 from the Romford and Brentwood districts of Essex. [. . .]

The traffic expanded in the late 1840s and 1850s at a time when London's urban cowkeepers were finding it difficult to keep pace with the requirements of a rapidly growing population, but it was confined to distances of under 20–25 miles [. . .] By the mid-1850s over 5 per cent of London's total milk supply was being transported by rail, and this proportion continued to increase gradually [. . .]

In 1865–6 a serious outbreak of rinderpest in London's cowsheds created shortages, and a great deal of accommodation milk was therefore imported by rail. But in 1867, as the emergency receded and demand for this extraordinary supply declined, it seemed that events might bear out a prediction made by J. C. Clutterbuck at the height of the epidemic that 'the carriage by railway from distant dairies will be superseded, and the original condition of the metropolitan milk trade re-established'.[1] This recession proved to be short-lived, for in the following year imports increased once more [. . .] By 1870, 'railway milk' was London's principal single source of supply [. . .]

When the Great Western Railway [. . .] took its first regular consignment of milk from Maidenhead to London in 1860, the project was ridiculed. [. . .] several important technical, logistical and psychological factors did make it difficult to visualize a trade sufficiently well organized to ensure the rapid and regular transportation of fresh milk from distant farms to the urban dweller's breakfast table.

Among the factors which made the milk trade 'peculiarly resistant to the railways'[2] was a lack of railway vehicles suitable for conveying milk over distances of more than a few miles. The goods wagons and parcel vans in general use for this sort of traffic were poorly ventilated, difficult to clean, and badly sprung. Moreover, the milk itself had the inconvenient characteristic of being both bulky and perishable; indeed, for most railway companies it was a minor item in their turnover, yielding little profit and therefore not justifying the expense of building special wagons. [. . .] The introduction of specially designed wagons was deferred until the 1880s and 1890s, and was restricted to lines where consignments were of sufficient scale to guarantee full loads of cans to London and empties back to the country. [. . .]

A second technical problem was the primitive design of early railway cans. Design competitions were organized in 1868 and 1879 to improve their durability and construction, but no fundamental changes were made to the traditional conical shape until the early twentieth century. This shape gave the can a low centre of gravity and made it comparatively easy to roll, but [. . .] it was very cumbersome to load into a railway van. These large vessels were rarely full [. . .] and the milk was therefore subject to considerable churning on a long journey, no doubt giving rise to the can's ironic misnomer of 'churn'.

No matter how perfect the container or means of transport, it would not have been possible to sustain a long-distance traffic without some means of prolonging the life of milk produced in the insanitary conditions of country sheds. The total absence of practical methods of preserving milk in the early years of the railway trade meant that in hot weather there was a high risk of souring on the shortest of journeys. But with the introduction of cooling apparatus in 1872 and chemical preservatives a few years later, it became possible for milk to be brought from further afield in even the most unsuitable weather.

The spread of the milk-cooler or 'refrigerator' was a slow process, but its effect upon the milk trade was important. According to one writer in 1889, 'the milk must still have been produced in the immediate neighbourhood of London, had not the method of reducing the temperature . . . by means of the refrigerator been discovered'.[3] One major difficulty was the initial cost of the apparatus, which many farmers were reluctant to incur, but a more important drawback in many areas was the absence of cold well or spring water, because the milk could only be cooled to the temperature of the water available. [. . .]

Chemicals had been added to milk from the earliest times in the hope of retarding the souring process, or simply to mask an adulteration. But it was not until the mid-1870s that patent preserving compounds were employed on a large scale in the dairy trade. Borax and boracic acid were popular, particularly for their cheapness, and they dominated the market until the 1890s when the more effective formalin was first used. By 1900 there were about 20 brands of chemical milk preservative on sale. About half the dairymen in London used them. [. . .]

The negative attitude of most railway companies to the milk traffic in its early stages may be illustrated in three ways. First [. . .] several companies remained reluctant to grant special facilities for the handling of milk at their stations. A major exception was the G.W.R., which became heavily committed to encouraging the milk trade. Drawing much of its milk from the hinterland of Swindon in Wiltshire, this company found it worthwhile in the 1880s to construct a special milk-halt nearby at Wootton Bassett; in addition, at the London end it spent a considerable amount on special milk platforms at Paddington in 1881, 1890. [. . .]

[. . .] The chaos which accompanied the arrival of milk trains in London's other stations, with crowds of wholesalers attempting to identify and claim the cans of their respective suppliers, was aggravated in the early decades of the country milk trade by the establishment of an *ad hoc* platform market, where small lots were decanted from can to can and sold to retailers and itinerant traders. This practice increased the risk of contamination and reduced the efficiency of the platform as a collection point.

A second disadvantage of the early railway milk traffic was the reluctance of the railway companies to make special provision for it in their timetables. Milk cans were forwarded in ordinary passenger trains which of course had to stop at many stations on their journey, and this did not enhance the chances of the milk reaching its destination unsoured. [. . .]

[. . .] Some milk specials were put on at periods of peak passenger traffic and were given priority over other trains in order to maintain average speeds of 30–40 m.p.h.; and on the G.W.R., the journey from Swindon Junction to London, which had at times taken over ten hours before 1865, was covered in less than two in 1902.

Third, the early extent of the 'milkshed' (the area from which milk was drawn) was partly limited by the great expense of sending milk further than 40 miles to London. In 1852, the freight rate of 1.75d. (0.78p) per imperial gallon levied on distances of greater than 40 miles must have been a discouragement to the dairying districts served by the G.W.R. [. . .]

We may conclude that in the early decades of the trade the absence of suitable arrangements for special trains, the lack of timetable adjustments, and terminal handling facilities, and an inappropriate freight rate structure, were all factors in the slow growth of London's railway milk traffic. [. . .]

The gradual removal of these [. . .] constraints upon the transport of perishable agricultural commodities by rail did not, of course, represent a sufficient condition for the growth of the railway traffic. There were a number of other factors, both economic and psychological, which influenced the decisions of those farmers, dealers and consumers who were potential participants in the trade. [. . .]

The financial circumstances and the attitude of the dairy farmer in the second half of the nineteenth century were important elements in his decision whether or not to produce liquid milk for the urban market opened up by the railways. Both were influenced by the medium-term fluctuations in prosperity epitomized by the so-called Great Agricultural Depression *c.* 1873–96. [. . .]

[. . .] The luxury of selling liquid milk was open only to those located within easy twice-daily reach of a railway station, but it did account for the rapid decline in the 1880s and 1890s of farmhouse cheese-making in the more accessible parts of Wiltshire, Derbyshire and Leicestershire. In particular, the clay soils of the Vale of White Horse in Berkshire, with their extension in Wiltshire from Swindon to Melksham, carried both pasture suited to grass-based dairying and a G.W.R. line with direct and rapid access to London. The area was therefore in the forefront of the movement toward liquid milk production for the London market from the 1860s. [. . .]

Liquid milk production was not inherently a more profitable enterprise than existing farming systems [. . .] but it did require less working capital. It was therefore attractive by

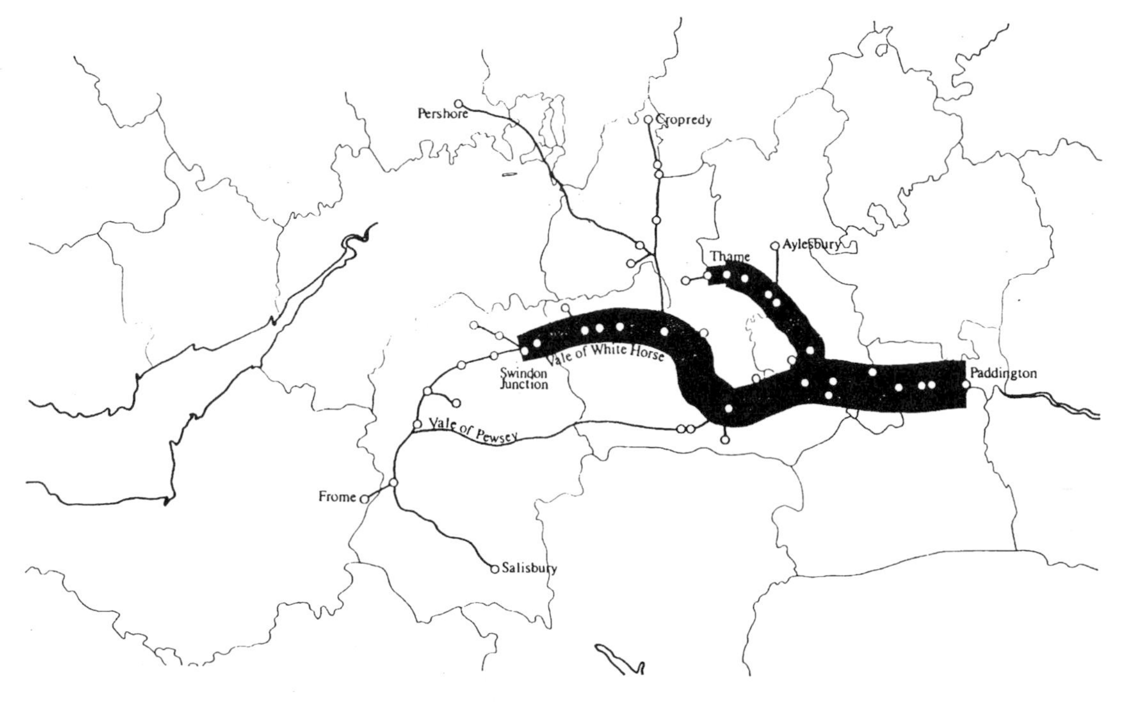

Figure 15.1 Cumulative flow map showing the milk sent by rail to London on the G.W.R. network in 1870

comparison with, for example, the sheep and corn husbandry of southern Wiltshire. [. . .]

These incentives were insufficient on their own to convince many farmers of the advantages of the milk trade over their traditional agricultural practice. In predominantly arable areas such a change would have required a psychological and technical leap of insurmountable proportions for land-owners and tenants alike. [. . .] First, the lack of suitable farm buildings for zero-grazing or winter production was not easy to overcome when even the landowner's capital was fully stretched [. . .] Second, there was a lack of the necessary skill and experience in a branch of farming which in some districts had been considered of a technically and socially inferior status. [. . .] In some arable areas farmers were ignorant of how to lay their land down to permanent grass. [. . .]

The condition, or apparent condition, of the milk delivered to households in London had a great bearing upon the consumers' willingness to repeat the purchase. Sourness was a complaint frequently heard about the pre-1865 country supply which came largely from Essex. This was partly because a scarcity of ground water in that county meant that the cooling of cans in a stream or pond was rarely possible. [. . .] Suspicious retail customers preferred to pay $\frac{1}{2}$–$\frac{3}{4}$ *d.* [. . .] more per quart for the urban-produced article and railway milk had the notoriety, which lasted until the mid-1870s, of being a poor man's beverage. [. . .]

Conclusion

The growth of country milk importation into London was a complex process. In this paper it has been possible to identify a number of stimuli to change; [. . .] unravelling the Gordian knot of cause and effect is exceptionally difficult in circumstances where the necessary and

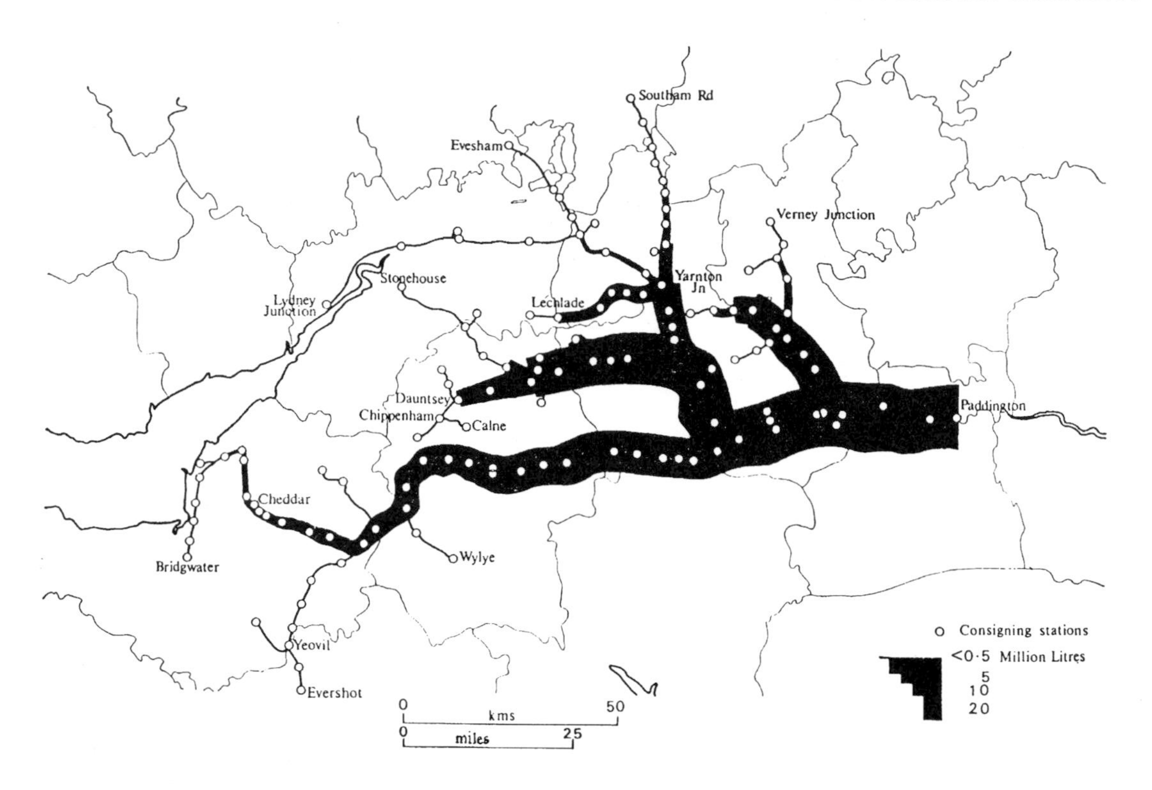

Figure 15.2 Cumulative flow map showing the milk sent by rail to London on the G.W.R. network in 1884

sufficient conditions of change, whether social, economic, or purely technological, cannot easily be distinguished. [. . .]

[. . .] It seems likely that the direction, nature and pace of change were moulded not only by the positive thrust of innovation and technological development, but also by the passive attitude of many farmers to the prevailing economic climate, and by a gradual erosion of economic, psychological and other barriers to change. [. . .] If rural suppliers had continued to find their dairy manufacturing or other agricultural activities more profitable than liquid milk production, it is conceivable that London's railway trade would have remained limited in volume and geographical extent well into the late nineteenth century. [. . .]

By 1914 the railways were providing London with upward of 96 per cent of her liquid milk supply. The 'milkshed' was extended from about 10 miles before 1840, to 20–25 miles in 1860, and to 200 miles by 1900, [. . .] The last foodstuff in which the capital city had been self-sufficient had therefore entered the national transportation system, although considerably later than might have been expected in view of the availability of railway freight movement since the late 1830s. [. . .]

Notes

1 J. C. Clutterbuck, 'On the Farming of Middlesex', *Journal of the Royal Agricultural Society of England* 5, 1869, p. 24.

2 H. J. Dyos and D. H. Aldcroft, *British Transport: an economic survey from the Seventeenth Century to the Twentieth*, Leicester University Press, 1969, p. 215.

3 G. Barham, 'The milk trade', in J. P. Sheldon, *The Farm and the Dairy*, London, 1889, p. 128.

16

MANUFACTURING IN THE METROPOLIS: THE DYNAMISM AND DYNAMICS OF PARISIAN INDUSTRY AT THE MID-NINETEENTH CENTURY

Barrie M. Ratcliffe

Source: *Journal of European Economic History*, 23 (1994), pp. 263–5, 267–70, 279, 281, 283, 285–6, 288–304, 318–22, 324–6

'We are a City of Philosophers,' Samuel Johnson wrote of London in 1776, 'we work with our Heads, and make the Boobies of Birmingham work for us with their hands'. Johnson has not been alone in underestimating the importance of industries in metropolitan economies. Since these industries apparently catered essentially to the local, and above all elite, market, historians have long treated manufacturing there less as a motor of growth than as a derivative of it. The kind of goods manufactured by many of the industries in nineteenth-century metropolises such as London or Paris is a further reason why, until very recently at least, industries there have not been treated as important. It may also be that scholars have been little inclined to examine such industries given their doubly discrete character: like hermit crabs they unobtrusively insert themselves in buildings and courtyards which previously served other purposes and, given the small size and high turnover of most enterprises, they leave but rare direct traces in archives. Their reluctance appeared all the more justified since the more heavily capitalized industries once found in city centres have always been subject to [centrifugal] forces which pull them to the periphery and, eventually, beyond city walls [. . .]

It is unlikely, though, that these are powerful enough reasons to explain why the historiography of metropolitan manufacturing is the poor relation of the rich work that continues to be done on nineteenth-century industrialization processes in general. It is unlikely because just a cursory glance at occupational censuses or even at some of the literature on large cities is enough to reveal some of the importance of metropolitan manufacturing in urban, regional and even national economies. Thus in the mid-nineteenth century half of New York's workforce was engaged in industrial work, as was just under half of London's. A closer examination reveals that a vast gamut of products was made [. . .]

[. . .] in every large city there were two types of industry [. . .] On the one hand, there were face-to-face industries, lightly capitalized, labour-intensive, producing goods generally small in bulk, little dependent on coal for fuel or power but subject to [centripetal] forces and thus highly clustered in city centres. On the other, large cities also sheltered more capital-intensive industries, especially engineering, shipbuilding, chemicals, and sugar refining, where firm size was often, [. . .] large and production processes more mechanized and were more subject to [centrifugal] forces, as high urban rents and surface transport costs pushed them from the core to the periphery with its open spaces, lower costs and transport nodes.

There must be other reasons, then, why students of the city have too often failed to accord manufacturing the attention it deserves. One of these is undoubtedly the kind of problematics that have been predominant in urban studies and which have led scholars to pay more attention to residential and neighbourhood patterns, and immigrant and urban cultures, to emphasize, in short, [. . .] lived space rather than production space in general, and industrial space in particular. Studies of nineteenth-century Paris, for instance, [. . .] have given much greater attention to demographic than to structural urbanization and, when the changing use of space has been examined, it has generally been through the residential and investment patterns of property-owners or the public works policies of the city rather than through the ways in which industries colonized urban space. [. . .]

The case of Paris in the first half of the nineteenth century illustrates these methodological and historiographical problems. The study of manufacturing there is hindered, first of all, by the difficulty of determining the spatial limits to the capital's industries. Manufacturing did not stop at city gates. Already from the 1820s onwards, an accelerating industrialization, that was linked to the needs of Paris, was taking place in the inner suburbs, especially those just to the north of city boundaries. At the same time, many Parisian manufacturers also employed outworkers in the provinces and, more importantly, manufacturing in the capital was part of a complex division of labour wherein Paris specialized in finishing processes that required the skills and flair which were one of the city's great strengths but upstream processes were usually carried out in the provinces. In many industries, then, earlier processes were carried out either in the Paris basin or in centres even further away. Our understanding is hampered, secondly, by current orthodoxy on the importance of industries in the capital at this time. Many of those who have discussed manufactures there have argued that the first half of the nineteenth century constitutes a hiatus, preceded by a period during the First Empire when 'modern' industry, such as cotton spinning, was implanted only to wither after the fall of Napoleon, and succeeded by the Second Empire when, with the completion of the rail network and massive public works in the city, industries such as ready-made clothes and shoes and engineering again began to expand, as they would do still more in the last decades of the century as the knowledge-centred Second Industrial Revolution reputedly called upon the kind of skills available in large cities such as Paris. There are two principal reasons why many historians believe that in the first half of the nineteenth century Parisian industries marked time. One is that they have adopted what we will argue are inappropriate criteria – firm size, mechanization, use of mineral fuels and production for mass markets – to determine the progress of industry and treated face-to-face industries as the «traditional» or at least secondary element in a so-called dual industrial economy. The clearest instance of the use of such measures is in the well-known study of the industrial geography of Paris published under the direction of Maurice Daumas and Jacques Payen in 1976. To carry out their detailed industrial census they adopted criterion for inclusion of a minimum of fifty workers before 1850 and of a hundred thereafter, thereby excluding altogether the great bulk of firms and industries in the capital. A further reason is that, influenced by the fact that in the nineteenth century «modern» industries grew up on or near coalfields or at transport hubs, scholars have emphasized physical resource endowment. [. . .]

We propose [. . .] to attempt three things: first, to determine the weight and dynamism of manufacturing in the French capital, but only, for evident reasons of simplifying our task, those industries within the city itself; second, to establish at least something of their importance in wider processes and, finally, to uncover the dynamics of change within the sector. [. . .]

We have used the industrial censuses that were carried out for the first time ever in France, or indeed elsewhere, in the 1840s, and then again in the 1860s, as one indicator of the importance of Parisian manufacturing in the spatial distribution of French industry [. . .] Second, we have used occupational censuses, available for the first time in the 1850s, as a further measure of the weight of manufacturing in Paris and the capital's importance in national structures. Most importantly, though, we have used the rich occupational censuses taken in London in 1851 and in Paris in 1856 to compare two metropolises which, by their size, stand out from other large cities in nineteenth-century Europe and which are comparable in functions and professional structures. [. . .]

[. . .] The first result of our endeavours is to show that Paris, with only three percent of France's population, was responsible for just under a quarter of total industrial production by value. [. . .]

[. . .] Parisian goods dominated most leading manufactures on France's export list: clothing, shoes, leather goods, jewellery and silverware, as well as those typical Parisian products, *articles de Paris.* [. . .]

[. . .] because of the advantages that might accrue from a more thorough comparative analysis of the French capital with another metropolis, we have adjusted the thrust of our analysis and made a detailed examination of manufacturing occupations in Paris by reworking available statistics and by establishing parallels and contrasts with London. This comparison was possible because for the first time in 1851 British census returns presented occupational tables. [. . .]

The results of our endeavours, presented in detail in Table 16.1 and in schematic form in Table 16.2 and Figures 16.1 and 16.2 are revealing in two ways. They permit us to confirm parallels and divergences in overall occupational structures in the two capitals. [. . .]

[. . .] Building and retailing, for example, are of comparable importance in each economy. It also shows differences. Domestic service was a more important occupation for Londoners, and particularly women, than it was for Parisians. It also indicates that banking and insurance, as well as transport (and especially, of course, dock workers), occupied a larger proportion of the active population in London than they did in Paris. Recent historians of the British capital are right, then, to stress that what set London apart within the British economy was the importance of its banking and insurance and domestic service sectors. Paris, on the other hand, had a larger share of its active population in public service, obviously reflecting the more centralized French political system. But the most significant difference in occupational structures between the two capitals is the proportion of the active population in manufacturing: 15 percent more of the occupied population in Paris were in industry than was the case in London. Manufacturing, indeed, occupied half the active population there. [. . .] four Parisian women in ten were in manufacturing as against only one woman in ten in London. There were actually more women in manufacturing in Paris despite the fact that London's population was twice as numerous.

Though a comparison of the distribution of the occupied population across sectors in manufacturing (Figure 16.3 and Table 16.3)

Table 16.1 1856 Census: Occupations in Paris

	Male	Number Female	Total
Agriculture	1,410	422	1,832
Mining	221	11	232
Building	55,666	912	56,578
Manufacturing	217,848	133,209	351,057
Transport	16,847	420	17,267
Retailing (Dealing)			
Coal and firewood	2,683	1,258	3,941
Raw materials	508	195	703
Foodstuffs	10,574	5,846	16,420
Tobacco	356	532	888
Wines, spirits, hotels	10,097	4,524	14,621
Lodging and coffee houses	8,247	2,830	11,077
Stationery and publications	3,802	1,117	4,919
Household utensils and ornaments	3,507	1,022	4,529
General dealers	8,083	7,311	15,394
Unspecified	2,990	96	3,086
Retailing Total	50,847	24,731	75,578
Banking, insurance, accounting	3,394	42	3,436
Public service and professions			
Government service	18,265	925	19,190
Sanitation	1,280	336	1,616
Armed forces	47,885	2,993	50,878
Police and prisons			
Law	4,204	13	4,217
Medicine	2,309	581	2,890
Art, amusement, literature, science	5,063	1,722	6,785
Education	2612	2,057	4,669
Religion	1,660	2,961	4,621
Other liberal professions	977	62	1,039
Total Public service and professions	84,255	11,650	95,905
Domestic service			
Indoor and outdoor	19,070	64,756	83,826
Extra service	15,982	26,154	42,136
Total Domestic service	35,052	90,910	125,962
Total Occupied Population	465,540	262,307	727,847

Source: 'Population générale de Paris et des arrondissements de Saint-Denis et de Sceaux par séries de professions,' *Recherches statistiques sur la ville de Paris et le département de la Seine*, VI (Paris: Dupont, 1860) pp. 625–54.

obviously cannot tell us anything about crucial facets of metropolitan manufacturing such as size of firms, production methods, products or their markets, it does reveal something of the patterns of production in each city. On the one hand, there are obvious, if not surprising, parallels: the clothing and textile industries were of comparable importance, with 54 per cent of the occupied population in manufacturing in London and 49 per cent in Paris. On the other, there are significant differences. Shipbuilding and the iron and steel industries occupied 7.6 per cent of those in manufacturing in London but only 0.8 per cent in Paris. The boot and

Table 16.2 Distribution of the active population of Paris (1856) and London (1851) across occupations (in percentages)

	Male	*Female*	*Both sexes*
Paris			
Agriculture	0.30	0.16	0.25
Mining	0.05	0.004	0.03
Building	11.96	0.35	7.77
Manufacturing	46.79	50.78	48.23
Transport	3.62	0.16	2.37
Retailing	10.92	9.43	10.38
Banking and insurance	0.73	0.02	0.47
Public service and professions	18.10	4.44	13.18
Domestic service	7.53	34.66	17.31
London			
Agriculture	3.23	0.37	2.20
Mining	0.59	0.02	0.38
Building	9.92	0.02	6.34
Manufacturing	33.86	32.83	33.48
Transport	11.6	0.17	7.46
General labourer	7.06	0.10	4.45
Retailing	14.13	8.00	11.91
Banking and insurance	2.56	0.00	1.63
Public service and professions	11.91	5.62	9.64
Domestic service	5.16	52.82	22.42

shoe industry of London, which was to prove so vulnerable to provincial competition later in the century, was also more important than shoemaking in Paris, as was millinery. The French capital's strengths, however, clearly lay elsewhere: Paris had somewhat greater concentrations of its active population in luxury industries such as jewellery and furniture, and in skilled trades such as printing and publishing and clock- and scientific instrument-making. [. . .]

As is well known, metropolitan manufactures have always been subject to two forces which push industries out of city centres: displacement, as some industrial processes move away from the city altogether, and decentralization, as some types of industry and enterprise move out from city centres to the urban periphery. Both of these forces were at work in Paris at this time.

As was the case with other large cities, Paris could not compete with other areas in products and processes which did not depend on the advantages economies of agglomeration conferred and which could profit from using cheaper labour and introducing machinery. Thus cutlery could prosper in the capital by producing speciality items, such as penknives and razors, and luxury ware that needed elaborate design and craft skills. Many of the other processes and the making of cheaper items, though, were increasingly located away from Paris. Hosiery, too, could survive in the capital by producing specialized and fashion items but could not meet provincial competition in ordinary articles, particularly after the introduction of the circular knitting frame. The cut-crystal industry that had once been found only in Paris also declined in the first half of the century, as mechanical cutters were introduced in the works that supplied glassware to the capital. [. . .] Even high-quality leather glove manufacture began to suffer from competition from the Dauphiné, and its labour force in Paris fell by a

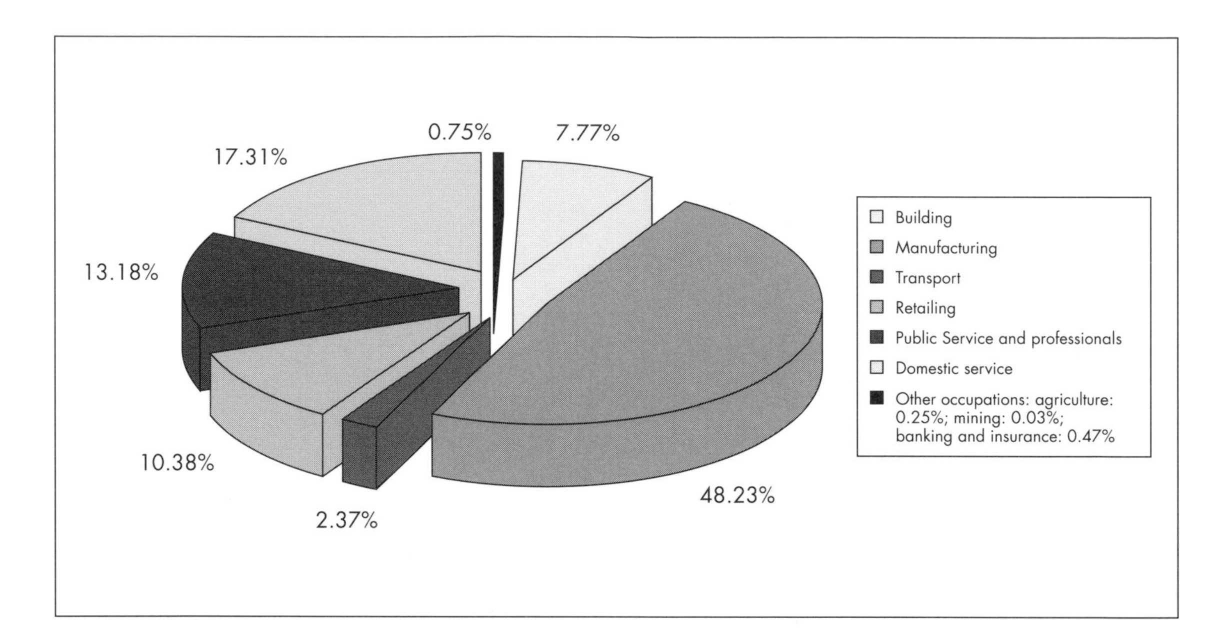

Figure 16.1 Distribution of the active population of Paris across occupations, 1856

half in the 1850s. The decline of industries such as these illustrates the competitive disadvantages which resulted from the high rents and cost of living in the capital.

[Centrifugal] forces that pulled to the periphery nuisance industries which polluted the environment and, above all, capital-intensive and land-hungry enterprises also led to declines in some sectors. The impact these shifts had on the industrial geography of Paris at this time can be roughly determined by an analysis of the 1856 occupation census. This shows that the industries in the Seine department which were most heavily concentrated in Paris itself were those most dependent on the competitive advantages offered by city-centre locations, while those which did not were more heavily concentrated in the suburbs. Those with the highest degree of concentration in the suburbs, therefore, were chemicals, carriage- and saddle-making and building. Other sources permit us to trace this decline across time. In 1821 Paris still had sixty-seven spinning mills and was a major French producer of spun cotton. But mechanized spinning is a classic case of an industry which cannot tolerate the high-cost urban environment and by 1847 there remained only twelve mills with a total workforce that was only a quarter of what the industry had employed at its zenith. In chemicals, too, the more heavily capitalized enterprises increasingly abandoned the city and even peripheral quarters for the suburbs, as did sugar refining. However, we should not exaggerate the speed of this process. [. . .] Significant numbers of capital-intensive and nuisance industries not only remained firmly planted within the city but new ones continued to be established there. This was the case with printing, some of whose processes were increasingly mechanized. It was also true of the dynamic high-precision machine-tool industry: half those dependent on this sector in the Seine department still lived in the capital at the mid-century. Firms in these and other industries remained in the city partly because of the inertia that results from solid implantation. More importantly, they remained there because they

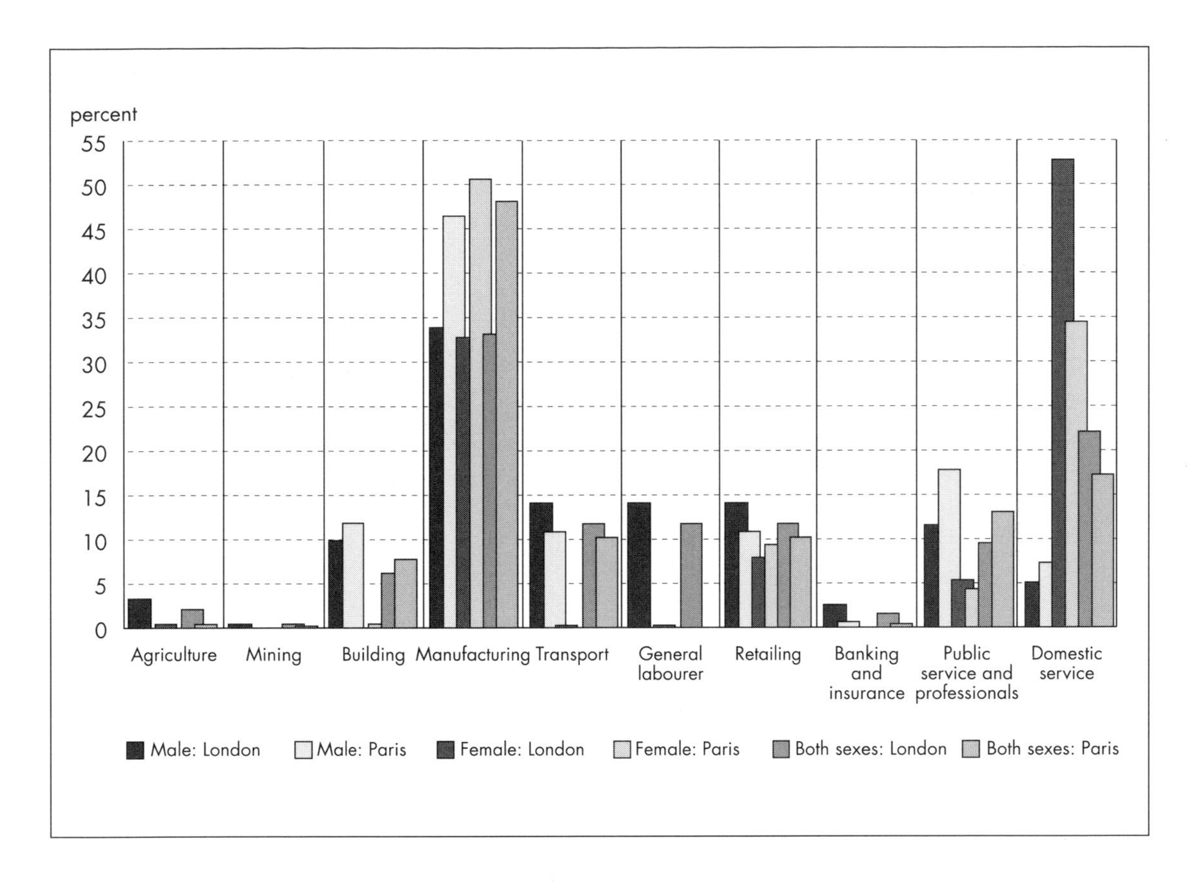

Figure 16.2 Distribution of the active population at mid-century: Paris and London compared

gained benefits from economies of agglomeration which compensated, at least for the moment, for the price they had to pay for an urban location. They stayed, too, because the powers of city government to evict nuisance industries were more limited than has often been supposed.

Proof of the dynamism of manufacturing, however, is not to be found in the changing industrial geography of the capital. It is to be found, instead, in three other kinds of change: the introduction of new products; technological innovation and, above all, organizational shifts. [. . .] The pressure to innovate and restructure was endemic to the system in Paris as in other giant cities, both during boom periods and downswings. [. . .]

These product changes were made across the vast range of manufactures, whether it be the introduction of iron bed frames or spring mattresses, mechanically-produced envelopes and fancy papers and cardboard, or new leather goods, such as money purses and wallets, or cigar and cigarette cases. The most obvious example, however, is ready-made clothes. Though the beginnings of ready-made are difficult to date, the industry took off in Paris in the 1820s and by mid-century was already responsible for over two-fifths of production in tailoring alone. [. . .]

During the first half of the nineteenth century the Seine was far ahead of any other department in the number of patents granted. This is but one indication of the importance of

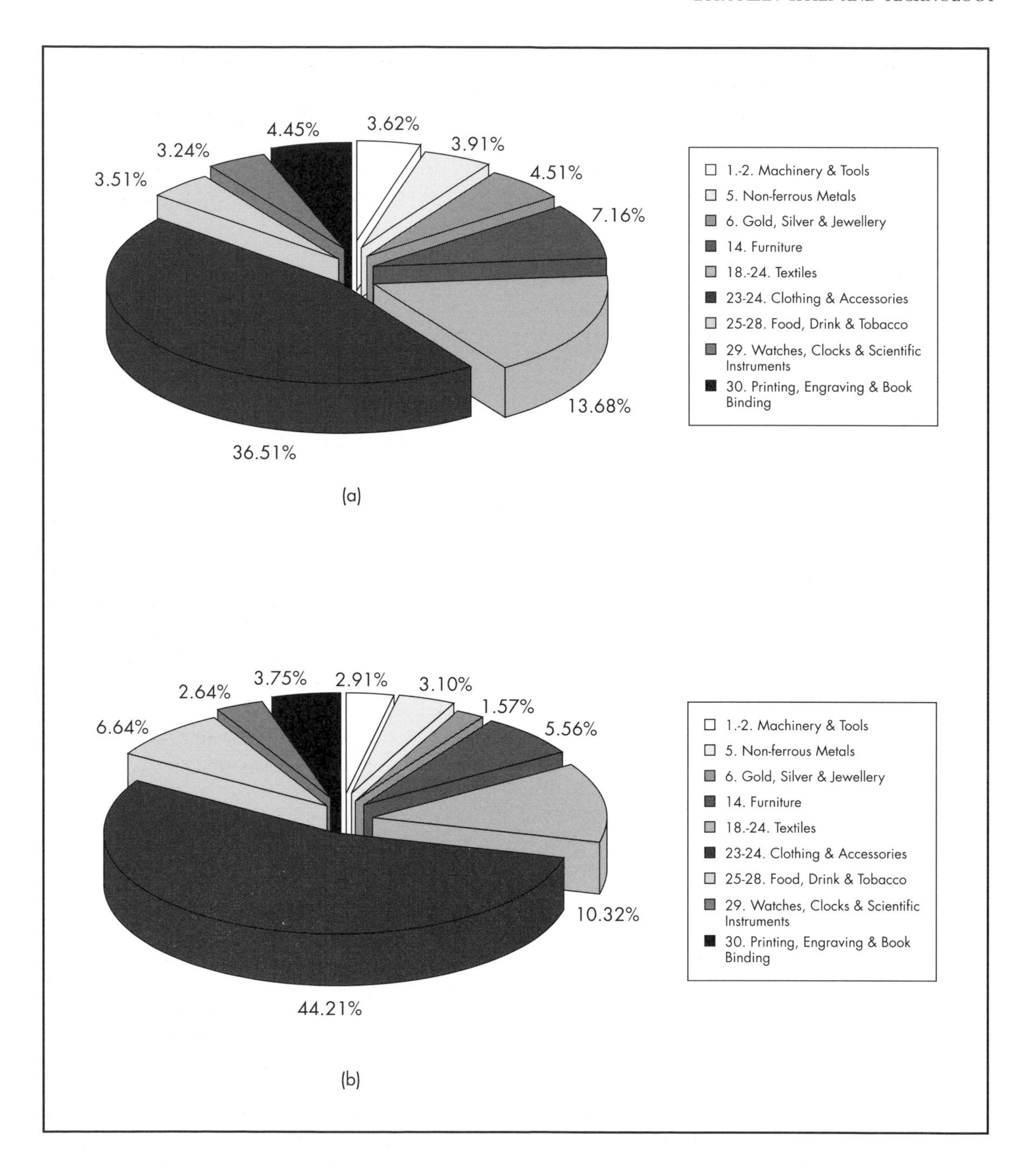

Figure 16.3 (a) Leading industries in Paris: percentages of occupied population
(b) Leading industries in London: percentages of occupied population

technical innovation. [. . .] There are others. There were those visible hardware innovations: machinery in printing, in wallpaper and tobacco manufacture. There was also a series of new processes which were introduced and perfected at this time and which enabled industrialists to extend their markets. In cutlery and kitchen- and dinner-ware, for instance, the introduction of nickelsilver, zinc and tin and, in the 1840s, of electrolytic metal plating, which meant that Christofle and his rivals could produce cheaper but still elegant imitations of fine silverware, all extended the range of lower-cost products. But change resulted less from such major breakthroughs than from small innovations across the board. Thus umbrella-making certainly profited from the invention of folding and self-opening umbrellas but it gained much more from a series of improvements in the frames, in the materials used for covering, and in the production of more elegant and varied handles. [. . .]

The most important facet of change in Paris, as in other metropolitan economies, is also the most elusive in surviving documents. This consists of organizational shifts [including] increasing recourse to the division of labour. An industrialist declared in 1860 that, because of the division of labour, the prices of Parisian precision instruments were one-third those asked by foreign competitors. [. . .] The remarkable expansion in ready-made clothes and clothing accessories [. . .] was made possible, not by the introduction of the sewing machine, which was only successfully adopted after the mid-century, but by recourse to seasonal labour [. . .] and to cheap labour that could be hired and let go as variations in production demanded. [. . .]

On the supply side, Paris enjoyed one advantage above all others, its human capital. [. . .] by far the most important explanations as to why Parisian industrialists were able to secure greater labour inputs and to divide labour tasks are to be found, on the one part, in migration flows to the capital and, on the other, in an 'industrial reserve army' composed not of what British scholars term casual labour but of women.

By their numbers, age structure, and the skills they brought with them, immigrants were unquestionably a major reason for the dynamism of manufacturing in the French capital and a much more important one for Paris, where newcomers made up two-thirds of the population, than for London, only two in five of whose inhabitants had not been born there. [. . .] The work that has been done on other cities also suggests that such transfers of human capital may have had positive effects on urban industries through the skills immigrants brought, as by their impact on the semi-skilled and unskilled labour market where some of them clustered [. . .]

We have advanced a series of arguments in order to establish the dynamism and dynamics of manufacturing in Paris in the first half of the nineteenth century. We have argued, first, that what previous work has been done on Paris was mainly [. . .] imprisoned in global interpretations of industrialization processes and the role urban manufacturing played in them, which prevented scholars from according Parisian industries their full importance [. . .]

The most dynamic sectors were those face-to-face industries – clothing, leather, precision engineering, furniture, jewellery, printing and publishing, specialized metals – many of which would continue to flourish in this as in other metropolitan centres. We cannot claim, however, that we have been able fully to determine how the capital's industries inserted into wider processes and still less that change in Parisian manufacturing was an autonomous process. The capital's industries grew in part because [. . .] they profited [. . .] from agricultural growth, from increasing internal and international trade, as from the choices individuals made to respond positively to signals from the capital by migrating there. Industrial

Table 16.3 Occupied population in manufacturing in Paris and London in the mid-nineteenth century

A: Paris		*Numbers*			*Percentage of occupied population*		
		Male	*Female*	*Total*	*Male %*	*Female %*	*Total %*
1	Machinery (1)	8,377	117	8,494	3.84	0.09	2.42
2	Tools (2)	2,615	178	2,793	1.20	0.13	0.80
3	Shipbuilding	6	1	7	0	0	0
4	Iron and steel (3)	2,665	134	2,799	1.22	0.10	0.80
5	Non-ferrous metals (4)	12,883	847	13,730	5.91	0.64	3.91
6	Gold, silver, jewellery (5)	11,340	4,593	15,933	5.20	3.45	4.54
7	Earthenware (6)	1,677	341	2,018	0.77	0.26	0.57
8	Coal and gas	469	12	481	0.21	0.01	0.14
9	Chemicals (7)	1,574	273	1,847	0.72	0.20	0.53
10	Furs and leather	5,087	552	5,639	2.34	0.41	1.61
11	Glue and tallow (8)	411	90	501	0.19	0.07	0.14
12	Hair brushes, etc. (9)	1,653	1,035	2,688	0.76	0.78	0.77
13	Woodworking, etc. (10)	6,690	998	7,688	3.07	0.75	2.19
14	Furniture (11)	20,922	4,229	25,151	9.60	3.17	7.16
15	Carriagemaking and harnesses (12)	7,400	650	8,050	3.40	0.49	2.29
16	Paper (13)	1,660	1,543	3,203	0.76	1.16	0.91
17	Floorcloths (14)	258	159	417	0.12	0.12	0.12
18–21	Textiles (15)	24,097	23,915	48,012	11.06	17.95	13.68
22	Dyeing	1,409	610	2,019	0.65	0.46	0.58
23	Clothing						
	(a) boot and shoemaking	16,089	9,010	25,099	7.39	6.76	7.15
	(b) hatting (16)	2,880	1,688	4,568	1.32	1.27	1.30
	(c) bonnet and capmaking	1,886	2,433	4,319	0.87	1.83	1.23
	(d) millinery	197	5,404	5,601	0.09	4.06	1.60
	(e) seamstresses	54	34,643	34,697	0.02	26.01	9.88
	(f) staymaking	152	1,854	2,006	0.07	1.39	0.57
	(g) linen good	471	19,101	19,572	0.22	14.34	5.58
	(h) tailoring (17)	13,695	7,065	20,760	6.29	5.30	5.91
	(i) others	2,868	4,070	6,938	1.32	3.06	1.98
Total clothing		38,292	85,268	123,560	17.58	64.01	35.20
24	Sundries connected with clothing (18)	2,697	1,915	4,612	1.24	1.44	1.31
25	Food preparation (19)	1,549	587	2,136	0.71	0.44	0.61
26	Baking and confectionery	5,151	1,487	6,638	2.36	1.12	1.89
27	Drink preparation	1,533	399	1,932	0.70	0.30	0.55
28	Tobacco products	365	1,266	1,631	0.17	0.95	0.46
29	Watches, instruments, etc.						
	(a) toys	905	608	1,513	0.42	0.46	0.43
	(b) watches and clocks (20)	2,909	227	3,136	1.34	0.17	0.89
	(c) musical instruments	1,418	73	1,491	0.65	0.05	0.42
	(d) scientific instruments (21)	2,032	394	2,426	0.93	0.30	0.69
	(e) perfumes	583	396	979	0.27	0.30	0.28
	(f) articles de Paris	1,186	509	1,695	0.54	0.38	0.48
	(g) hunting and fishing equipment	96	35	131	0.04	0.03	0.04
Total watches, instruments, etc.		9,129	2,242	11,371	4.19	1.68	3.24
30	Printing, engraving and bookbinding	12,029	3,598	15,627	5.52	2.70	4.45
31	Unspecified (22)	22,187	9,893	32,080	10.18	7.43	9.14
Total for manufacturing		217,848	133,209	351,057	100	100	100

Table 16.3 *continued*

B: London		*Male*	*Numbers* *Female*	*Total*	*Percentage of occupied population* *Male %*	*Female %*	*Total %*
1	Machinery (1)	8,321		8,321	3.45		2.23
2	Tools (2)	2,557		2,557	1.06		0.68
3	Shipbuilding	8,342		8,342	3.46		2.23
4	Iron and steel (3)	19,486	176	19,662	8.09	0.13	5.27
5	Non-ferrous metals (4)	11,274	292	11,566	4.68	0.22	3.10
6	Gold, silver, jewellery (5)	5,706	144	5,850	2.37	0.11	1.57
7	Earthenware (6)	1,372	160	1,532	0.57	0.12	0.41
8	Coal and gas	1,702		1,702	0.71		0.46
9	Chemicals (7)	2,630	376	3,006	1.09	0.28	0.80
10	Furs and leather	6,263	1,600	7,863	2.60	1.21	2.11
11	Glue and tallow (8)	4,311	332	4,643	1.79	0.25	1.24
12	Hair brushes, etc. (9)	3,055	1,712	4,767	1.27	1.29	1.28
13	Woodworking, etc. (10)	14,202	27	14,229	5.89	0.02	3.81
14	Furniture (11)	18,658	2,431	21,089	7.74	1.84	5.65
15	Carriagemaking and harnesses (12)	9,605	175	9,780	3.99	0.13	2.62
16	Paper (13)	2,648	1,267	3,915	1.10	0.96	1.05
17	Floorcloths (14)						
18–21	Textiles (15)	21,625	16,894	38,519	8.98	12.75	10.32
22	Dyeing	1,735		1,735	0.72		0.46
23	Clothing						
	(a) boot and shoemaking	30,855	19,774	50,629	12.81	14.92	13.56
	(b) hatting (16)	2,494	2,434	4,928	1.03	1.84	1.32
	(c) bonnet and capmaking	3,046	3,046			2.30	0.82
	(d) millinery	43,928	43,928			33.16	11.76
	(e) seamstresses	21,210	21,210			16.01	5.68
	(f) staymaking	2,466	2,466			1.86	0.66
	(g) linen goods						
	(h) tailoring (17)	22,479	8,292	30,771	9.33	6.26	8.24
	(i) others						
Total clothing		59,181	103,835	163,016	24.56	78.37	43.66
24	Sundries connected with clothing (18)	841	1,221	2,062	0.35	0.92	0.55
25	Food preparation (19)	1,200	3	1,203	0.50		0.32
26	Baking and confectionery	13,762	1,150	14,912	5.71	0.87	3.99
27	Drink preparation	7,604		7,604	3.16		2.04
28	Tobacco products	901	193	1,094	0.37	0.15	0.29
29	Watches, instruments, etc.						
	(a) toys						
	(b) watches and clocks (20)	4,847		4,847	2.01		1.30
	(c) musical instruments	2,929		2,929	1.22		0.78
	(d) scientific instruments (21)	1,578	93	1,671	0.65	0.07	0.45
	(e) perfumes						
	(f) articles de Paris						
	(g) hunting and fishing equipment						
Total watches, instruments, etc.		9,354	517	9,871	3.88	0.39	2.64
30	Printing, engraving and bookbinding	14,002		14,002	5.81		3.75
31	Unspecified (22)	1,662		1,662	0.69		0.44
Total for manufacturing		240,919	132,486	373,405	100	100	100

Notes

1 Machine-building also includes scale and pump making.
2 Comprises makers of edge-tools, cutlers, armourers, medal and coin makers, pen makers and typefounders.
3 Also includes blacksmiths.
4 Manufacturing of copper, brass, tin and lead objects.
5 Comprises jewellery (fine and costume) and gold and silverware.
6 Includes porcelain, crystal and glassware.
7 Includes matchmaking, fireworks, colour and ink manufacturing, as well as chemical products.
8 Also includes wax and soap.
9 Comprises brushes and combs.
10 Also includes cork and basket making.
11 Along with furniture, this sector includes dressing-case making, inlaid work, upholstering and wallpaper manufacturing.
12 Included are wheelwrights and makers of carriages, military equipment and luggage.
13 Comprises paper, cardboard and playing cards.
14 Includes floorcloths and rubberized cloths.
15 Also included are ropemaking and artificial flowers. The latter occupied 8,424 persons in Paris (94 percent of them women).
16 Straw hat making has been added to hats.
17 The 'others' category comprises gloves (not leather), clogs, ready-made clothes for women and breeches making.
18 Comprises umbrellas, walking sticks, buttons and fans.
19 Conserves, noodles, jams, sugar refining, vinegar and mustard and chocolate making.
20 Paris did not manufacture significant quantities of watches.
21 Trussmakers have been included with surgical and scientific instrument makers.
22 'Unspecified' is a more significant category for Paris than for London not just because it includes a larger group of industrial occupations which the compilers of the 1856 census lumped together as 'unclassified' but also because we have included here the *commissionnaires-négociants* (5,006 persons) who were of increasing significance in many branches of Parisian industry, who were all grouped together in the census.

growth in Paris, then, possessed a dynamism of its own but it also depended on wider forces. All the difficulty, as in every effort to understand urbanization processes, is to establish the relative weight of each. [. . .]

We have argued, finally, that the explanation for the dynamism of Parisian industries at this time is to be found in changes taking place in the city and in wider processes and that these can best be understood if we take into account not only supply factors, such as product and process innovation, changing labour inputs and productivity, but also demand. [. . .]

We might also suggest that our preliminary findings should serve to stimulate further research on the complex [. . .] linkages that Parisian manufacturing had with regional and national economies [. . .]. It should also encourage much-needed work on industry's changing occupation of urban space. Though manufacturing was certainly not the only factor modelling and remodelling space, too much attention has so far been given to the ways in which urban planners and elites, as property owners and speculators, moulded urban forms. [. . .] We are beginning to realize, for instance, that immigrants were less innocents flooding the city, to become 'dangerous classes', than workers bringing labour and skills in response to signals emitted from the capital, and that rising population densities in central areas were symptoms less of urban pathology [. . .] than of the residential patterns of labour drawn to the employment nodes of industries which were growing and changing. [. . .] Our research already suggests [. . .] that the Paris of the first half of the nineteenth century was undergoing not just the demographic change, which other scholars have already stressed, but structural change and that manufacturing was one of its principal motors.

17

A NEW SEWER-SYSTEM FOR PARIS

David H. Pinkney

Source: David H. Pinkney, *Napoleon III and the Rebuilding of Paris,* Princeton, 1958, pp. 127–9, 132–5, 137–42, 148–9

Napoleon III once asked Edwin Chadwick, Britain's great public health crusader, what he thought of his improvements in Paris.

'Sir,' he replied, 'it was said of Augustus that he found Rome brick and left it marble. May it be said of you that you found Paris stinking and left it sweet'. [. . .] At mid-century, Paris was, as Chadwick affirmed, a smelly city, and much of the trouble lay in its drainage. Most streets lacked underground drains. Those in existence were too small and frequently overflowed and flooded low-lying areas. The Seine itself was an open sewer flowing through the center of the city. [. . .] Haussmann and the engineer Belgrand gave virtually every street in old Paris its own underground drain and substantially extended the network of street drains in the annexed zone, and they built a comprehensive system of collectors that carried waste water to the Seine well below Paris and ended pollution of the river within the city.

This revolution in Paris underground was the conception of Haussmann and Belgrand. [. . .]

[. . .] Haussmann's and Belgrand's predecessors in Paris had accomplished important improvements in sewerage that suggested the directions their program might take. In the twenties a municipal engineer had developed a low cost method of building sewers of cement and rough stone, and the city had used it in constructing some sixty miles of new lines during the two following decades. Another engineer had introduced a rational pattern into the location of new street drains and had prepared the first map and statistics of the existing lines. To carry all waste water to the river below the city, the Prefect Berger and his Director of Public Works, Dupuit, had started construction of the first collectors, one under the Rue de Rivoli and another under the quais of the Left Bank. The design of these galleries was revolutionary, and Belgrand copied it in all the collectors he built during the Empire.

These were useful beginnings. [. . .] But the actual accomplishment [. . .] was small and uncoordinated in any comprehensive plan. The 260 miles of streets still had fewer than 100 miles of sewers, and they lacked the capacity to handle heavy rains. Their dimensions had been determined by the size of a man, facility in cleaning being the designer's first consideration; Haussmann could find no indication that any effort had ever been made to compute in advance of construction how much water a gallery might be required to carry. [. . .]

[. . .] The galleries built after 1855 were egg-shaped, made of hydraulic cement, and their walls were but half as thick as those of the old-style sewers. A given amount of building material used in a gallery of the new design produced a cross-section about 50 per cent larger than it would in an old style gallery. In these larger vaults was space for water mains, and later they provided protected and accessible positions for electric, telegraph, and telephone wires and for the pneumatic tubes that carried letters about the city at high speed. For the first time, moreover, the dimensions of sewers were related to the volume of water they would be required to carry. Belgrand computed the cross-section required for each 250 acres of area that a sewer must drain, and the new lines were built and most of the old rebuilt according to that formula.

Haussmann's extension of the network of street sewers and individual house connections was an accelerated continuation of the work begun before his time. [. . .] Previously, engineers had planned one collector on each bank parallel with the river. Haussmann and Belgrand proposed a system of eleven collectors covering the whole city, and they had already begun construction of its major component, the great general collector that would receive the flow of all the others and carry it to the Seine at Asnières, below Paris.

The Collector of Asnières was a solution to the problem raised by the discovery that no collector following the river could have sufficient incline and sufficient length to carry it beyond the city and still emerge above flood stage. It took advantage of the peculiar course of the Seine below Paris, where the river doubles back on itself in a great bend around Boulogne and flows northeastward for a few miles, coming at Asnières within two miles of the city's northwestern limits. [. . .]

The execution of the plans for the collectors began in the spring of 1857 when Belgrand asked for bids on the construction of the Collector of Asnières, the *Cloaca Maxima* of the system. (Haussmann frequently used the Roman name and often referred to the Roman sewers in his reports, in part because he had to go back that far to find a precedent for what he was attempting, but also, certainly, because he hoped the Roman comparison would heighten the appeal of his program to the Emperor.) Belgrand's specifications called for a larger gallery than any previously built for a sewer. It was elliptical in shape with a maximum interior width of about eighteen feet and a maximum height of nearly fourteen-and-a-half feet. [. . .] The gallery presented some formidable construction problems. The water table under much of the Right Bank of Paris, including the area through which the collector would pass, was unusually high and turned the strata of sand and gravel a few yards below the surface into quicksand. The walls of a trench or tunnel cut through it would not stand up, and excavations caused shifts in the soil that endangered foundations of nearby buildings. [. . .]

[. . .] Belgrand undertook to eliminate the problem. He dug a series of pits along the course of the collector, installed a steam engine and pump at each, and pumped out the water, lowering the water level by eight feet. The sand and gravel layer, once drained of water, had enough consistency to permit construction of the gallery with no more than ordinary hazard. [. . .] Two years later Haussmann reported to the Municipal Council that the Collector of Asnières was completed and in use over its entire length. [. . .]

For the Left Bank Haussmann proposed three collectors radiating like spokes of a wheel from the southern end of the Pont de la Concorde opposite the General Collector of Asnières. One was to drain the eastern quarters, the second the central quarters below the Hôtel des Invalides, and the third the western end of the Left Bank. A siphon across the river would connect them with the Collector of Asnières. [. . .]

[. . .] Belgrand's proposed collectors to drain the southwest quarters on both sides of the river remained unbuilt at the end of the Empire, and provisionally the waters of these districts continued to run directly into the Seine.

The new collector under the Place de l'Etoile was called the Collector of the Bièvre, being regarded as a continuation of the principal collector of the Left Bank, which began at the Bièvre River on the east side of the city. [. . .]

In 1868 Belgrand began work on the siphon under the Seine that was to connect the two sections of the Collector of the Bièvre. Its principal elements were two sheet-iron tubes thirty-nine inches in diameter and 492 feet long, each composed of sections that were assembled on the river bank above the Pont de l'Alma and there joined together. Belgrand planned to lower the hermetically sealed tubes into the river, float them into position across the stream just above the bridge, and then sink them onto a prepared bed. Maneuvering more than 110 tons of iron stretched over nearly 500 feet was no simple operation. Preparations for lowering the first tube thirteen feet down an incline from the river bank to the water took up the better part of two weeks at the end of July and early August, 1868, and even with those careful preparations workmen let it drop more than three feet. Happily the tube emerged undamaged. A week later on August 18 they lowered the second tube safely, and the two were anchored side by side in midstream and permanently fixed together with iron beams. They were then moved into position above the prepared bed, an operation complicated by the exceptionally low water in the river at the end of a dry summer. On August 26 before hundreds of spectators who came to watch the operations, the engineers in charge made the first attempt to submerge the huge structure, using iron weights to force it under water. Just below the surface it tipped, dumped some of the weights to the bottom of the river, and re-emerged. Operations had to be suspended until divers recovered the lost weights. Not until September 1, after they had blocked all river traffic for more than a week, were the tubes successfully sunk. Divers fixed them in place with iron braces, and later two feet of concrete were poured over them. On November 2, 1868, municipal engineers turned the waters of the Left Bank section of the Collector of the Bièvre into the siphon.

With the completion of this siphon and the Right Bank section of the Collector of the Bièvre in the same year Haussmann and Belgrand had finished the principal elements of their network of collectors, and they had come close to achieving their aim of freeing the Seine in Paris of all sewer water. Only the extreme southeast and southwest quarters and the two islands emptied their storm sewers directly into the river, and Belgrand had plans to bring those areas into the system. The new collectors with their large capacities and emergency flood-gates also cut sharply into the number of days when flood waters of the Seine interrupted sewer service. [. . .]

During the Franco-Prussian War the collectors had some days of unforeseen disfavor with the public. When Parisians learned that the French armies were being routed in the east, rumor spread that Germans infiltrating the collectors might suddenly spring out of the earth in the center of Paris. Later the encirclement of the city by the besieging enemy changed the fantasy to a real threat, and the authorities built walls across the General Collector of Asnières to close it to potential invaders.

An unmixed blessing for Paris at peace, the collector system was a menace for the riverside communities downstream from the outfall. It concentrated at a single point all the evils of congestion and contamination that were formerly spread along several miles of the river in the city. An analysis of the water flowing from the general collector showed that in a year it carried 154,000 tons of solids and 77,000 tons of dissolved matter into the river. The solids included mud, sand, and gravel, coming especially from macadam streets, and organic matter, chiefly the leavings of curb-side garbage collections and the less than perfect system of removing horse droppings. At the mouth of the collector the solids formed a delta that had to be dredged away periodically at the

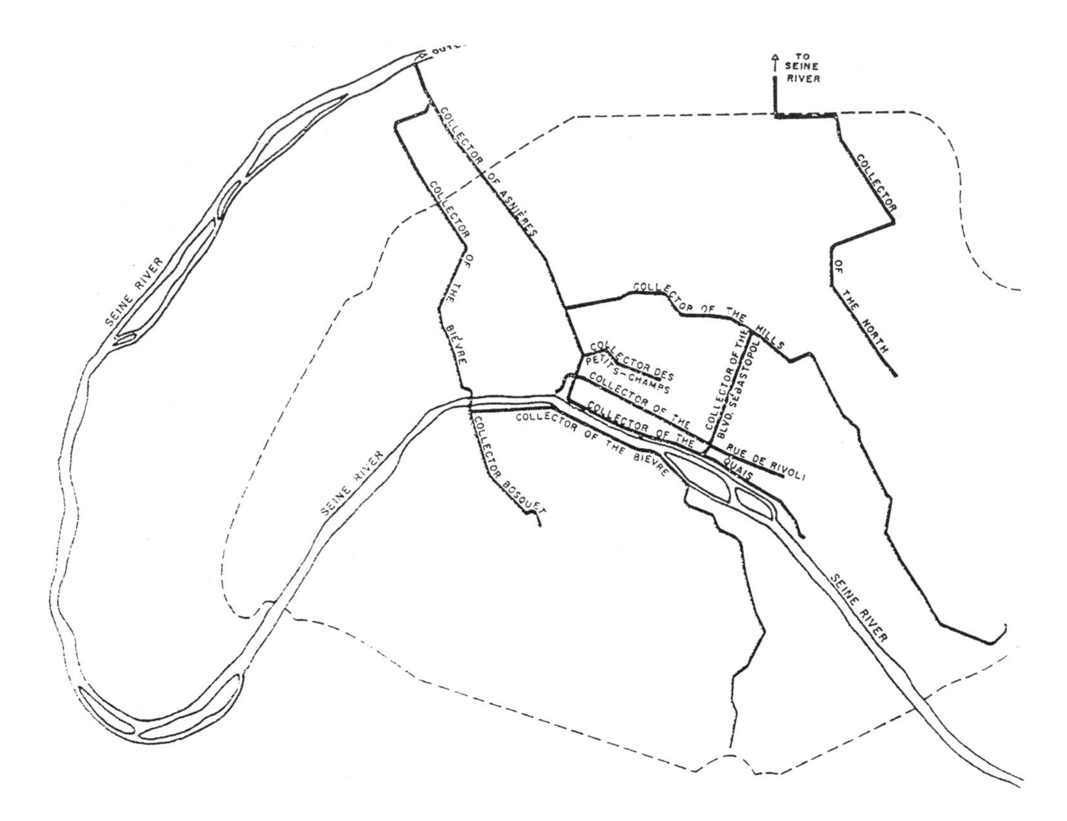

Figure 17.1 Collector sewers of Paris, 1870

city's expense, and other foreign matter made a dark streak in the river visible miles downstream.

Valuable fertilizers were being lost, and in the last years of the Empire, the engineer Mille, the city's expert on solid sewage disposal, experimented with the use of the sewer water for irrigation and fertilization of gardens. In 1868 he obtained good results on a small plot at Clichy near the outlet of the general collector, and the following year the city pumped 1,300,000 gallons daily to the plain of Gennevilliers, a short distance downstream, and made it available without charge to any proprietor who would use it. By October, 1869, ninety acres in this area were being watered, and the formerly inhospitable soil began to produce a high yield of vegetables. Before the end of that year, Mille reported, rents of land had multiplied as much as five-fold. The experiment pointed the way toward a possible solution of the problem of river contamination, but several decades passed before more than a small fraction of the sewer water was used as Mille proposed.

Haussmann [. . .] had laid down as one of the essential requirements of an acceptable system of sewers that the principal galleries be designed to permit the use of mechanical cleaning devices. Heretofore the sewers had been cleaned by hand. Men working with scrapers and using movable sluice-gates to produce a strong rush of water pushed mud, gravel, and other debris into piles under manholes; there it was hauled to the surface in buckets and loaded into carts for removal. When the Rue de Rivoli sewer was built. Dupuit, the engineer in charge, installed a rail on each of the footways so that cars straddling the channel could be run in the gallery itself to carry debris, and he also worked on the idea of a car with a kind of sluice-gate projecting below it to obstruct the flow in the channel and produce a water action that would clean the channel. [. . .] Later Belgrand took up the idea, and in 1858 a newly built sluice-car proved capable of an efficient job of cleaning. The device was a platform measuring about four by six feet. mounted on four wheels with a movable sluice gate suspended below it. The gate was slightly smaller than the cross-section of the sewer channel, and near the lower end it had two small openings. Lowered into the channel it blocked the flow of water. When the difference in levels on opposite sides of the gate reached six or eight inches the water pouring through the two openings and around the edges of the gate produced an agitation that stirred up the sand and silt on the channel floor. The weight of the impounded water on the gate moved the car forward, and the gate and water together pushed the mass of stirred-up solids ahead as a spring wind blows dry sand along a beach. A bank of moving sand and sludge might extend 300 or more feet ahead of the car. In the Collector of Asnières the greater width of the channel precluded the use of straddling cars, but engineers applied the same technique of hydraulic cleaning there, using a boat instead of a car to carry the sluice-gate. A single boat operating around the clock could traverse the three mile length of the general collector in from eight to twenty days. For more frequent cleaning several boats were operated simultaneously. [. . .] Sluice-boats were used in the Collector of Asnières and the Collector of the Bièvre and in a section of the Collector of the Quais, sluice-cars in all others. In the smaller sewers cleaning crews continued to use the old manual methods. [. . .]

[. . .] At the end of the Empire sewage no longer ran in the streets of the old city, no longer coated the pavement with ice in the winter; low areas and cellars were rarely flooded; and the Seine passed through Paris largely uncontaminated.

Part 2
EUROPEAN CITIES SINCE 1870

Historians have seen the 1870s as the start of the 'Second Industrial Revolution', a second phase of industrialization that, extending into the twentieth century, brought new forms of power, new materials, new forms of transport and automation. Reading 18 considers the effects on the city of Glasgow of the new public transport: electric trams (at the close of the nineteenth century) and an underground railway (operated by cable at first, and electrified in 1935). Here there is analysis of how laying the new lines aroused local politics. And the consequences for the class structure of Glasgow's West End district are revealed.

Reading 19 focuses on the great changes in twentieth-century industrial buildings resulting from the widespread use of structural steel and reinforced concrete, the development of production machinery and the adoption of techniques of scientific management in the production process. The Shredded Wheat factory, opened in 1925 in the new Welwyn Garden City, is the main subject for analysis. It illustrates vividly the way the adoption of rapidly changing technology affected the layout and working conditions in factories of the time.

Changes to the structure of Britain's domestic buildings came later. Reading 20 explains why a movement of opinion grew in the 1940s for taller blocks of flats. That was followed in the 1950s by intensive experimentation with new materials and more scientific methods of construction. It all culminated in the 1960s in a multitude of new building forms that have changed the landscape of British cities.

London's traffic congestion in the late 1890s moved a leading civil engineer, Sir John Wolfe Barry, to complain of decades of neglect in urban planning – a deplorable failure to provide the capital with arterial thoroughfares. It was, in his view, a disgrace to the world's greatest metropolis which, in this respect, had fallen far behind Paris and Vienna. His call for immediate action, recorded in reading 21, has a present-day ring with its comparative assessment of the merits of trams, underground and overground railways as systems of urban transit, and the account taken of the financial losses to businessmen caused by delayed traffic.

The year 1906 was a landmark in another aspect of London's development. It saw the arrival of ultra-modern American construction and on an American scale. The steel-frame and reinforced concrete style was brought to Oxford Street through the enterprise of Gordon Selfridge, a businessman from Chicago (the leading centre of this new structural engineering) intent on establishing a giant department store in the world's largest city. Reading 22 tells the fascinating story of its completion, which depended on the technical expertise of American architects. Here the most interesting point of all is the obstacle presented by London's building regulations, whose specifications had been devised in an era when no one conceived of shops on the scale of Selfridge's. All of which again brings to our attention the paramount importance in urban history of considering not only technological innovations but the prevailing conditions affecting their implementation in individual cities.

While reading 21 looked across the Channel with envy at the wide boulevards of late-nineteenth-century Paris, the reality was very different. Instead of an efficient flow of fast-moving traffic, Paris' fine boulevards were badly congested and subject to messing by horses. This is conveyed by reading 23, which then concentrates on the still-worsening state of Paris streets in the twentieth century, due to the increasing number of cars. These unforeseen problems and growing recognition of the inadequacy of Parisian streets stimulated decades of urban planning from the rejected visionary schemes of Le Corbusier in the 1920s to the adoption, in the 1950s, of the idea of a peripheral urban motorway. In 1960 there were a million cars in Paris; five years later, that number had doubled. Many of the trees that had been an essential feature of Haussmann's mid-nineteenth-century transformation of Paris now fell victim to street widening, and parked cars invaded pedestrians' pavements. By 1970, the new president of France, Georges Pompidou, was telling Parisians they must accept the dominance of the car as a symbol of progress, regardless of traditional urban aesthetics. This text presents the controversies generated within Paris by the various plans and assesses the consequences for the city. And, here again, we see the importance of political power in influencing technological change in cities.

Politics certainly has to be taken into account in the creation of the Paris Métro. That is apparent in the analytical study of reading 24. The planning dates from the 1870s. While some Victorians wished London was endowed with Parisian boulevards, in the case of underground railways the direction of interested gaze was reversed: Parisians looked to what had been achieved in London, though not always with approval. This extract from Norma Evenson's book on a century of change in Paris, brings to life the contemporary climate of opinion – the fears, the hopes – on the projected innovation, and examines the contending schemes that ranged from the ingenious to the bizarre and impracticable. We learn of initial apprehension in Paris, very similar to the first reactions to the proposal for the London Tube, that citizens would shun the depressing or dangerous experience of underground travel. That was soon proved to be baseless pessimism. But, for various reasons, construction was long delayed; there was, in the author's words, a 'long debate'. Work on the Métro did not begin until 1898. The decision to start building it was fired by circumstances that are today familiar to us. Plans for the vast international exhibitions of the 1990s in Seville and Lisbon, with predictions of an influx of millions of tourists, included elaborate improvements in transport systems in the host cities. So it was in Paris in the late 1890s, with its fast-approaching Exposition in 1900. Paris would not cope with the expected huge crowds of visitors with its existing transport facilities; that was the immediate cause of implementing plans for the Métro. Once completed, the Métro soon became an integral part of everyday city life: by 1960, the text informs us, Parisians at the age of retirement had, on average, spent two years of their life underground. Several general issues emerge in this text: the role of aesthetics in designing the new building forms of Métro station entrances, the effects of the Métro on surface traffic, and the role of the Métro and its later complement, the RER, in the developing relationship between the city and its outlying suburbs.

Berlin's development into one of the world's great cities began later than London and Paris, in the 1870s. But its rate of growth was faster than either: from a population of some 400,000 in 1850, by 1900 it had grown to almost two million. Readings 25 and 26 each consider how the decisions to adopt new technology influenced (or were influenced by) the fabric of the expanding city. Hans Kollhoff highlights the replacement of Berlin's primitive cesspools in the 1870s by the most up-to-date sewage system in the world. And he brings out the importance of the new railway system in shaping the appearance of the city. But the main feature for Kollhoff of the developing German capital seems to have been the tension between utilitarian and aesthetic considerations in the new building types. His discussion of one of these new building types – the Panorama, a giant structure containing within it a panoramic picture on public display – is revealing: the original propaganda piece depicting the greatness of Prussia in history was replaced by a huge image portraying the technological wonders of New York.

These allusions to the inspiration of American technology, occasional in Kollhoff's piece, are developed in reading 26, by Thomas Hughes, into a full-blown comparison between Berlin and New York (the two texts originally appeared in the same volume devoted to that comparison). Here Hughes categorizes each city as a 'technological metropolis' and a product of the Second Industrial Revolution, a revolution largely engendered by applications of the new electrical science. Berlin, by the 1880s, was the site of the world's largest electrical, manufacturing companies, which adopted the system of the American inventor, Thomas Edison. As Hughes shows, other American developments also strongly influenced the structure of Berlin: scientific management and mass production, adopted by German architects after the First World War to solve Berlin's dreadful housing conditions. Perhaps the most interesting conclusion deducible from this text is that there was no uniform progressive march of modern technology. Hughes rightly points to the entirely different traditions of Berlin and New York - distinctive historically, politically and economically - to explain why the same modern technology was guided along different channels in the two cities.

Bombing in the Second World War brought devastation to many European cities. Even before the war was over, detailed plans were devised to rebuild and repair unprecedented levels of damage to the urban fabric. Readings 27 and 28 are a sample of this reconstruction in Britain and Germany. Reading 27 deals with the programme of the 1940s for prefabricated houses, an emergency measure for rectifying the acute housing shortage in Britain's bombed-out cities. Production was organized by collaboration between government and industry. Techniques of mass-production and the assembly-line allowed a house to be completed within 12 minutes. The detailed account here shows how it was achieved and how the product, originally intended as 'temporary' accommodation, continued to function for many years. Especially interesting is the transfer of war-industry factories and materials for peacetime civilian use. Aluminium, the material of aircraft, was now used to build prefabricated bungalows, and the factories and workers that once turned out fighter planes were now, in peacetime, converted to the rapid manufacture of houses.

In the following reading 28, by Jeffry Diefendorf, the setting is post-war Germany and the focus is traffic planning for the projected rebuilding of destroyed areas at the centres of historic cities. The bombing created new opportunities for a radical re-planning of street layout with attention concentrated on traffic, redirecting or channelling traffic to relieve congestion and so to bring new life to these destroyed cities. Hannover, Hamburg, Stuttgart and Cologne are all briefly examined, investigating the sources of the new plans, their reception and what was achieved. The solution was not the same in every case. The views of Cologne's planner-in-chief stand out for their resistance to projects that gave priority to traffic, seen as an unacceptable path that amounted to the dehumanization of cities by technology. That recurring issue in urban history of the expropriation of private property for sweeping change again arises here, though Diefendorf plays down its importance in Germany's reconstruction. What mattered most in the end was that the planners' expectations were rapidly overtaken by wholly unforeseen levels of car ownership. The eventual result was to give back the centres of German cities to pedestrians.

18

URBAN TRANSPORT AND THE DEVELOPMENT OF GLASGOW'S WEST END, 1830–1914

Michael Simpson

Source: *Journal of Transport History*, 2nd series, 1 (1972), pp. 146–60

Between 1800 and 1914, Glasgow grew from a town of under 100,000 to a city of a million, with almost a million more people living in the rest of the Clydeside conurbation.[1] This extensive urbanization was accompanied by the growth of a complex transport network. In the case of the western suburbs of the city, most forms of transport followed, rather than led, suburban development. The experience of Glasgow's West End is similar to that of the suburbs investigated by Dyos and Kellett.[2] Of the four major forms of transport provided in the Victorian West End, only the first, the Great Western Road, a turnpike, could be said to have contributed greatly to urban growth. The trams and commuter trains arrived after 1870, by which time the main lines of the suburb's development had been laid down and at least half of the present number of houses and flats had been occupied. The Glasgow District Subway, opened in 1896, made no discernible contribution to urban growth and the most that can be said for the tramways and railways is that they aided the filling-in of gaps in the development pattern around their routes. In no case did either of them build lines and stations ahead of substantial urban growth and termini were always located behind the western edge of the developed area.

The West End proper is an area about three miles in length and up to a mile in width, which lies west of Charing Cross and extends as far as Anniesland Cross. During its period of development as a suburb, it was inhabited exclusively by the upper middle class. Householders, listed in the annual Post Office directories were the leading mercantile, industrial, banking and professional people in the city. [. . .] Many of these people also owned properties at the seaside or in the country. West End houses were never advertised at less than £600 and the average asking price was always around £2,000, with some houses rising as high as £7,000. Many had stables and most families had servants living in. In the Burgh of Hillhead, in the heart of the West End, there was one servant for every six people in 1890. High-class tenements were of at least five rooms and often rose to ten, while the average house consisted of between ten and 14 rooms. The operation of market forces, the absence of competitors for most of the nineteenth century and restrictive covenants on development maintained the exclusiveness of the West End almost intact down to the First World War.

Before 1800, Glasgow's middle classes had lived above their counting houses, which were mainly located west of Glasgow Cross, then the city centre. Industrialization rendered the old town crowded and dirty. Fear of the masses, distaste for the polluted atmosphere of the inner city, and the demands of business for more office space, combined with rising real incomes and changing fashions which decreed the separation of home and work, led to the wholesale movement of the middle class to a new suburb, Blythswood. This was one mile west of the central business district and adjacent to it. Extensive speculation in these lands began in 1802 and by 1830 it had resulted in a fine suburb of classical terraces laid out on a grid pattern, the streets climbing steeply over the drumlins (low, conical mounds of boulder clay which are features of the north shore of the lower Clyde). Even before Blythswood was fully built up, its earliest parts were being converted to business use. Attempts to develop other districts of Glasgow for middle-class suburbs were frustrated by pollution, the rapid growth of industry and railways and the flood of working-class immigrants. [. . .] Since long-distance commuting was out of the question in 1830, there was only one area to which the upper middle class could turn, the West End. At this time, it was a wedge of country lying between the already burgeoning industrial and proletarian suburbs of Partick and Maryhill. It was divided into a score of small and medium-sized estates, occupied by *ersatz* country gentlemen - city lawyers and merchants.

The West End had many advantages for the upper middle class. It abutted onto Blythswood and the Yoker Turnpike (now Argyle Street and Dumbarton Road) formed its southern border. The south-eastern corner therefore had good communications from the first. The early terraces were provided with stabling and it is probable that many business and professional families travelled about the city by private carriage or hansom cab. Horse buses plied on the city's principal streets by the 'forties. The West End was upwind of existing industry. It stretched across five drumlins, which gave splendid views, and their steepness discouraged industry and working-class housing from locating there. A high standard of architecture and layout was adopted by all the estates in their developments. The feu contracts (equivalent to English long leases), by which the ground was leased for building, strictly regulated the development of the area.

Although the south-eastern corner was quickly built up between 1830 and 1860, the bulk of the West End was more or less inaccessible from town. One contemporary described such roads as existed as being unfit for driving cattle.[3] The major part of the West End could not be developed for either industrial or urban uses until it had direct all-weather communications with Glasgow. The need was met by the Great Western Road, a turnpike authorized by an Act of 1836. [. . .]

The Great Western Road, which ran straight as an arrow from St George's Cross, where the city streets ended, through the heart of the West End to Anniesland Cross, a distance of just over three miles, was opened in the winter of 1840-1, together with two feeder roads, Byres Road and St George's Road South, which ran south to link up with the Yoker Road, and another connection was made via the Yoker Turnpike's branch, Crow Road, which ran north to Anniesland Cross. [. . .]

The history of the Kelvinside estate provides an example of the importance attached to the Great Western Road by West End landowners. This 576-acre property was by far the largest in the West End. It was purchased in 1839 by two solicitors, Matthew Montgomerie and John Park Fleming, and the inventor, James Beaumont Neilson. The bulk of the property lay either side of the proposed Great Western Road. Montgomerie, the most dynamic of the three, had 'an implicit belief that the best class of houses in the rapidly-expanding suburbs of Glasgow must ultimately come to be built on the line of Great Western Road'. Hence, when the road ran into financial difficulties in its early years, Montgomerie led the rescue operation and became the most active of its Trustees.

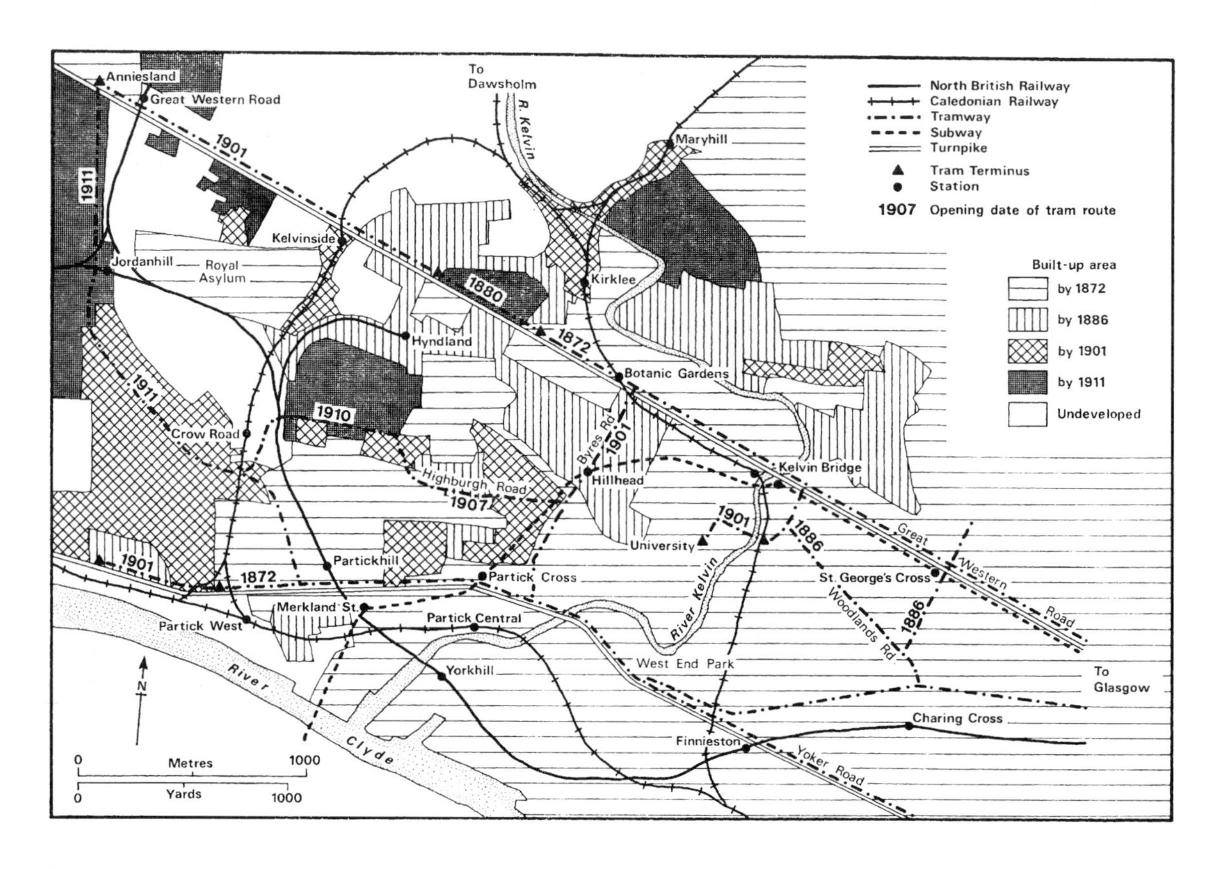

Figure 18.1 Transport and urban growth in the West End of Glasgow, 1830–1914

He realized the crucial importance to his ambitions for Kelvinside of this broad, straight, metalled shaft leading straight from the city centre to his property. [. . .]

The Great Western Road struggled for a decade from its opening but by the 'fifties Glasgow was experiencing its great mid-Victorian boom with a consequent rapid growth of its suburbs. The Great Western Road therefore profited from a substantial rise in both traffic and income. By the time of its demise in 1883 (it was taken over by the City of Glasgow and, outside the city, by a County Road Board), it had just about cleared its capital debts and interest charges. Although not an outstanding success considered as a business enterprise, it fulfilled a major role in the urbanization of the West End. Unlike the Yoker Road, which only tapped the southern edge of the area, it neatly bisected the West End and brought it within reach of the middle-class business and professional men. Without it, it is difficult to conceive of the West End being built up so quickly or so attractively, since it acted as a great axis round which the West End estates were planned.

The Great Western Road made the urbanization of the West End possible. It could not determine the character of urbanization, for that was the decision of the developing landlords. Moreover, while it is probable that its excellence as an artery to and from town quickened the pace of growth, its ability to influence the rate of development was limited. Even after it opened, Kelvinside was still considered 'quite in the country'. Apart from communications, two other major factors influencing the rate of growth were the extent of middle-class demand for housing and the advance of the Glasgow built-up area to the eastern edge of the West End. Demand began to rise steeply from the 'fifties and the built-up area reached Hillhead at about the same time. The Glaswegian bourgeoisie did not appear willing to leapfrog far into the countryside at that time. Many West End estates had building land on offer for 20 years or more before it was taken up. The experience of the Kelvinside Estate Company illustrates this well. It used many devices to stimulate feuing. Perhaps its most direct inducement to prospective settlers was the horse bus service which it subsidised from 1847. Operated by one of the principal local liverymen, John Ewing Walker, it ran from Botanic Gardens to Glasgow Cross. At first a two-hourly, one-horse bus, it became in the 'fifties a half-hourly three-horse car. Wylie Lochhead, another well-known livery stable, ran a bus along Great Western Road from about the same date as Walker's but at the instigation of the Road's Trustees. Despite the Company's energetic attempts to hasten the feuing of its lands, growth was very slow. It was not until Hillhead was well developed - in 1869 it was created a Burgh and had over 3,600 inhabitants - that Kelvinside began to grow with any rapidity; from 1868, feuing went ahead at a steady pace. The West End as a whole grew most rapidly between 1870 and 1890. In 1830 its population had been about 600, at a crude estimate. By 1850 it stood at only about 1,350 but in 1870 it had risen to around 8,000. By 1890 it was up to 20,000 and thereafter rose slowly to 26,000 or so by the outbreak of war in 1914. If Partick, Maryhill, Jordanhill and Anniesland, the other western suburbs, are added, the 1870 population was about 38,000, rising to well above 50,000 by 1880 and about 65,000 in 1890. The final pre-war population was around 90,000.[4] This was the area taken into consideration by the planners of the tramways and railways.

The period 1870–1914 was the great age of tramway building and Glasgow was one of the first cities to take up the new form of urban transport. The Corporation promoted an Act in 1870 which allowed it to construct and lease lines to a private operating company. In 1871, it signed a 23-year lease with a London-based company, the Glasgow Tramways Company. The first lines began working in 1872, using horse-drawn cars. During the 'Company era' (1872–94), routes stuck to the principal roads and never ventured beyond the edge of the built-up area. Although owned by the Corporation of Glasgow, lines were laid down both inside and outside the city. In the west, lines ran along the Yoker Road to Partick via both Argyle Street and Sauchiehall Street. The principal West End route was naturally along Great Western Road, from Glasgow Cross to Botanic Gardens, which was then about half a mile east of the limit of the built-up area. The trams averaged 6 m.p.h. as against the horse buses' 4 m.p.h. but were just as draughty and uncomfortable. However, they did provide a more frequent and efficient service to town, running on a $7\frac{1}{2}$-minute headway for 16 hours a day. Fares to the city centre from the West End were *2d.* or *3d.*

Throughout the 'seventies, the Company claimed that it could not make even its most popular routes pay and it could not afford to extend existing services. New communities desired tramways, 'the advantages of Tramway communication having now been thoroughly recognised'. But both the Corporation and the Company were reluctant to invest in new routes which did not promise considerable traffic from the outset. An early example of this caution can be seen in their response to repeated petitions from Kelvinside residents, supported by James Brown Montgomerie-Fleming, their feudal superior, calling for an extension of the Great Western Road route from Botanic

Gardens to the Royal Asylum, the entrance to which formed the western edge of the built-up area from about 1875. The Corporation, having decided in 1877 that this two-thirds of a mile addition would cost £3,750 to construct and would not be justified by the resulting traffic, declined to make it. After further petitions, a quarter-mile spur was added to the line in 1880, leaving it somewhat short of the fringe of development. The new services which were provided in the 'Company era' all ran in fully-developed areas which were already densely settled long before trams came to Glasgow. For example, in 1886, St George's Road and Woodlands Road received lines which linked Great Western Road with Sauchiehall Street and a short branch from Woodlands Road served the University. Through services to eastern and southern parts of the city were gradually provided in the 'eighties. By this time, the Company was earning a net profit of $9\frac{1}{2}$ per cent.

From the late 'eighties, the question of the renewal of the Company's lease dominated tramway affairs until municipalization took place in 1894 and the debate caused a hiatus in the provision of new services. Even after municipal operation began, no new routes were built until it had been decided which form of mechanical traction should be adopted. Electrification was determined upon in 1898 and accomplished by 1901. The new electric 'caurs' which 'shoodlied' (shook from side to side) down the road became the objects of the citizens' appreciation, affection and pride for 60 years.

The immediate benefits of municipalization for the West End were new through routes and more frequent services. As early as 1896, the Tramways Committee set up a sub-committee on extensions. The first request made to it came from a group of Hillhead citizens who wanted an addition to the University line. This was not undertaken until 1901 and then only as far as Bank Street, at the foot of University Avenue. Although Parliamentary sanction was obtained on several occasions to carry this line along University Avenue to Byres Road, objections were lodged by the Senate of the University which feared damage to the delicate instruments in the laboratories from the vibration of the cars. Residents of University Avenue also protested against the scheme, on the grounds that street noise would be raised to an intolerable level and that traffic would be seriously dislocated. As a result of these objections the line was never laid. In 1907, residents in Highburgh Road raised similar objections to a proposed extension along their road from Byres Road to Hyndland. Everyone wanted trams, it seemed, so long as they were routed along the next street where they would still be convenient without destroying one's amenity. In the Highburgh Road case, the objectors were defeated. Partick Burgh Council and residents and proprietors in Hyndland badgered the Corporation to provide a line from Byres Road to Crow Road via Highburgh Road, Hyndland Road and Clarence Drive. Pressure was put upon the Corporation for three years before it finally agreed to make part of the desired addition, as far as Hyndland Road. Local interests remained unsatisfied and agitation for the full extension continued. They achieved success in 1910 and the new addition linked up with a line along Crow Road from Partick to Anniesland via Jordanhill, the last piece of tramway building in the western suburbs, opened in 1911. Trams ran the length of the Great Western Road from 1901, terminating at Anniesland Cross. A route was also opened in 1901 along Byres Road and Church Street, connecting Dumbarton Road with Great Western Road.

Although the Corporation ultimately proved more amenable to requests for new tramlines after municipalization, it is evident that it did little on its own initiative; most of the West End extensions were the result of persistent pressure by Partick Burgh Council, ward committees, developers and residents. The Corporation was still unwilling to consider new routes which did not guarantee a good return from the outset. For example, authorization was obtained for a line from Kelvinside to Maryhill but it was never built, the Tramways Committee deciding that it was an uneconomic proposition. The General Manager always had to show that a proposed extension would not incur losses. Discussing the suggested Crow Road line in 1908, for instance, he expected 'a fair traffic . . . which would at once yield a revenue which would at least pay working expenses, and would undoubtedly improve.' Nevertheless, this line was not opened until 1911. One of the factors behind municipalization had been the belief that the Corporation could make as big a profit out of operating the trams as the Company. This proved to be true, for the annual net surplus before 1914 never fell below £50,000. Since municipal undertakings were a valuable source of income to the city, public profit was an important consideration in assessing proposals for extensions. Although numerous additions were made to the tramway network of the western suburbs between 1901 and 1911, they nowhere ran ahead of settlement. In fact termini were always located behind the building frontier. A good example of this was the line built from Byres Road to Hyndland Road in 1907. This stopped about a mile to the east of the edge of the built-up area in Partick, despite local demands that it should terminate half-a-mile further west. The role of the trams in the development of the West End appears to have been a limited one. Their chief function was to fill in gaps in residential development. They always followed sizeable populations and appeared in a locality only when all or most of it had been built up. Demands for new tramway services were either totally rejected or satisfied only in part or after several years of campaigning. In the end, trams did the 'aristocratic' West End no good, for they undermined its exclusiveness. The cheap tram and its great rival, the suburban train, were a boon to the speculative tenement builders who dominated West End construction after 1890. They allowed the building of smart, easily run apartments with all modern conveniences for the lower middle class. The tall red sandstone tenement blocks which arose in Hillhead, Dowanhill and Hyndland, and the small semi-detached houses of Jordanhill, allied to fast and convenient transportation, permitted the *petite bourgeoisie* to share the privilege which had for so long been the preserve of the *haute bourgeoisie,* life in a pleasant and easily accessible suburb. By the time that the last rails were being laid on Crow Road in 1911, residential building in the western suburbs had come to a virtual halt. The lack of space for further expansion, together with the loss of exclusiveness, helped to drive the upper middle class out to an outer ring of suburbs – Lenzie, Uddingston, Helensburgh, Milngavie and Bearsden – which were still separated from Glasgow by a green belt which at the same time enjoyed new rail and even tram connections with the city, enabling them to act as commuter suburbs in the modern fashion. By 1914, in fact, many Glasgow tram routes ran for upwards of ten miles, the trams themselves averaging a modest 8 m.p.h., often on 3-minute headways and carrying passengers an average of 2.32 miles for 1*d.*

The railways were curiously slow to move into the western suburbs. [. . .]

The press was apparently quicker to see the traffic potential of this area than the railways. A Glasgow newspaper reckoned that suburban passenger lines would not only be welcomed by residents but would also prove profitable to whichever company seized the initiative. When a Bill to promote a railway of this nature was announced in 1881, the *Glasgow Herald* declared: 'It has been somewhat remarkable that while other districts of the city are accommodated by railways the western district, which contains the greater part of the wealth and population, and especially of the travelling public, should have been overlooked.' The *Herald* estimated that over 200,000 people lived west of Glasgow Cross.[5] [. . .]

All of the city's suburban services, claimed *The Bailie,* a local humorous journal, in 1886, were promoted by local men unconnected with any of the big companies. The main line companies had often come in later with financial assistance, Parliamentary support, directors and equipment, sometimes eventually absorbing the smaller company.[6] An example of this procedure was the history of the Glasgow City & District Railway, promoted in 1881 by several local businessmen and landed proprietors. [. . .]

[. . .] The C.D.R. was a boon to the middle-class commuters of the West End. [. . .] It provided through lines from east to west across the whole of the Glasgow region on the northern shore of the Clyde. [. . .] A writer on the eve of the First World War described the Kelvinside and Dowanhill district as 'a fashionable residential neighbourhood' brought 'within ten minutes of the city' by the C.D.R. 'and for this reason popular with businessmen'. Not only was the city brought closer to the suburbs in point of time but 'the seaside is brought still nearer our doors'.[7] The C.D.R. quickly built up traffic to a level of seven million passengers a year but the Tramways Company's passengers also increased. Rising population in the west and the absence of direct railway competition there except at Hyndland and Partickhill led to this continuing success of the trams. Down to 1914, rail fares remained competitive with those of the trams. From Hyndland to Queen Street, the third-class single fare in 1914 was $1\frac{1}{2}$*d.*, a penny less than the tram. The first-class single fare was 3*d.* and the return $4\frac{1}{2}$*d.*, while the third-class return cost $2\frac{1}{2}$*d.* [. . .]

In the late 'eighties, plans were submitted to Parliament for a Glasgow District Subways Company. After overcoming objections from the railways and other bodies likely to be affected by tunnelling the scheme was authorized in 1890 and the subway opened in 1896. A $6\frac{1}{2}$-mile circular line, swinging first westward then south under the Clyde from the city centre, it was worked until electrification in 1935 by a continuous cable system. Entirely underground and running solely through long built-up districts, it had West End stations at St George's Cross, Kelvin Bridge, Hillhead (Byres Road), Partick Cross and Merkland Street (South Partick). Its chief merit was that it provided an alternative connection with the city and also linked the West End to the South Side. However, it was never a great favourite with the travelling public, who found its stations dingy, its cars uncomfortable and its service unreliable because of frequent breakdowns in the cable haulage. The Company rarely managed to pay 1 per cent on its capital and the service was mediocre until the Corporation purchased the undertaking in 1922. The electrification of the trams and the opening of competitive new routes after 1901 cut off its growth prospects. Because it ran through already densely settled areas, it made no contribution to urbanization.

The contribution of the railways to suburban development was even more marginal than that of the tramways. They catered deliberately to areas already well developed, though in some cases – at Hyndland, for example – their presence led to the filling-in of gaps in the residential pattern. The first commuter line in the west opened in 1886, by which time the suburbs west of Glasgow were so built-up that it could tap a population of around a quarter of a million. Railway building was completed by 1901, when the rival trams were electrified and entered on a final spate of extensions, thus inhibiting the railways from undertaking further construction. [The railway companies] adopted the tactic of letting an associate or subsidiary test the capital market and the traffic potential before being willing to plunge in themselves. Given such extreme caution, it would have been wholly out of character for the main line companies to have laid lines and built stations in rural areas beyond the suburban frontier, or to have sponsored urban development themselves. There was no equivalent in Victorian Glasgow of the Metropolitan Railway Country Estates. The chief advantage of the railways was that they provided the fastest form of travel between the suburbs and the central business district - 10 to 15 minutes from western stations. They were cheap and efficient though smoky and not particularly comfortable. They almost certainly suffered from electric tram competition, which was just as cheap and efficient and was more comfortable, cleaner and had more routes and boarding points. Furthermore, as Kellett has pointed out, the western suburbs were not far enough from Glasgow for the railway's greatest asset - the ability to carry large numbers of people for low fares at relatively high speeds over lengthy distances - to show to the best effect.[8] However, at the same time that lines were being built in the inner suburbs, services were being improved on routes to towns lying beyond the contemporary 'green belt'. In most cases these towns had rail connections with Glasgow before 1880 but not commuter services. It was not only the West End which obtained businessmen's trains in the 'eighties and 'nineties; so too did Helensburgh, Bearsden, Milngavie, Bishopbriggs, Kirkintilloch, Lenzie and Uddingston. The numerous stone-built villas in these towns bear dates after 1880, testimony to their growth as commuter suburbs from that time. By 1888, upper-middle-class residence in these places was common enough for it to be referred to in the hearings of the Glasgow Boundaries' Commission, when it was even claimed that Glasgow businessmen were also willing to live as far away as Edinburgh (45 miles).[9] Not only did the new rail services of the late nineteenth century enable the upper middle class to live further afield than the West End, they also permitted the lower middle class to infiltrate the West End. As in the case of the trams, they encouraged the building of tenements for the *petite bourgeoisie* on vacant plots round about their stations. The best example of this railway- (and tram-) inspired gap-filling is the Hyndland estate, which lies on the south side of the original Hyndland station and which was feued out exclusively for tenements after the coming of the railway in 1886. Like the trams, the trains were a mixed blessing for the high-class West End, for they, too, helped to destroy its exclusiveness.

In assessing the contribution of transport and communications to the development of the West End, one is struck by the immense importance of the Great Western Road. While this could not decide the class of building and could only partially influence its timing, it did make urbanization feasible and probably brought it forward by perhaps 15 or 20 years. Given the availability of land, capital and a potential market, the only other factor necessary for suburban growth was direct, all-weather communications with the central business district. The Great Western Road enabled developers to exploit the other advantages of the West End - clean air, scenic beauty and convenience. Finally, the road acted as a point around which estates along its length could be handsomely laid out. The contribution of the horse buses is more difficult to analyse, owing to the lack of evidence. However, it is significant that the owners of the largest estate, Kelvinside, thought that a horse bus service would induce prospective feuars to reside there; they actually subsidized one as early as 1847. At that time, judging by the presence of stables behind nearly all West End houses, most people journeyed to work by private carriage or by hansom cab. The Great Western Road, a metalled road, 60 feet wide and dead straight, was admirably suited to carriage-drawing.

In the case of both tramways and railways, promoters of lines were attracted by the prospect of an immediately lucrative traffic, since substantial development had already taken place in the West End and neighbouring suburbs. Neither the tramways nor the railways showed any desire to move ahead of sizeable populations; indeed, tramway officials were only persuaded with great difficulty to extend their lines up to points near the urban frontier. The railways trod very gingerly and came to the West End only when it was two-thirds complete. Nevertheless, although neither trams nor trains did anything to advance the boundaries of the built-up area, they did facilitate the feuing of isolated vacant plots within developed districts. They probably hastened this process and they certainly brought a suburban residence within the reach of the lower middle class. Their arrival and the new opportunities afforded by electric trams and particularly trains to live at a quite considerable distance from Glasgow (7–20 miles) led the upper middle class to leapfrog beyond the existing built-up area to old villages and small towns. Although the upper middle class made full use of trams and trains once they were provided, they were sufficiently independent of them because of their wealth to use more exclusive forms of transport such as the private carriage and the hire cab; this ability to afford expensive and private transportation allowed them to settle in the West End well before the trams and suburban trains were thought about. Landowners and builders, too, were prepared to undertake development throughout the period in advance of tramways and railways. Apart from limited areas within the developed region which they helped to fill in, both tramways and railways were stimulated by existing growth rather than themselves initiating it. Indeed, it is unlikely that they could do otherwise, for whether they were in private or public ownership, both forms of transport were primarily concerned with making profits. It is evident from the history of the tramways and railways that transport managers considered traffic potential very carefully before committing large amounts of capital to serving the western suburbs. If cheap mass transport was of dubious benefit to the upper middle class of the West End, for the working-class populations of Partick, Maryhill and Anniesland it was largely irrelevant. Considerable settlements grew up there well before the coming of the trams. The inhabitants of these mining, quarrying and industrial towns lived close to their employment and had no great need of urban transport. Until at least 1890, they were all distinct communities, self-sufficient and independent of Glasgow. Apart from the speculators who built tenements in the left-over bits of Hyndland, Partickhill, Anniesland Cross, Hillhead and the North Kelvinside quarter of Maryhill, the people who really benefited from cheap mass transportation were the members of the rapidly growing army of white collar workers who, from the late 'eighties, were able to afford new flats at 10 to 15 shillings a week and a threepenny ride to and from town each day, provided that it took no longer than half-an-hour.

Notes

1 B. R. Mitchell, *British Historical Statistics* (1962), 24–7.

2 H. J. Dyos, *Victorian Suburb* (1961), 63, 70–2, 76, 79–80; J. R. Kellett, *The Impact of Railways Upon Victorian Cities* (1969), 354–82.

3 W. Simpson, *Glasgow in the Forties* (1899), notes to plate 43.

4 Estimates of population for such a composite area are necessarily crude but the basis is: *Post Office Directory* maps; Burgh of Hillhead, *Official Lists, 1869-91; Hillhead and Kelvinside (Annexation to Glasgow) Bill* (1886), vol. I, 602; vol. II, 421, 473, 533–41.

5 *Glasgow Herald*, 18 November 1881.

6 *The Bailie*, 17 March 1886.

7 *Railway Times*, 23 October 1886.

8 Kellett, *op. cit.*, 354.

9 Glasgow Boundaries Commission, *Report*, vol. II (Evidence) (1888), 350.

19

THE SHREDDED WHEAT FACTORY AT WELWYN GARDEN CITY

Richard J. Butterfield

Source: R. J. Butterfield, 'The Industrial Archaeology of the Twentieth Century: the Shredded Wheat Factory at Welwyn Garden City', *Industrial Archaeology Review* 16 (1994), pp. 196, 198–9, 201–15

Introduction

The record of Industrial Archaeology is impressive. From obscure amateur origins in the quite recent past, it has recorded innumerable industrial sites, developed techniques of survey and analysis, influenced planning decisions and raised public awareness of Britain's industrial inheritance. However, these successes almost all relate to the period of the Industrial Revolution and its aftermath. Industrial Archaeology remains wedded to the agenda of iron, steam, canals and textiles set during the 1950s and 1960s, and still fails seriously to address the problem of understanding industry as it has developed in this century.

To some extent, this is due to a simple dislike of the twentieth-century landscape. In W. G. Hoskins' opinion, since the Industrial Revolution and 'especially since the year 1914, every single change in the English landscape has either uglified it or destroyed its meaning or both.'[1] J. B. Priestley, writing in the early 1930s could not take the new factories on the Great West Road out of London seriously:

> These decorative little buildings, all glass and concrete and chromium plate, seem to my barbaric mind to be merely playing at being factories. You could go up to any one of the charming little fellows, I feel, and safely order an ice cream or select a few picture postcards. But as for industry, real industry with double entry and bills of lading, I cannot believe them capable of it.[2]

Priestley welcomed the changing nature of industry, but could not help thinking that the new England of the 1930s was lacking in the 'character, zest, gusto, flavour, bite, drive and originality of the old.'[3] Respect for hard physical labour, dirt and smoke touches the hearts of many industrial archaeologists, but the machismo of those qualities is not to be found on the sanitised factory floor producing delicate cosmetics and fancy goods.

Nevertheless, the 1980s saw a growing appreciation for twentieth-century styles as urban redevelopment and changing business organisation placed relatively recent additions to the landscape under threat. The Firestone building for example (among the 'pleasing facades' derided by Priestley in the 1930s), was demolished in 1980 to make way for a retail park. This is not an isolated example, but is symptomatic of the changing nature of business activity.

Variously described as 'New Industries'[4] and the 'Second Industrial Revolution',[5] twentieth century industry has usually been identified with the manufacture of motor vehicles, electrical engineering, precision instruments, artificial fibres, petro-chemicals and consumer durables. However, the distinction between

Figure 19.1 Thomas Ambler's Ardsley Mill dates from 1912. A state of the art ferro-concrete building constructed on the Hennébique system, it was purpose-built to house the traditional industry of worsted spinning

'old' and 'new' industries is unhelpful in highlighting changes in the nature of site evidence. More useful, is to identify new and old forms. The twentieth century can then be identified as a period of change characterised by the use of highly specialised production machinery, employing electric power to manufacture large volumes of increasingly differentiated products in scientifically laid-out plant, constructed using structural steel and concrete (Figure 19.1). All of these stylised features were available in the nineteenth century, but it was only in the 1920s that they began to see widespread application in Britain in both 'new' *and* 'traditional' industries.

Figure 19.2 A publicity photograph of the Shredded Wheat factory showing the initial form of the plant

Welwyn Shredded Wheat factory

The Shredded Wheat breakfast cereal was produced in the U.S.A. from 1892, after the discovery of the shredding process by Denver lawyer Henry Perky. Exports to the U.K. began in 1908, and this trade soon became large enough to justify building a British factory, which opened in 1925 in the new town of Welwyn Garden City (Figure 19.2). The factory, its products and the organisation of production make this an archetypal twentieth century industry. Most of the original structure remains, even after huge changes in the nature of production and the building of major extensions in 1939 and 1959.

Today, cleaned and pressure cooked grain is still shredded between pairs of smooth and grooved rotating rollers on the same pattern as in 1925 (Figures 19.4 and 19.5). The grain is squeezed into the grooves and sheared into strands through the differential speed of the rollers, before being cleared by a steel comb and formed into a bed on the conveyor which moves below. At the end of each line of rollers, a rotary cutter pressed the bed into the familiar Shredded Wheat shape ready for the ovens.

Packing technology presents a very different and discontinuous story. The latest machinery is fully automated with site examination

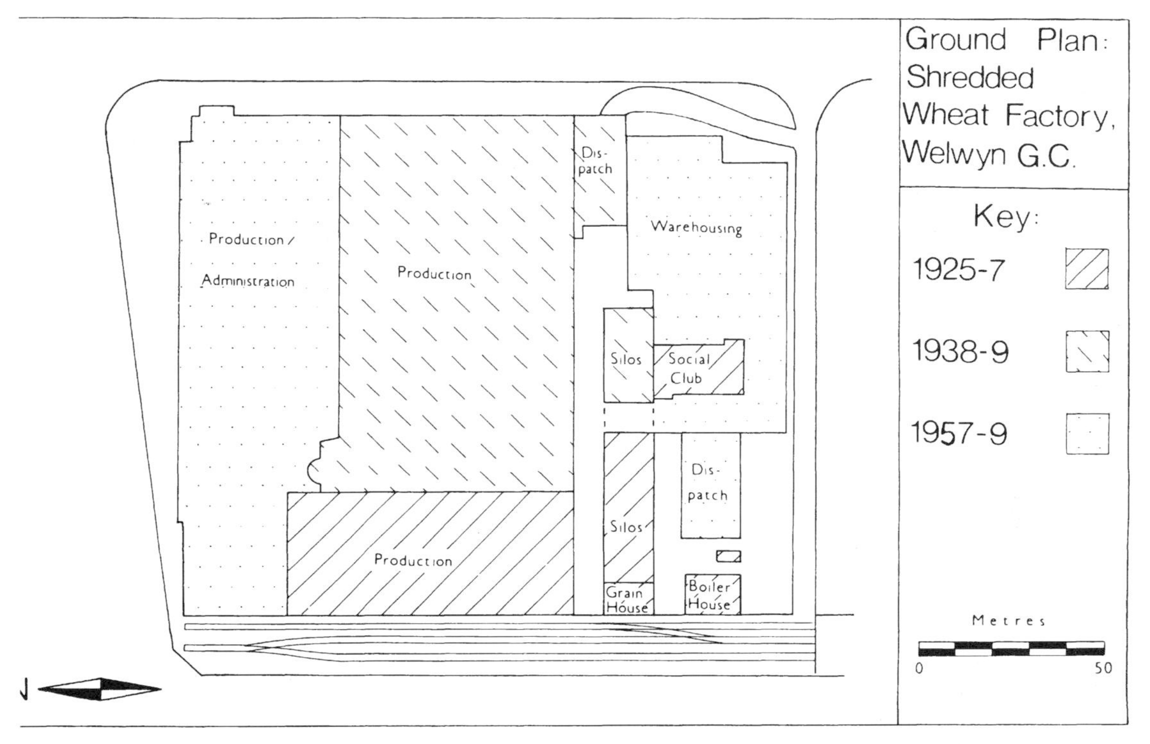

Figure 19.3 Ground plan of the Shredded Wheat factory, showing site development from 1925 to 1959

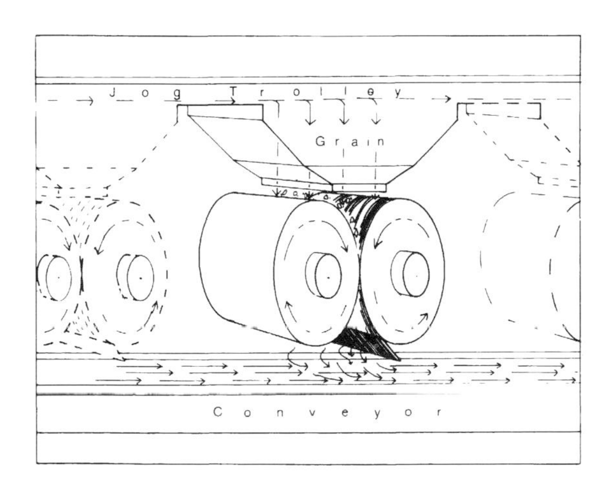

Figure 19.4 Diagram showing the operation of the grain-shredding rollers. (Not to scale)

Figure 19.5 An early view of the shredding machinery

showing it to be based around a number of clearly defined operations. As the biscuits come out of the ovens onto a conveyor, they are automatically divided into groups of three for wrapping. Flat packs are automatically lifted by suction and formed into packets before the wrapped biscuits are pushed inside. The sell-by date is stamped on the lid of each packet before being sealed. Quality control consists of each packet being weighed and passed through a metal detector. Six people are needed to supervise the entire operation.

The original packing equipment has long been scrapped so oral history accounts, newspaper reports and archive photographs must be used to reconstruct the packing process. These sources tell us that Shredded Wheat was arranged in trays of forty-eight biscuits and manhandled into electric conveyors which took them through one of three packing lines. Packers sat eight on either side of each line and put the biscuits into packets as they passed (Figure 19.6). Remarkably, this labour-intensive system employing sixteen packers on each line survived until 1960 when automatic wrapping and packing machinery was installed (Figure 19.7).

When the factory opened, it was producing in the region of 100 million Shredded Wheat biscuits per week. By 1990, this had risen to 500 million with greatly increased productivity and with large volumes of other products being manufactured as well. A combination of sources is needed to assess the contribution of capital equipment to this process. Technical drawings and archival photographs show that the shredding machinery has been little modified, but tell us nothing about how the machinery was operated in practice. Newspaper reports of the opening of the plant are needed to show that only seventeen or eighteen of the pairs of rollers in each bank of twenty-nine rollers were used at any one time.[6] Consultation with production staff reveals that this same ratio is employed today to produce the desired thickness of shredded grains and to prevent overheating. This reduces the scope for increases in throughput, and output at this stage of production has been raised by the construction of additional shredding lines.

Figure 19.6 Hand-packing tables, around 1930

Figure 19.7 The first automated packing line to be used at the Shredded Wheat factory, newly installed in 1960

Throughput in packing was also increased by the multiplication of existing technology. From 1960, however, the automation incorporated in investment in new packing lines was aimed less at increasing Shredded Wheat production (which was already reaching a peak) than improving productivity.

In both these cases, machinery (and hence its remains) is only one component in productivity. The intensity with which machinery is used is crucial. An extension of the working week from five to seven days brought huge increases in output, but had no identifiable effects on the nature of site remains. If one of the early packing tables had survived, it would be unlikely to offer any significant new information. This is because the dexterity needed for packing lay with the packers; little complexity was built into the machinery itself. Documentary and oral history sources are needed to fill in the detail of how the packers actually worked. Modern automated packing machinery is more complex, with all of this information programmed into its design. This gives it great potential as a historical source, shedding light on working practices as well as the state of technology.

Power systems

[. . .] In the twentieth century, electricity, oil and gas have become the dominant modes of artificial power in Western economies. The flexibility of these sources makes power a relatively minor constraint on location, outside industries with specialised requirements such as aluminium manufacturing which uses vast quantities of electricity and is invariably located near generating plant.

Data from the Censii of Production allows movements in power use to be quantified in the twentieth century. This shows the early part of the century to have been a key period of transition with the proportion of power applied electrically in U.K. factories rising from 24.4 per cent in 1907 to 66.2 per cent in 1930.[7,8] [. . .]

The Shredded Wheat factory has never had a power generating facility on site, with electricity and gas being supplied from power stations and gasworks. [. . .] Nevertheless, power has to be applied to machinery, and this involves supply and transmission systems. Technical drawings and archive photographs show that a single electric motor powered a whole line of shredding rollers until at least the late-1950s (Figure 19.8), whereas small electric motors are now fitted to each pair of rollers.

Figure 19.8 A rank of shredding rollers, powered by a single electric motor in the late 1950s

This yields clear advantages in operation. Individual pairs of rollers can be switched on and off at will so power is not wasted and maintenance can be carried out without interrupting production. Indeed, the improbability of multiple motor or roller failure means that reliability and productivity is increased as well. These points are admittedly modest, but they should be balanced against the limited nature of transmission systems and the absence of a generating facility on site. However, the factory buildings provide additional evidence of power usage.

The initial form of the factory incorporated multi-storey buildings and vertical flows of materials. The grainhouse is still located in the same part of the original complex, raising grain by elevators and passing it through a series of cleaning and cooking processes with the aid of gravity. Gravity could not possibly be incorporated into the statistics, and yet site evidence reveals it to be essential to the character of this section of plant.

Shredded Wheat production is based around grain processing, and initially the production lines were housed in a multi-storey building. This carried with it no advantage in the transmission or application of power due to the use of direct-drive electric motors and in 1939 Shredded Wheat manufacture moved to a new single-storey extension to the factory. [. . .]

The organisation of production

Rationality in the organisation of production is not a new concept. Adam Smith drew attention to the increases in productivity that are possible from specialisation and division of labour as early as the eighteenth century. In his famous description of a pin-factory, he showed how division of labour resulted in levels of worker productivity 240 times greater than was possible by workers operating in isolation. None the less, it is only in the twentieth century that management has been elevated to the level of a science.

Most of the early attempts to analyse production scientifically emanated from the United States of America. F. W. Taylor writing in 1911, put forward a case for what he termed 'scientific management', whereby factory organisation would be the result of careful workplace study.[9] He argued that managers should divide the production process into simple, co-ordinated operations and by laying down rules confine 'brain work' to the planning department. Henry Ford is thought to be responsible for first applying such principles to mass production, and developed moving production in lines in 1913 at his Highland Park car plant in Michigan. Like Taylorism, Fordism places more control in the hands of management who regulate work through the speed of the production line. [. . .]

At the Shredded Wheat factory, the evidence of oral history, archive photographs and the buildings themselves show that the principles of scientific management were applied from the outset. Production planners made a careful study of the parent plant in Niagara, U.S.A., and this formed the basis for the classic 'U'-shaped layout of the factory floor at Welwyn Garden City (Figure 19.9). This layout represents a logical sequence of processes organised to manufacture a single type of product. Each area of

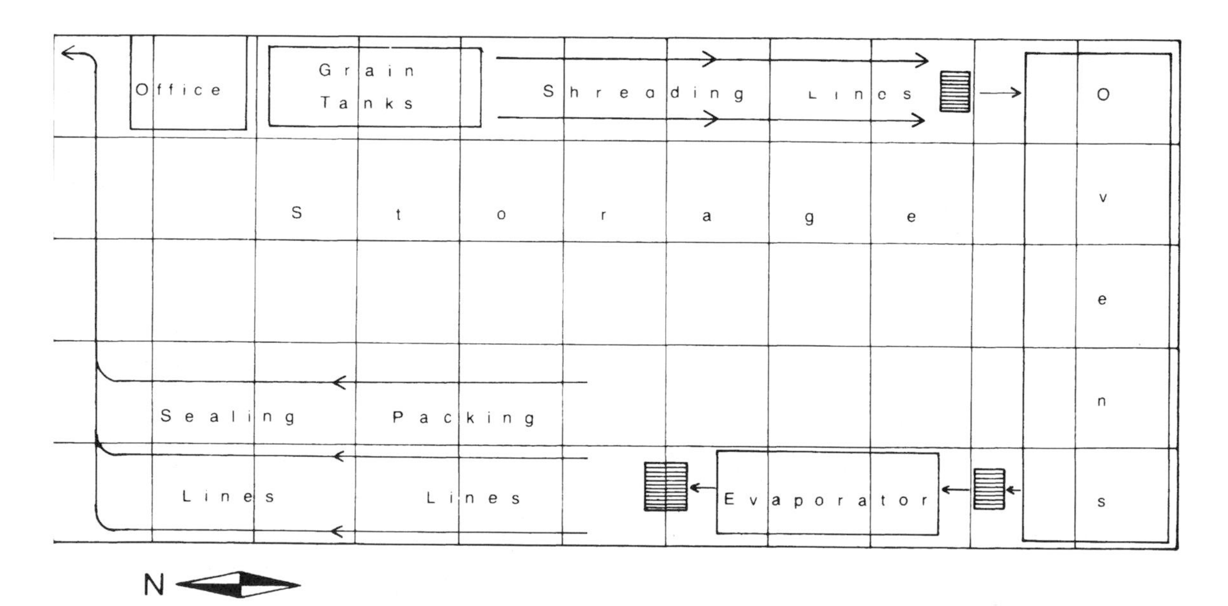

Figure 19.9 The organisation of production on the first floor of the multi-storey section of the Shredded Wheat factory, around 1930. (Not to scale)

the factory floor was specialised around a particular function, though this was compromised by the provision of storage space within the production area. Material handling was the aspect of production furthest from Taylor's ideals, with the throughput of the factory dependent upon the speed of the men loading and unloading the trays of biscuits (Figure 19.10). Little provision for flexibility or development was built into the scheme.

Examination of photographic evidence of the revised production flow installed in the 1939 extension highlights material handling and new product lines to be the major developments (Figure 19.11). A conveyor system took the biscuits to and from continuous ovens before delivering them to the packing lines, dispensing with the need for manual handling.

In 1939 'Small Shredded Wheat' manufacture began and in 1953 'Shreddies' were added to the range. This spread the risk of product failure in the market place, built on existing skills and technology and offered the flexibility to utilise any surplus capacity for the shredded product which was in greatest demand. More specialised functions for managers and workers were developed with a separate building for warehousing and dispatch. The viewing platforms which were constructed to monitor production can still be seen. These developments were on Ford and Taylor's lines with tightly organised production of a narrow range of products for the mass market, bringing economies of scale.

No direct physical evidence survives of the layout of the factory floor before 1959. Instead, oral testimony, machine analysis and documentary sources must be used. However, the buildings remain and add important insights into the nature of the production processes which they contained over the entire history of the factory.

The initial location of the production lines was on the first floor of the multi-storey section of the factory. The numerous thick columns needed to support the machinery clearly represented a major obstacle to the layout of the production lines. Indeed, the weight and bulk of the shredders and ovens provided a further constraint to reorganisation and no substantial changes were possible. By way of contrast, the north-lit extension has a large open floor only infrequently punctuated by the 'I' section rolled steel columns which support a light steel roof. These features provided the extra floorspace needed to accommodate automatic conveyors and new product lines in 1939. High roof space was also provided and this was filled by new continuous ovens. These innovations could not have been incorporated into the

Figure 19.10 Trays of Shredded Wheat biscuits being manhandled in and out of the ovens soon after the opening of the plant

Figure 19.11 A view of the single-storey range around 1939, showing the continuous ovens and conveyors

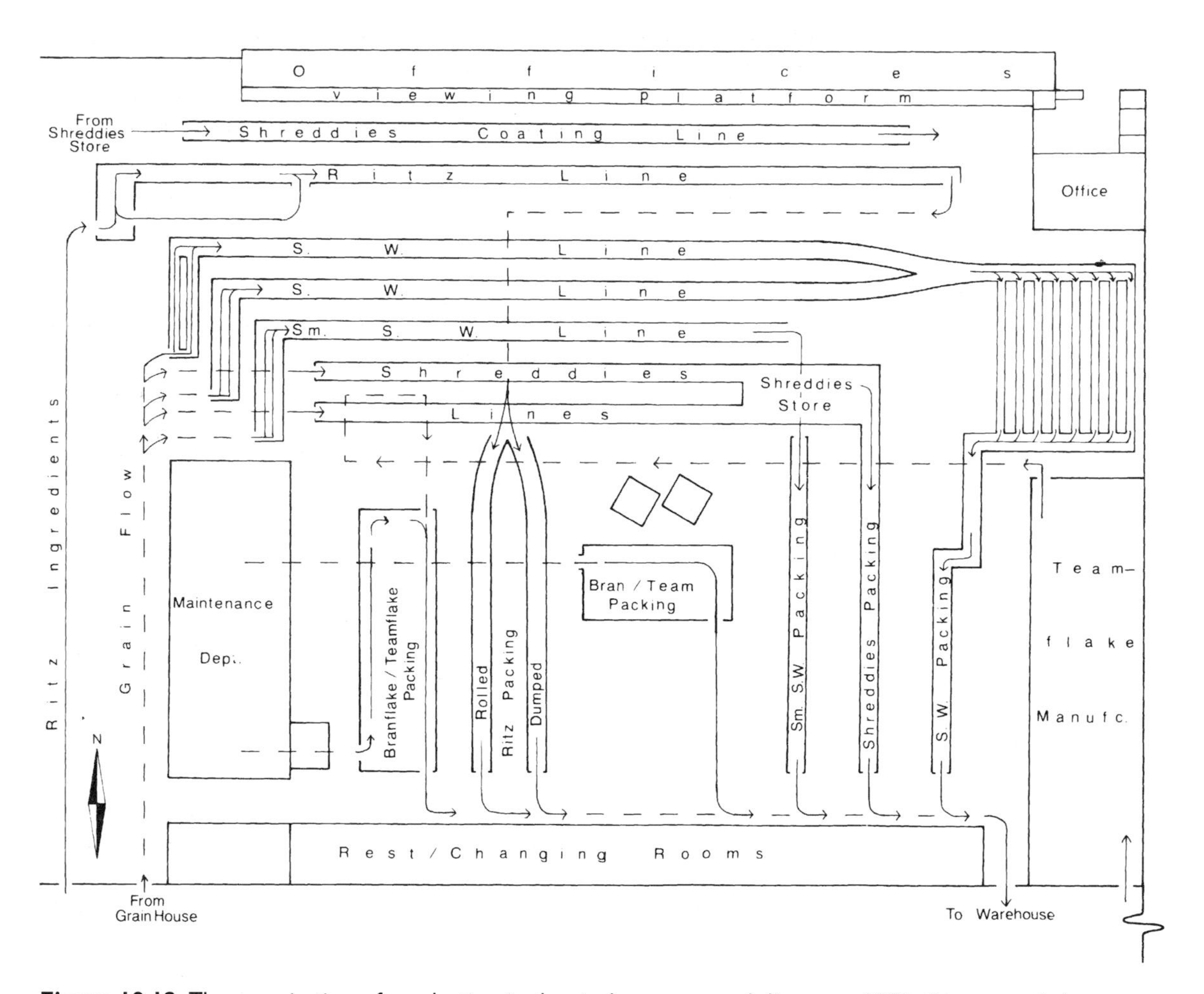

Figure 19.12 The organisation of production in the single-storey north-lit range, 1990. (Not to scale.)

original production flow or the original buildings, so the form of the extension was crucial in facilitating reorganisation.

In 1959 the factory was extended once again. The factory floor has not seen a thorough reorganisation since that time, so it is possible to analyse the existing layout as evidence of production management at the plant over the last thirty years (Figure 19.12). All cereal products at the plant are now either shredded or flaked in nature. The core operations of the factory are the shredded products and these capital-intensive lines are well laid out across the centre of the factory floor. Grain cleaning, shredding and cooking equipment is common to all the shredded products and the parallel siting of the shredders means that material handling is integrated. The flaked products of Teamflakes and Branflakes similarly see integration through common ingredients, with production and packing processes which can be flexibly used for either product.

The recent history of the factory has also seen a trend away from integration as efforts have been made to capture a larger share of the increasingly sophisticated post-war market. In

1955, diversification began with the introduction of Mary Baker cake mixes and in 1961 Ritz Crackers joined the range. Protein Wispflakes, Branflakes and Teamflakes modernised the range of adult breakfast foods and in the 1970s innovatory products such as Golden Nuggets were developed for the children's market.

All of these products used new ingredients and new production techniques. Golden Nuggets required product development, specialised production machinery, staff training and marketing. Though the new product sold well, production could not be integrated with existing lines, costs became prohibitive and along with the cake mixes, production ended. [. . .]

Production solutions to alter the form of shredded grains leave rather more physical evidence, beginning as early as 1939 with the introduction of Small Shredded Wheat. In 1990, this idea was revived by coating Shreddies with cocoa and sugar. Flexibly located on the edge of the production area, an experimental Shreddie-coating line was installed, incorporating hand packing and served with Shreddies by forklift truck. At this stage, flexibility was seen as more important than high levels of productivity or integration until the response of the market could be assessed.

This approach to product development replicates the experience of the existing Ritz line. Ritz Crackers had been imported from U.S. factories from the 1930s, so when home production began in 1961 the company could rely on an existing market base. Nevertheless, the line remains in its original position on the fringes of the production area and since the demise of cake mix manufacture there is little scope for integration with other lines.

A dual system, therefore, seems to have emerged at the plant. On the one hand, well established products are produced on well laid-out, capital-intensive lines with high levels of integration. On the other, lower-tech experimental lines with high levels of added value have been located on the fringes of the factory floor. These lines cross and weave, bearing witness to the thirty years of piecemeal development which produced them. The resulting network provides a stratigraphy with as much potential as any found in the ancient world. [. . .]

The 'master tool' of production

Buildings are probably the type of site remains which are most studied by industrial archaeologists. [. . .]

Study of surviving textile mills has been particularly rewarding. [. . .] The leading role played by mills in eighteenth and nineteenth century industrial building design is related to the pioneering role of the textile industry as the first to combine mechanisation, a central power source and factory organisation for a mass market.

In the twentieth century, these features have come to characterise a much wider range of industries. In the United States, scientific management demanded a similarly scientific approach to building design. Quantity surveying was refined to meet these needs. The A. G. W. Shaw Company of Chicago was one of the pioneers of accurate quotations and accounting to the building trade and referred to factory buildings as the 'master tool' of production.[10] Their publications allowed planners to compare the relative costs of materials and services and dictate precise requirements to architects whose work could be assessed on production-related criteria.

New materials and construction techniques assisted architects in meeting the demands of industrialists. The properties of structural steel had been well explored in a variety of late nineteenth century American civil engineering projects. Bridges and train sheds provided the experience and public confidence to develop wider spans and increasing floor-loads for factories. Structural concrete was developed in

France and heavily promoted from the turn of the century by François Hennébique and Albert Kahn.

The application of science to factory planning and the use of structural steel and reinforced concrete was limited in Britain before the First World War. However, during the 1920s and 1930s a rash of American-owned factories was built in the London area to satisfy the growing British demand for consumer goods. Heinz at Harlesden (1922), Hoover at Perivale (1931) and the Ford Motor Company at Dagenham (1931) were all initiated by American parent companies after careful design based on the experience of existing plants. Their mere presence influenced British architects and designs, and competition from these well-designed and efficient plants forced the pace of change in British manufacturing companies.

Thinking on factory design soon moved on. The lifespan of an industrial building will normally be longer than that of the production equipment which it is designed to house, so it became clear that factory design must incorporate the potential for change and not provide too tight a fit for 'Taylor-made' production flows.

This context for decision-making marks a distinct contrast with the norm for the nineteenth century. In many industries the optimum form of organisation and building structure had been long established. The slow rate of growth and innovation meant that little provision for flexibility was needed. In the twentieth century, the rapid pace of technical innovation means that this is rarely possible. Buildings can therefore be linked with a range of processes during their lifetime, even while remaining in the same industry.

Company records give a picture of the materials and construction techniques employed at the Shredded Wheat factory in the post-war period. Building drawings, plans of the steel reinforcements for the concrete structure and

Figure 19.13 Construction work on the octagonal columns of the Shredded Wheat factory in 1925

some photographs of work in progress record aspects of the construction of the factory in 1925 (Figure 19.13). Records for other phases of development vary greatly, but in any period the buildings stand as a primary source to be consulted to check documents and provide raw data.

The survival of sections of plant dating from a number of different periods allows approaches to particular design problems to be compared. Natural light provides a useful case study. The design of the original multistorey building incorporates mushroom columns which give uninterupted ceiling space, reduce the structural load taken by the outside walls and allow large areas of wallspace to be given over to windows. The problems of variability in light and heat become clear on a visit to this section of the factory and in 1939 north-facing roof lights were adopted for the new extension. The final major extension made to the factory in the late 1950s shows no evidence of the use of natural light for production at all. The roof is unglazed and the factory floor is bound to the north and east by office space. In theory, this path of development indicates that building form has created increasingly controllable environments. In practice, site visits testify that while the impact of the external environment is minimal, problems of heat, noise and dust created by the processes themselves are the unresolved root of the problem.

Each phase of factory development produced a notably different form which can be correlated with the planning processes which created them. The first phase of the Shredded Wheat factory was the product of scientific study of the existing plant at Niagara, U.S.A. This gave valuable experience of the suitability of the buildings and of the optimal organisation of production. The planners were thus able to draw-up a highly detailed design brief which gave Welwyn Garden City Limited's architect Louis de Soissons very little freedom over the form which the plant would take.

The multi-storey design which they adopted was well suited to the initial organisation of production. Some provision was also made for expansion, with the specifications of the foundations and columns capable of supporting a further two storeys. This would have resulted in a factory even more akin to its American parent. In practice, after only fifteen years, not only was this option rejected, but Shredded Wheat production was actually transferred from this section of the factory and relocated in a single-storey extension.

This casts some doubt on the degree of rationality involved in those early decisions. The land and building technology were both available in 1925 to build a flexible, single storey factory. Nevertheless, quantity surveys pointed out to clear cost advantages for a multi-storey building. According to Taylor and Thompson's figures, the multi-storey format offered building costs that were up to 20 per cent cheaper than for a single-storey building.[11] Furthermore, planners could not have foreseen the developments in material handling which were to make a single-storey open plan factory essential in the 1930s.

Shredded Wheat was by no means unusual in making these choices when it did. The Wander Foods factory, purpose-built at Kings Langley in 1913, and the Weetabix factory, centred on a re-used flour mill at Burton Latimer from 1932, are both examples of factories using cereals whose early phases share a multi-storey configuration when very few of their activities actually involved 'milling'. Wallis, Gilbert and Partners built the Wrigley factory at North Wembley in 1928 to a multi-storey format similar to the Shredded Wheat factory, but three years later their designs for factories along the Great West Road were uniformly single storey sheds fronted by two-storey office ranges.

However, the initial design of the Shredded Wheat factory had to satisfy more than just production-related criteria. The start of production at Welwyn Garden City was accompanied by a marketing campaign aimed at exploiting public fears about food adulteration and poor hygiene in shops which weighed and packed cereals from sacks. This could only be effective if the Shredded Wheat factory itself was *seen* to be clean, so people were invited to 'come to Welwyn Garden City and see the wonderful new home of Shredded Wheat, where the "100% food" is made under the most hygienic conditions imaginable' (Figure 19.14).

This approach to product promotion was based upon techniques used in America

Figure 19.14 One of the vans which delivered complimentary packets of breakfast cereal to households throughout Britain in the 1920s. It features an idealised view of the factory on the side

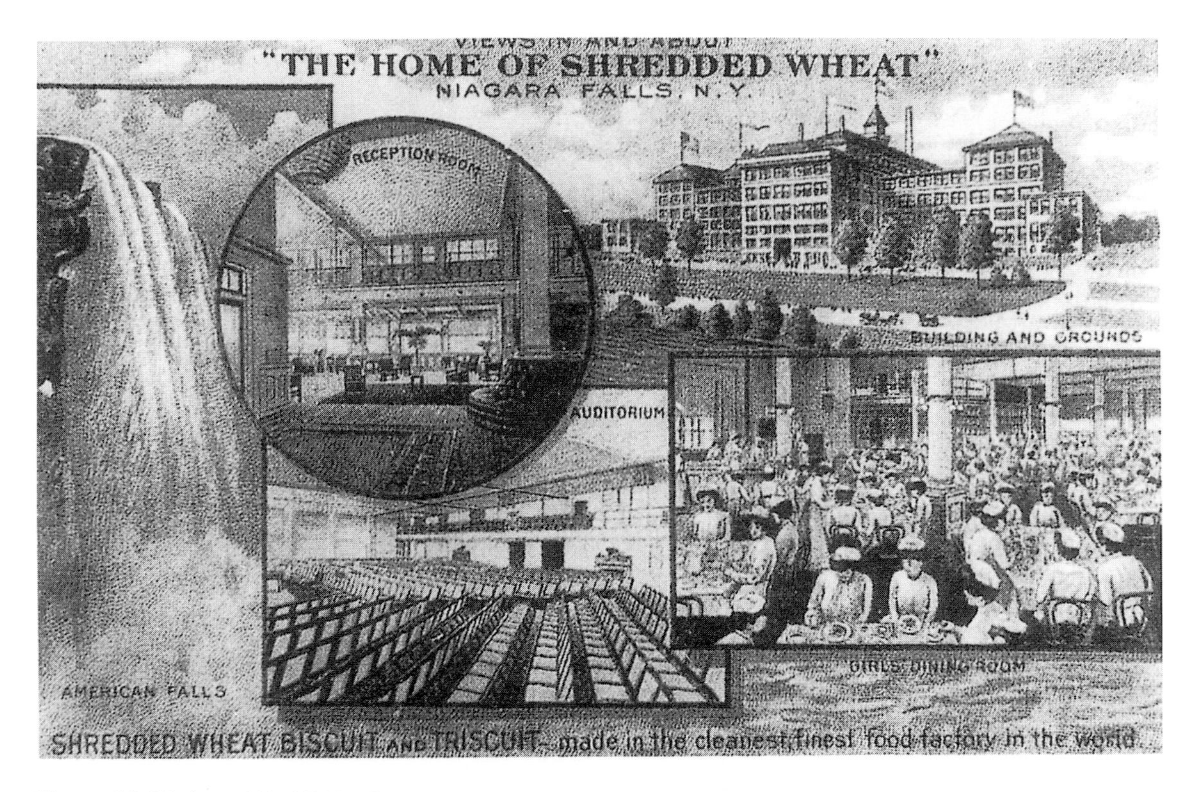

Figure 19.15 A pre-World War One poster promoting the newly opened Shredded Wheat factory at Niagara Falls, New York State, U.S.A.

twenty-five years earlier. The newly built Niagara factory was of pioneering steel frame construction and, under the title the 'Palace of Light', it symbolised the link between wholesome food and the fight against social deprivation[12] (Figure 19.15). The Welwyn Garden City factory too employed striking and innovative design with the whitewashed concrete, ceramic tiles and pollution-free location combining to present an image of hygiene and progress. The multi-storey format added prestige and openness to the site, and gave visual balance against the scale of the grain silos. The factory became the company's principal marketing tool and its image was printed on every packet of cereal down to 1960.

The working practices and social life of the factory were promoted as part of the Shredded Wheat image. At the opening of the plant, the Managing Director Joseph W. Bryce said, 'this is a co-operative family' and described the provision of rest rooms, showers, recreation facilities and a company restaurant.[13] A football pitch and a social club were built on the east side of the factory and they featured prominently in the illustrations printed on cereal packets.

Subsequent extensions have undermined the image of the factory. Architectural elevations and photographs make aesthetic comparisons difficult, so these sources need to be complemented by isometric reconstructions (Figure 19.16). These have the virtue of a common vantage point and reveal that the 1939 extension significantly altered the look of the factory and yet maintained visual balance and appeal. A clean image was preserved through the use of ceramic tiles and whitewashing, and the fac-

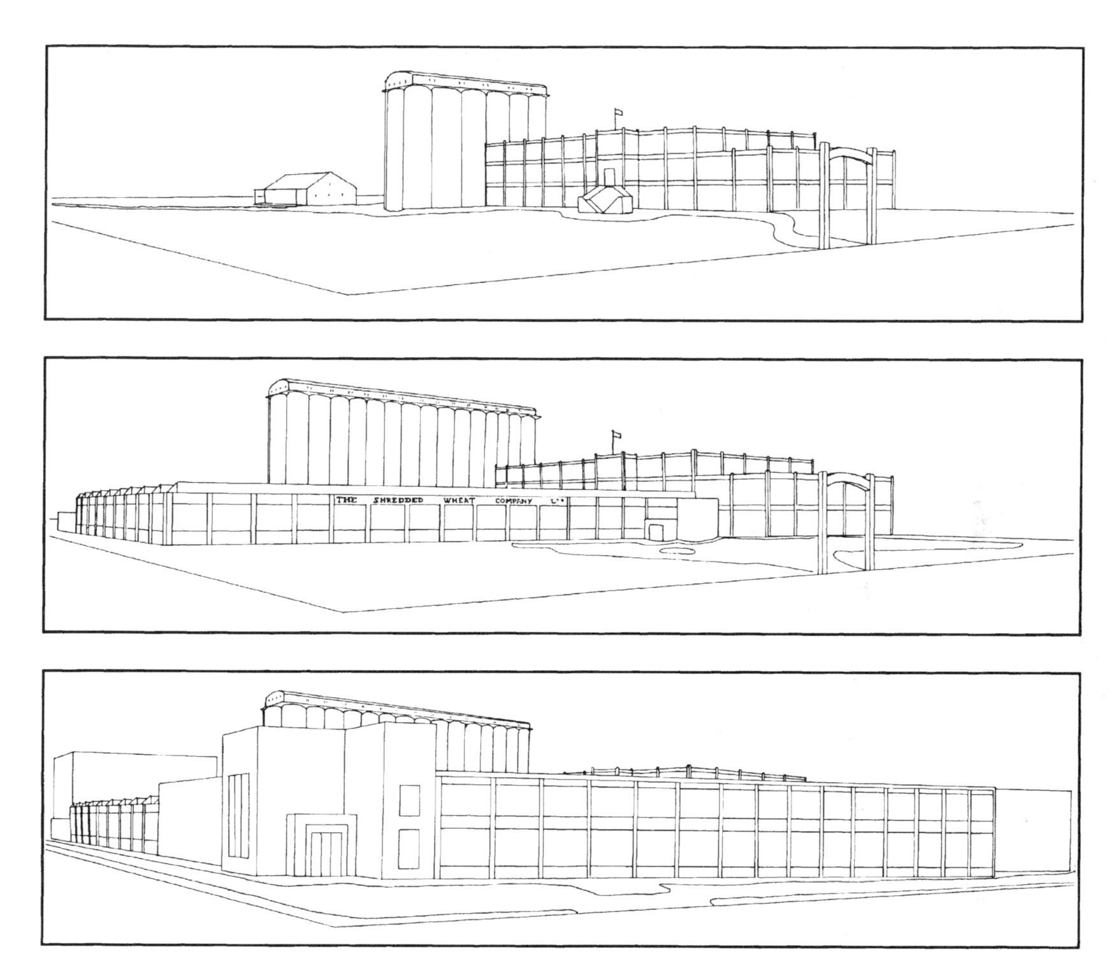

Figure 19.16 Isometric views of the factory from a common vantage point, recreating its form in 1930, 1940 and 1990

tory continued to be heavily promoted. Nevertheless, the on-site football pitch was sacrificed along with the balance between industry and nature suggested by a site fronted by greenery.

The extensions of the late 1950s covered the whole site with buildings and obscured the silos from the view of the road which passes the factory. This busy road is an element of the factory environment which is impossible to control and the row of poplars which now mark the perimeter of the site can do no more than pay lip-service to the garden city ideal. Such circumstances make active site promotion impossible. However, in the mature market which had developed for breakfast cereals, high profile products are generated from a low profile factory, increasingly specialising in production instead of being promoted for marketing purposes. Likewise, the social welfare facilities which were provided to attract the first workers to the new town of Welwyn Garden City became superfluous as alternative entertainments developed in the town and company paternalism itself went into decline.

The Welwyn Garden City factory justifies itself as the 'master tool' of production, but also acts as the 'master tool' of Shredded Wheat archaeology. It provides a record of its construction and the development of services such as lighting. It gives context to machinery, power systems and production flows as they have been applied and reapplied to the buildings. Above all, it bears witness to the changing priorities of planners and highlights the role of marketing alongside the functional needs of production. [. . .]

Notes

1 W. Hoskins, *The Making of the English Landscape*, London, 1955, 298.

2 J. B. Priestley, *English Journey*, London, 1934, 10, 11.

3 *Ibid.*, 379.

4 H. W. Richardson, *Economic Recovery in Britain, 1932–9*, London, 1967.

5 K. Hudson, *Food, Clothes & Shelter, Twentieth Century Industrial Archaeology* (1978), 4.

6 *Hertfordshire Advertiser* (13/3/1926).

7 HMSO, 'First Census of Production, 1907', *Final Report* (1912).

8 HMSO, 'Fourth Census of Production, 1930', *Final Report* (1935).

9 F. W. Taylor, *Scientific Management* (New York, 1974 edn).

10 A. G. W. Shaw Company, *The Library of Factory Management: Volume F Buildings and Upkeep* (Chicago, 1915).

11 Taylor and Thompson, *Concrete Costs*, quoted in Shaw.

12 'The Making of Shredded Wheat', *Courier* (21/7/1910), 171–6.

13 'Factory in Fields. Industry's New Trend. Hertfordshire Enterprise', *Daily Telegraph* (13/3/1926).

20

THE BRITISH POST-WAR TOWER BLOCK: TECHNOLOGY AND POLITICS

Miles Glendinning and Stefan Muthesius

Source: Miles Glendinning and Stefan Muthesius, *Tower Block: modern public housing in England, Scotland, Wales and Northern Ireland,* Yale University Press, 1994, pp. 73–84, 86–7, 89, 153–6

The most important single issue with high blocks was, of course, construction. We can begin with the simple realisation that, for buildings of more than five storeys, the old methods would not do any more. We shall deal here with the major structural innovations from the 1940s into the early 1960s. [. . .]

Our main concern [. . .] is thus the subject of building as engineering. Public housing formed a major constituent part of the postwar history of structural engineering: the most eminent names of the profession, such as Samuely and Arup, recur frequently. At the same time, architects usually remained in control of the overall design process [. . .]

Structural experiment was something that had, up to 1930, been conspicuously absent from the design of flats. There simply did not seem any reason, economically, to change the old methods. In the thirties, a few architectural firms and Modern groups, such as Tecton and MARS, spearheaded the investigation of new kinds of building. It was not until the late forties that more widespread thinking on the construction of flats and a continuous series of innovations began. By 1954, a building scientist, W. A. Allen, and an architect, E. D. Mills, could speak of 'our inability to settle down [to an agreed method of modern building;] . . . we are in a highly creative period where we seek originality at every turn'[1] [. . .]. The first aim of this new 'architectural science' movement was to devise a programme for dealing with postwar economic problems: materials and labour shortages. But it also contributed to a fundamental reappraisal of the methods of building construction, in which economics play only a subsidiary role.

Although later commentators, during the fifties and sixties, tended to belittle some of the results of the 'architectural physics' of the wartime and immediate postwar years, for their lack of immediate applicability, the change in general approach proved to be extremely influential. It was already evident in *House Construction* (1944: the first Burt Committee report), and it appeared more prominently in the third edition of the much-valued Building Research Station handbook, *Principles of Modern Building* (1959).[2] The handbook's first edition, in 1938, had divided the subject into walls, floors and other basic parts of a structure. This was a treatment firmly in the tradition of all builders' manuals: one dealt with methods of construction, with the trades, and with materials. The 1959 edition, however, proceeded in a completely different way. Its subject-matter was divided primarily into the chief demands made upon a structure: namely, that water, wind and cold should be excluded, fire prevented, and daylight admitted; it then explained how various methods of construction could help to fulfil these demands. Primary importance was now attached to what was somewhat later called 'functional performance', and to its scientific measurement; we have already dealt with one of the novel branches of this study, the 'provision' of daylight and sunlight.

Having thus established the priorities, and the logical procedure as a whole, new kinds of scientific precision were likewise crucial for the new methods of construction. Structural design in accordance with these principles, whether with old or new materials, involved various subtle but pervasive changes of practice. 'Measurability' now stood at the beginning of each undertaking. Instead of the old, vaguely 'empirical', 'margin of safety' and ad-hoc provision of additional thickness, one now prescribed exact performance criteria of stability and protection, and allocated the material and devised the structure accordingly.[3] The new principles were most fully developed in the well-known formula which separated 'structural' and 'functional' demands. This, in turn, fitted most closely the favourite constructional model of those decades, that of frame plus infill: the frame to hold up the building, and the cladding to keep out the weather.

One of the results of this new thinking was that many existing building regulations were considered to be a hindrance to efficiency. It was claimed that, because of their lack of a basis in measurement, they had erred unnecessarily on the side of caution. We saw, above, these developments at work in designing for daylight: the inflexible old blanket limitations on building height and proximity were now superfluous, as one could achieve greater density and daylight penetration by manipulating the shapes of buildings from the ground upwards. In the fifties, existing local byelaws were increasingly waived, and in the early sixties nationwide building regulations and standards were established in Scotland, England, and Wales. An epoch of unprecedented lightness in building had set in.

The new science of construction, the drive for economics (linked to the belief in the possibility of large-scale standardisation), fuelled the desire for the industrial prefabrication of building components. Wartime mass production seemed to provide direct models. [. . .]

A further organisational development which greatly facilitated structural experimentation in the construction of flats was the mechanisation of site processes. Buildings of six or fewer storeys could be dealt with by mobile cranes. For higher buildings, the tower crane was introduced to Britain about 1950, and by 1954 some two hundred were already in action. Such contracts were literally 'centred around the tower crane and its capacity'. At Picton Street (see Figure 20.3), the tower crane could lift weights of thirty tons and assist in the in-situ casting process as well. Another major advantage was the fact that it made possible the elimination of scaffolding. In the words of an engineer in the mid-fifties, such a modern building site was characterised by 'complete order . . . and an absence of workmen', in contrast to the traditional 'over-populated, semi-organised rubbish dump'. [. . .]

In the early 1950s, frame and infilling, or frame and cladding, became the almost universally accepted principle of multi-storey construction. As far as materials were concerned, the great majority of very high urban buildings to that date had relied on steel (Figure 20.1). [. . .] Among major London postwar experiments in steel-framing were the St Pancras Way and Cromer Street developments for St Pancras (1946 and 1947), and the Great Ormond Street (Dombey Street) scheme for Holborn of 1946; in Birmingham, the Duddeston and Nechells Unit 1 blocks (1950) were steel-framed. But steel was expensive, and was intermittently in very short supply around 1950. Its use in council housing became extremely rare – a few spectacular examples were constructed in the Clyde Valley steel belt in the early sixties, in Glasgow (Red Road) [. . .] and Motherwell

Figure 20.1 London: Holborn MBC, Great Ormond Street Area, Dombey Street, 1948–9; steel-framed flats. The frame is, of course, not visible but can just be felt in the thinness of part of the elevation.
Source: The Architectural Review (AR) 11-1949, p. 337

(Parkhead Street), and also in Paisley (George Street Stage 2).

The alternative was reinforced concrete (r.c.). The well-known Grindelberg 10- to 15-storey slab blocks in Hamburg of the late forties actually changed from steel to reinforced concrete frame during their construction, apparently saving 20 per cent of the construction costs. Reinforced concrete had progressed considerably in Britain after World War II. It was something that could be made available anywhere and most of its ingredients could be procured locally. By comparison, the design, fabrication and transport of steel structures could be cumbersome. There was now a new lightness in reinforced concrete construction, which contrasted with earlier ideas that concrete had to be something bulky and heavy (Figure 20.2). The use of shuttering, that is, the moulds required for concrete casting, had become rationalised into a number of flexible and precisely organised processes: there were continuous sliding shutters, intermittent climbing shutters and, especially important in our context, shutters standardised to the floor height and wall length of rooms. The concrete mix itself could now be drier than had usually been the case before, provided it was graded more carefully (in accordance with scientific measurement of the various properties of the ingredients), and this lighter mixture could be compacted into the moulds more efficiently by a process of vibration. Additional strength could be achieved, and the amount of steel used in the reinforcement could be greatly reduced, through the new technique of 'prestressing': that is, the wires embedded in the concrete are tensioned while the concrete is poured around them. Prestressing makes the r.c. beam better able to withstand tensions, and was much used for floor construction from World War II onwards. 'Poststressing', on the other hand, was a new method used in strengthening the vertical members of a frame after they had been put together on site.

The majority of 'non-traditional' blocks of council flats, at least until the mid-fifties, were of r.c. frame construction. A tremendous model for designers was the complicated frame of the Unité in Marseilles, where there had also been a midstream change from steel to concrete. In early postwar multi-storey construction in Britain, the frame method was used for an 8-storey block at Paddington MBC's Church Street development (from 1951); the 16-storey main block at the City's Golden Lane development (from 1954), the LCC's point blocks at Portsmouth Road (Alton East) and Roehampton Lane (Alton West) (both 11 storeys); and the Brandon Estate (17 storeys). [. . .] Internal dividing walls and external facing could, of course, be carried out in a great variety of materials, including a smooth brick, as in the

Figure 20.2 Coventry: Tile Hill, flats, diagrammatic view, demonstrating the regularity of reinforced concrete framework construction.
Source: AR 7-1956, p. 29

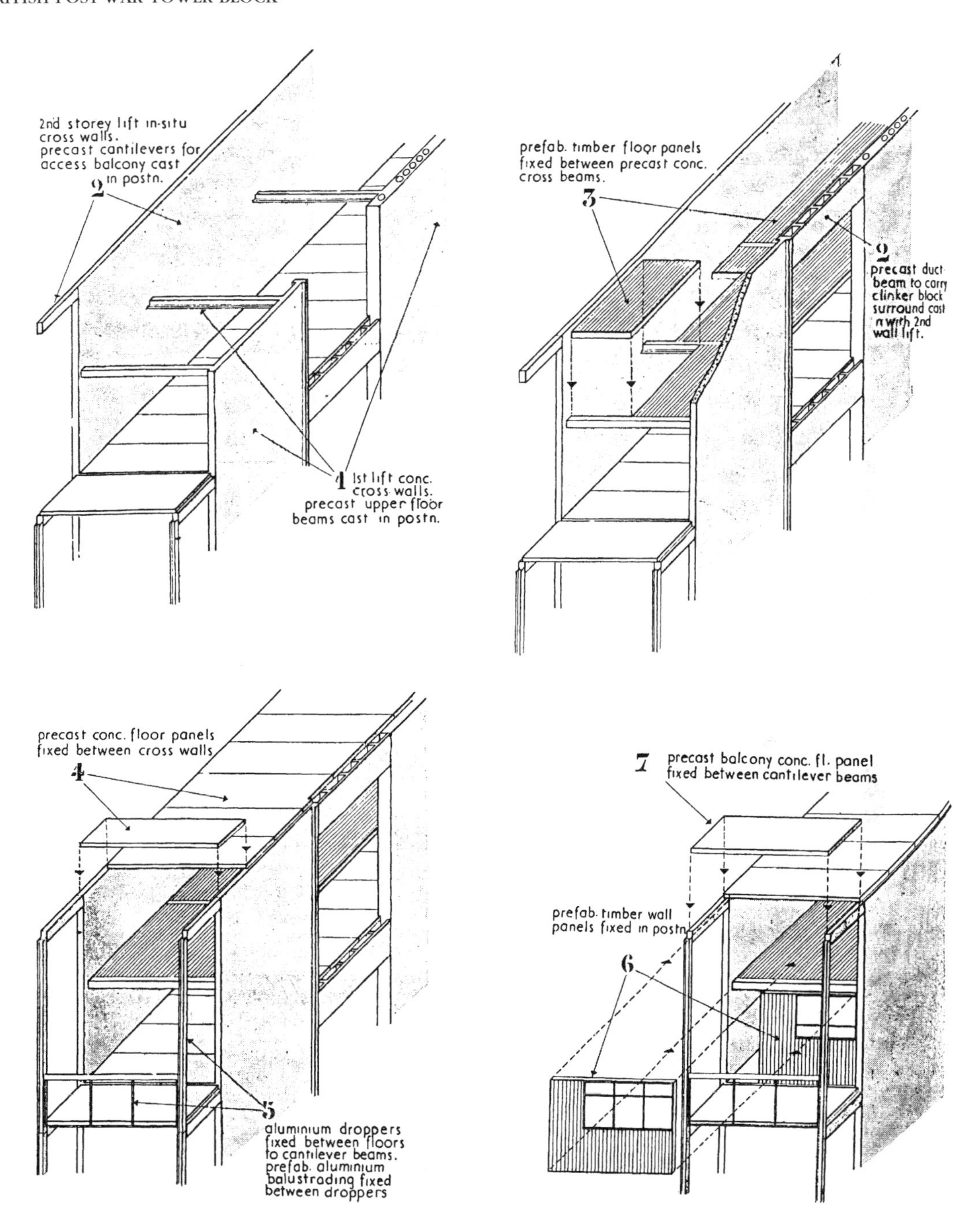

Figure 20.3 Cross-wall – box-frame reinforced concrete construction. London: LCC Picton Street, Southwark, designed from 1953 onwards by A. W. Cleeve Barr; contractor: Laing; engineer: Ove Arup & Ptns. Almost all parts are cast or made before they are assembled with the help of the tower crane and quasi-slotted together. The drawings show practically all the parts needed to form the basic structure of the block and the balconies attached to it.
Source: A. W. Cleeve Barr *Public Authority Housing*, 1958

case of Britain's first point block, at Harlow (designed by Gibberd). [. . .]

Reinforced concrete is simple in principle: what is complicated is the casting process. The more casting of small and narrow parts and corners, the more lengthy and costly the process of construction, because of the complications of the shuttering. Furthermore, there are the many problems of fitting the walls neatly into the frame. Hence a method was devised which involved a different way of looking at the construction of a block as consisting of a number of units or cells. This kind of thinking takes the internal dividing walls as the starting point together with the floors and other horizontal members. If we think of both of these elements as something relatively rigid, we can combine them into a kind of frame. This frame has the added advantage of providing readymade the main internal divisions at the same time – provided, of course, that these divisions do not need to be broken through. In other words, this kind of frame system suits buildings which are rigidly divided into small units – such as blocks of small flats. Two equally important terms, 'cross wall' and 'box frame', were used to describe this method; occasionally 'egg-crate' was also used. [. . .]

The first completed postwar example of an r.c. box frame in London (in 1948), was claimed to be a small block of maisonettes, Brett Manor, built in Hackney by a charitable trust to the designs of Edward D. Mills. Mills and Lubetkin both collaborated with Ove Arup, who also became involved with the Building Research Station's experimental block at Abbots Langley.

[. . .] The LCC then adopted the box frame as the constructional method for its new type of 11-storey maisonette slab block: the Zeilenbau ideal of a stack of dwellings in a slim block with maximum openings on either side. [. . .]

Peter Dunican (an Arup partner) and Cleeve Barr both fervently believed in the box frame. Constructionally, their own experiment at Picton Street represented the early culmination of this development of the LCC slab block (Figure 20.3). Reinforced and unreinforced cross walls alternate; there was particularly careful measurement of the strength of all respective materials: as strong as necessary and as light as possible. Longitudinal strength in the box frame 'house of cards' was provided at both ends by strong flanking walls, as well as by the grouping of stairs and lifts near the centre.

So far, we have discussed only the 'bare bones' of a block of dwellings. Earlier in this chapter, however, we hinted at a major general innovation in Modern building: the separation of frame and 'infill'. What we are concerned with here is the outside covering of the frame, or box frame – in the language of structural engineering, the cladding. This, too, needed much more rigorous scientific consideration. In 1955, Dunican and Cleeve Barr elaborated on the desired performance of cladding: light in weight and strong. Its chief functions were to keep out wet and cold, and to provide soundproofing and fire resistance. Lastly, the external surface of the cladding was subject to aesthetic scrutiny: the new freedom as regards the look of the building was often emphasised even in technical accounts. [. . .]

In the early fifties, there was still uncertainty about new materials, and brick still seemed the best proposition. [. . .] But a few years later, experiments abounded. No single material would do: there needed to be a combination of materials, each used according to its own special performance, as in the case of the insulating agents, woodwool and polystyrene. The external layer was frequently of metal (for example, corrugated aluminium), but there was also, surprisingly, much use of wood externally, for instance at Picton Street (see Figure 20.3). The cross-wall maisonette block of whatever height, with its varied, often brightly coloured infill cladding, was certainly one of the most characteristic sights of English urban housing of the late fifties and early sixties [. . .].

A separate, but related area of experiment was that of flat roofs, which were 'in the centre of violent controversies', we read in 1946. Such arguments were often of a semi-aesthetic nature. It was for general rationalist reasons of constructional consistency that one ought not to introduce timber roofs in buildings whose ceilings were of concrete. But to many engineers and designers, the flat roof's practical advantages also seemed proven: it was 'better at keeping the weather out than pitched roofs', it was cheaper, and it provided 'space for recreation'.[4]

The requirements of structural and functional performance of Modern blocks of flats converged in the design of the joint: the Modern joint was something completely 'without precedent'.[5] In traditional construction, the myriads of individual joints merge with the small 'building components', the bricks, into a quasi-monolithic whole. But the new emphasis on light cladding and on the efficient assembly of large cladding panels conferred a new special importance on the joint and the variety of its sealants (see Figure 20.7). [. . .]

We must return to the subject of prefabrication (Figure 20.4). Cleeve Barr claimed of his Picton Street scheme that it was 'perhaps the most rationalized yet . . . by maximising the use of precasting'.[6] The precasting of floors, for instance, obviated the cumbersome in-situ handling of horizontal shuttering. Apart from the cross walls, most other repetitive components of the Picton Street blocks were prefabricated (see Figure 20.3). These were dropped into position by the tower crane. Furthermore, the production of the in-situ parts was mechanically assisted by the crane, which could move the floor-high shuttering for the cross walls from one location to the next. There was also, we remember, no need for traditional scaffolding in these operations. The designers asserted that speed was greatly improved, as was the standard of the concrete finishes. The cross-wall system – or the box frame – Barr and

Figure 20.4 Reinforced concrete framework construction. London: LCC Bentham Road, Hackney, planned from 1952, built from 1955, designed by C. G. Weald, C. St J. Wilson, et al.; engineer: F. J. Samuely. Most parts are precast on site.
Source: Cleeve Barr

Dunican concluded, 'lends itself to repetition and standardisation of structural and non-structural elements and therefore to prefabrication and more rapid building methods'.[7] [. . .]

Finally, it was something of a paradox that another method of building was frequently mentioned in the context of prefabrication: namely, 'no-fines' construction, quantitatively the most successful of all the new postwar non-traditional house construction methods. The majority of non-traditional two-storey houses of this period were of no-fines construction, built by Wimpey, the Scottish Special Housing Association, and Laing (in the latter case under the name 'Easiform'). The opaque term 'no-fines' denotes, simply, the omission of fine sand from the concrete/aggregate mix, resulting in a cellular kind of concrete with reputedly good insulating properties. For low buildings, steel reinforcement is not needed. Moreover, the concrete is less wet and therefore lighter, and less expensive shuttering can thus be used. Here, too, the tower crane reduced costs and made it possible to go higher. In 1953, there began the building of 11-storey no-fines blocks at Tile Hill Neighbourhood (North), Coventry, and, by the end of the 1950s, the firm had commenced no-fines blocks of 20 storeys, at Royston 'A', Glasgow, by 1967, the company was building to 25 storeys at Townhead 'B' for the same authority. [. . .]

The 1960s

[. . .] In the early sixties, we enter the most important phase of the use of new construction methods in public housing in Britain. [. . .]

The essence of most of the construction methods of this period, the large precast concrete panel, can be seen as a continuation of the box frame of the fifties. [. . .]

The first major foreign-designed prefabrication construction methods introduced to Britain were those of the French building firm Camus and the Danish firm Larsen Nielsen. The former, under discussion in England and Scotland since 1959 and contracted by Liverpool CBC in 1963, incorporated panels 21 feet in length; its floors and panels, each at least 6 inches thick, together formed a 'multicellular structure rigidly braced in all directions' (Figures 20.5 and 20.6).[8] The Larsen-Nielsen method, adapted by Taylor Woodrow-Anglian and used, from 1963, by the LCC, incorporated a main structure of basic loadbearing cross walls, again 6 inches in thickness [. . .].

Concrete Ltd's 'Bison Wall-Frame', the most prolifically used 'system' in the UK, named the matter confusingly, yet aptly: there is no frame in the sense of frame plus infill; the walls themselves constitute the 'frame' – a series of all-round boxes (Figures 20.7 and 20.8). This was

Figure 20.5 Liverpool CBC: Boundary Street, Logan Towers, planned from 1963; Camus system.
Source: Architect and Building News 18-2-1966, p. 21

occasionally described as 'rigid box construction', but this must not be confused with the prefabrication of complete units of rooms or flats – an essentially Eastern European method not found in multi-storey flat building in the UK. In Bison blocks of twelve or more storeys, all walls are loadbearing, including the external walls; there was no such thing as a 'partition wall'. Here Bison went beyond the 1950s method of cross-wall construction which left two sides of the 'box' open. The wall and floor panels are up to 21 feet in length, and all internal walls have a thickness, again, of 6 inches. A two-bedroom flat could be assembled out of the incredibly small number of 21 pre-cast com-

Figure 20.6 'Camus System': plastic wall tiling for external walls is placed into position in the mould before the concrete is poured in; vertical battery casting for walls.
Source: R. M. E. Diamant, *Industrialised Building*, 1964

ponents, which included a completely prefabricated bathroom and toilet unit and elements of the structure comprising the stairs; the section enclosing the lifts and stairs was put together out of precast units three storeys in height. Despite the emphasis in Concrete's promotional literature on Bison's flexibility of planning, essentially it offered only two and three bedroom flats, and maisonettes were not possible. [. . .]

Virtually all parts of the new prefabricated blocks of flats were produced in specially built factories, situated at a greater or lesser distance from the site. Concrete Ltd, for instance, created a number of strategically placed factories throughout England and Scotland. This was, of course, a crucial development, a decisive step away from the situation in the fifties, when most of the prefabricated parts of the new advanced blocks were cast nearby on site. A major exception was Wates, who prided themselves that the setting-up of 'factories on site' meant no breakages or delivery delays [. . .].

[. . .] No-fines construction, of course, lay completely outside the methods just discussed - yet it accounted for the greatest number of cottages and high flats built in any 'system' over the postwar period (Figure 20.9). During the early and mid-fifties, multi-storey variants of no-fines were evolved by Wimpey and the Scottish Special Housing Association, and built, initially, in Birmingham, Kirkcaldy, Coventry and Glasgow. By the sixties, only Wimpey was left building large numbers of no-fines high blocks. Originally, everything was poured on site. But in later Wimpey multi-storey blocks, complete bathrooms and kitchens were prefabricated, and the '1001' point-block type of the early sixties incorporated also innovations in the field of finishing, such as dry lining. New mechanised techniques - the use of tower cranes and a 'cyclic programme programmed by a computer' - also achieved substantial time savings.

Despite the somewhat unfashionable roughcast finish with which it was associated, no-fines approximated the most closely of all the new methods to the universal or comprehensive 'open system' of Modern construction advocated by many architects, in that it could

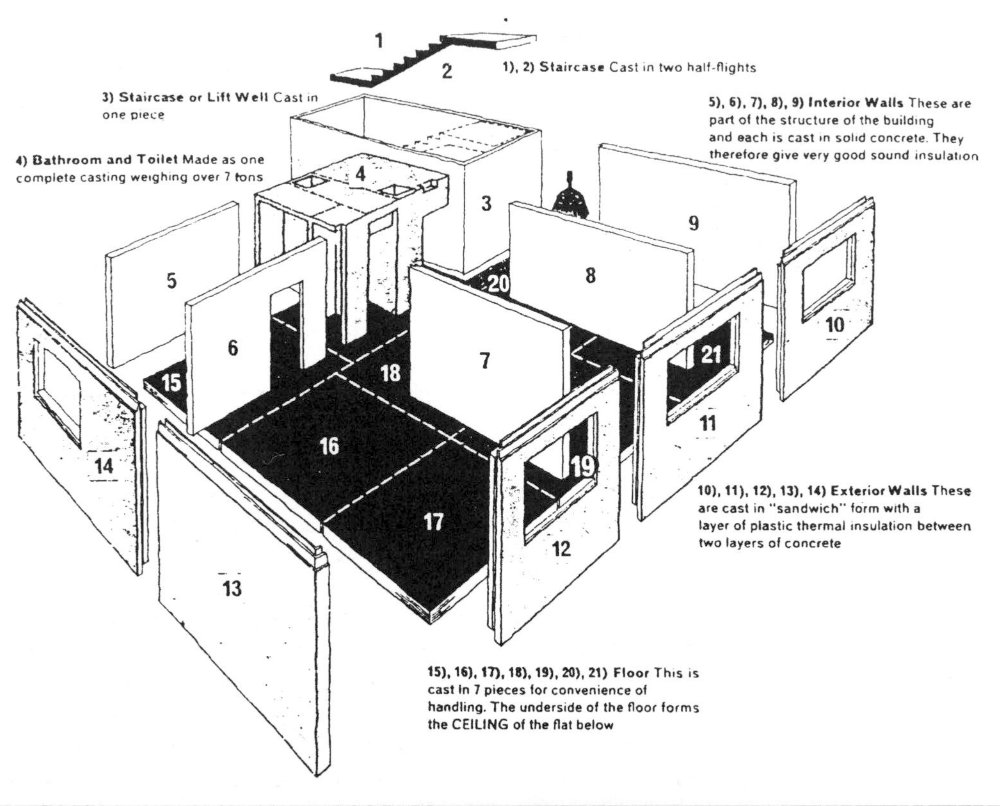

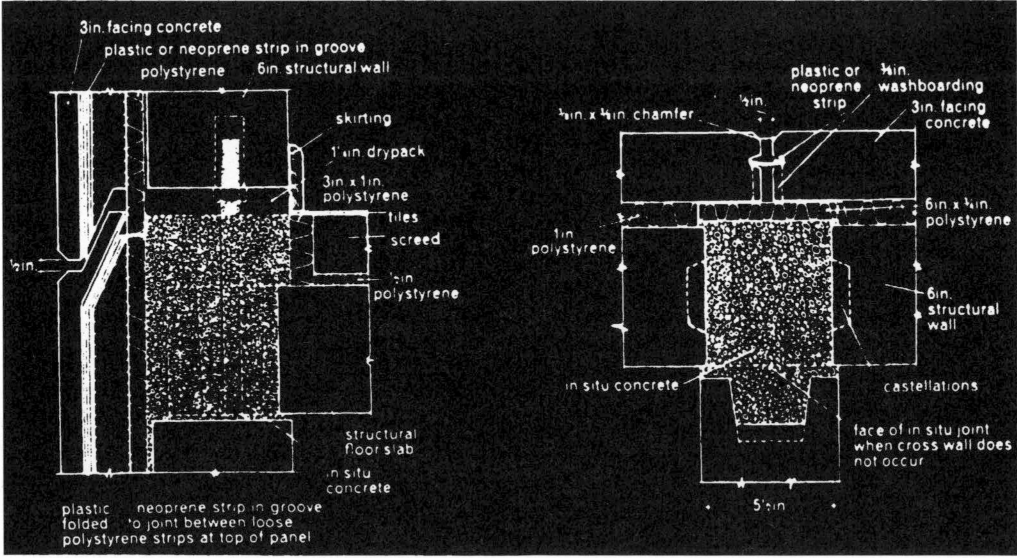

Figure 20.7 Concrete Ltd 'Bison Wall-Frame System' (1962). Only 21 parts were needed for one two-bedroom flat in a high block. External joints: left: vertical section through an external wall at the point of joining with an interior wall; right: horizontal section.

Source: *AJ* 1, 8, 1962, p. 262; A. F. L. Deeson, *The Comprehensive Industrialised Buildings Annual*, 1965

Figure 20.8 London: Tower Hamlets LBC, Mansfield Buildings Area Stage I, Charles Dickens House, Bethnal Green Road, 1967. 'Bison' block, showing the varied kinds of facing given to the prefabricated panels

be used with equal facility for all types of dwellings. [. . .]

It was an essential aim, even a raison d'être for all new methods of construction to provide as complete a range of services as possible. Great financial savings could be achieved by fitting these in at prefabrication stage. Occasionally this had been advocated in the 1950s. In prefabricated schemes of the sixties, many service conduits, such as electric under-floor heating ducts, could be included in the precast panels. Likewise, window and door frames were attached to the panels at an early stage and joinery work was reduced to a minimum. The more sophisticated and varied methods of casting now available also made it possible to eliminate plastering, by the provision of internal surfaces suitable for the direct application of wallpaper or paint. Large prefabricated panel building perpetuated the concern of fifties designers for the 'manipulation' of walls to provide insulation. In the case of sound insulation, the weight of the panels in itself provided the answer. [. . .]

[. . .] By the end of the 1950s, it was widely assumed that the constructional, servicing and architectural problems of high blocks had been solved. [. . .] So far, we have dealt with dwellings as devised by housing reformers and architects, and built as advanced showpiece developments, such as those of the LCC. Now we turn to the subject of dwelling provision pure and simple. Here we find a rather different set of issues and values: urgent local political demands, financial and organisational constraints and initiatives, lobbying, negotiating, monitoring, pressurising. These were subsumed under the single, all-encompassing aim of 'output', or 'production'; the latter word is here used not so much in its more general economic sense as to denote the demand for, and process of, large-scale building of new dwellings. Now the pace was set by groups different from the designers and reformers [. . .], above all by 'crusading' councillors, pursuing the local-political aim of 'housing their people' and safeguarding their patronage powers of housebuilding and letting, but also by engineers or production-minded architects of local authorities or contractors, serving their political or commercial masters through efficient pursuit of output. The relative simplicity of their aim was reflected in the layout of many developments built in the great 'production drive' of the late fifties and sixties. Now, high blocks were less likely to be built in the context of complicated architectural and planning proposals, but rather as one-off groups, placed on any kind of site, urban or suburban. [. . .]

It seems only too obvious that, during [the 1950s] the values of 'production' and most of those of 'Modern design' were closely bound up with one another. Designers and producers alike took it for granted that the new public housing should both be provided in large quantities, and also be Modern in appearance, construction and equipment. And yet there were, at the same time, emphatic divergences of interest. Many of the designers [. . .] already held, and continued to hold, a contemptuous attitude towards 'mere production'. In the sixties, in cities such as Glasgow, a different kind of 'designer' became much more prominent: one who acted chiefly as technical assistant to

Figure 20.9 Wimpey Type 1001/6 point block.
Source: Wimpey Rationalised planning in No-Fines Construction, [*c*.1963]

the driving councillors and organising officials. This new balance of power led to fresh conflicts between 'design' and 'production'.

Yet large-scale production was no novelty within the field of public housing: its establishment as one of the chief yardsticks of achievement for the new municipal housing organisations after World War I had been followed by several waves of high output in various parts of Britain. [. . .] during the fifties and sixties, this 'tradition' of urban housing production progressively became associated with Modern flats. Much of the groundwork for this had been accomplished in the interwar and early postwar years, by those authorities that erected large numbers of flats of broadly 'traditional' tenemental types.

In the building of flats between the wars, as in other areas of housing production, the most significant differences within Britain were centred around the Scotland–England fault line. In Scotland, the national tradition of monumental tenement housing remained an enduring influence, although increasingly stigmatised by the biased 'standards' introduced by Anglocentric housing reformers, and by the effects of rent control in pushing existing tenements into a cycle of decline. Many new tenements were built after 1919 by municipalities, but largely for the rehousing of the 'residuum' cleared from slums. As a result of this gradual internalisation of cultural 'inferiorism', even working-class opinion seemed to have turned against tenements by the forties. But the interwar nationalist 'Scots Renascence' also saw the start of a revaluation of the tenement: Edinburgh's City Architect, Ebenezer MacRae, repudiated cottages and called for a 'return to the old Scots traditional form of building', and the official Highton Report (1935) praised contemporary Continental flats. After the war, there would be a renewed boom in tenement building; the production-orientated characteristics of this phase would, in their turn, be perpetuated in the energetic campaigns of Modern flat construction during the sixties.

In England and Wales, on the other hand, interwar council housing production was dominated by suburban cottage estates: slum-clearance and large-scale flat-building only really took hold in the 1930s, in London and a few large provincial centres such as Liverpool. In the capital, responsibility for clearance and building was divided between the London County Council and the 'second-tier' authorities, the Metropolitan Boroughs. This arrangement was later to cause much friction, but, during the thirties, the autocratic Labour LCC administration of Herbert Morrison maintained an easy dominance. In slum-clearance areas, the Council built large schemes of balcony or staircase access 'block dwellings' up to five or even six storeys in height, and in styles (such as neo-Georgian) considered 'modern' under interwar criteria. Only a few of the more precocious Labour-controlled boroughs, such as Bermondsey MBC, attempted flat-building on a major scale.

Immediately following World War II, it seemed that the interwar balance of power and activity in London housing production would be maintained. The LCC's decision in January 1946, as a matter of expediency, to concentrate responsibility for design and erection of its new housing in the hands of its Director of Housing and Valuer, C. H. Walker, paid rapid dividends. Within four years, through administrative efficiency, the avoidance of 'flights of fancy' and the use of cautiously improved prewar patterns, 19,171 dwellings had been completed and a further 51,436 were approved or under construction. The flats which made up 44 per cent of this total were traditional 'block dwellings', usually of four or five storeys in height (although a few higher blocks, up to eight storeys, were built for the first time). Walker was also able to impose his production-orientated values and preferred housing patterns on London's lower-tier housing authorities, the Metropolitan Boroughs, as a result of the Ministry's delegation to the LCC, up to 1950, of responsibility for loan-sanction vetting of their projects [. . .] Walker's pressure and assistance accustomed all Metropolitan Boroughs to the idea of large-scale production of flats. [. . .] Block dwellings or tenements were also under construction in a few other English and Scots cities. But by this date [. . .], the LCC and several 'avant-garde' boroughs, and the New Towns, had already fallen under the influence of designers who emphatically rejected block dwellings, and pushed for the building of Modern flats. The 'heroic' early period of Modern housing architecture overlapped with late and vigorous bursts of production of pre-Modern dwelling types.

It was only a few years later, however – in the late 1950s – that Modern flats themselves were adopted for the same production-orientated purposes as Glasgow Corporation's tenements or Walker's block dwellings. [. . .] Now the types which the LCC designers had advocated as an antidote to Walker's production use of block dwellings were themselves embraced by Walker-like officials, and councillor 'housing crusaders', throughout Britain. Here they were used in ways which were unacceptable to, and vehemently criticised by, the LCC designers and their councillor supporters such as Evelyn Denington. The production-dominated building of Modern flats, in contexts such as the 'package-deal' contract (designed and built by the contractor's staff), required the intimate involvement of many 'designers' in the technical sense – municipal engineers to identify and lay out the sites, contractors' architects and engineers to design and superintend the erection of the blocks - yet this activity would not have been acknowledged as 'design' by the LCC architects.

It is at this point that high blocks began to play a crucial role: they turned out, perhaps unexpectedly for some, to be especially suitable for the rapid exploitation of small gap-sites, which were becoming increasingly prevalent in the fifties. We thus encounter a crucial moment in the history of the high Modern block: initiated by 'designers' in the context of new planning and architectural concepts, these values now became translated into pure 'production' advantages. It was in this gap-filling capacity that multi-storey blocks were much more suitable than the older tenements or block dwellings with their far greater horizontal spread. However, this coalescence between Modern design and production did not mean that the schism between the two factions was closed: on the contrary, it in many ways widened. The producers saw high blocks as gap-fillers; the designers rejected this argument as simplistic and dangerous. Eventually, fortified by a general collapse in the idea of 'production', the late sixties were to bring a complete turn against high flats within the design profession, the very profession which had introduced the type, but which now repudiated the many blocks built as only a parody or misuse of the original idea.

Most recent historians have argued that high flats were 'imposed' on local 'communities' by external forces. However, [. . .] the truth was very different - they were a key *expression* of local municipal power, rather than an attack on it. Indeed, one of the main reasons for the translation of Modern flats from design to production purposes was the urgent desire of some cities to defend their power against the perceived threat posed by one powerful 'national' group of designers, the town planning profession.

Our story starts with a powerful, Government-supported challenge to the established structure of local authority housing provision, during the 1940s. This challenge set up a 'land trap' for local authorities, by curbing their land supply, and thus threatening their autonomy in housing policy. It took two loosely associated forms, both trying to turn away from 'mere' local housebuilding to a town planning conception of housing as part of some wider process of reconstruction or modernisation. The less contentious of the two policies was an energetic resumption of slum-clearance from the mid-fifties; this was linked, in England and Wales, with a curb on municipal development of suburban sites. [. . .] Much the more threatening aspect of the 'land trap' for local authorities was the second policy: an attempt by town planners and others interested in 'rational reconstruction' to constrain the development of urban housing land, and to force cities to overspill outside their boundaries many of those displaced by slum-clearance.

This orchestrated 'British' challenge, during the forties and fifties, established broad constraints on the choices available to local authorities on housing policy. But it did not dictate or determine their response. There was much population loss from cities during the fifties and sixties, but this happened mostly in the form of 'unplanned overspill' of the middle classes. The officially preferred policy, a combination of slum-clearance and large-scale planned overspill of industry and working-class housing, was not, in most cases, seriously implemented. Instead, powerful local authorities determinedly set about the task of keeping control of their own housing destinies, by combining slum-clearance with a building policy at variance with Government-endorsed decentralism: the massed development of high flats on their own territory, using one element of the LCC mixed development formula in a manner akin to Walker's use of block dwellings for 'site cramming'. In England and Wales, this powerful movement cut across the grain of a popularly entrenched anti-urbanism and (in the

English case) individualism. In Scotland, by contrast, the high national-cultural status of urban settlements, as natural centres of order and authority, facilitated decisive collective action by cities and burghs. It was here, too, that the fewest mixed developments, and the most high blocks on their own, were built.

Thus the Modern housing which swept through Scotland and England during the fifties and sixties was not actually caused by the 'land trap' itself, but by the particular response chosen by the municipalities, in their opposition to the Government's pressure for 'nationally' planned and orchestrated reconstruction. Finally, once established, high-flat building was taken up for uses divorced from its original association with perceived land shortage – for instance in Dundee, where it allowed rapid exploitation of plentiful suburban land. [. . .]

Our focus on the great power wielded by councillor 'crusaders' and official 'organisers' in the largest authorities represents a sharp divergence from the theory of decision-making in housing provision put forward by recent historians, such as Dunleavy, Owens or Finnimore, which has emphasised the power exerted by other groups or 'actors' (the national architectural or planning professions, or building contractors), and has correspondingly portrayed local groups and individuals as weak and exploited.[9] Here, by contrast, the decentralised nature of decision-making in public housing is interpreted as a source of strength, enabling driving 'housing leaders', set on output expansion, to emerge at local level from among both members and officers. [. . .]

This story reaches a dramatic climax in the multi-storey crash-drive of Glasgow. Here, in the 'shock city of the Modern housing revolution', [. . .] no 'national British' group (planners, civil servants or contractors) was able to withstand the overwhelming negotiating and organising power wielded by two 'local' figures: the Corporation's Housing Committee Convener, Councillor David Gibson, and the engineer Lewis Cross, 'Housing Progress Officer' in charge of sites and contracts. The architectural results of Glasgow's 1962 multi-storey revolution were blocks of an unfettered monumentality, unparalleled at that date not only in these islands, but in Europe as a whole. [. . .]

The LCC's large mixed developments of the mid and late fifties, and Glasgow's multi-storey-only 'package-deals' of the mid and late sixties, employed physically similar Modern high blocks; and both represented the large-scale exercise of 'State power' in public housing provision. Yet the driving values of the LCC architects and the Glasgow 'crusaders', in some ways, could hardly have been more different!

Notes

1 *Journal of the Royal Institute of British Architects (JRIBA)* 6-1954, p. 319.

2 Dept. of Scientific and Industrial Research, Building Research Station, *Principles of Modern Building*, Vol. I, 3rd ed., 1959; cf. R. Fitzmaurice (Dept. of Scient. and Ind. Res., BRS) *Principles of Modern Building*, Vol. I, 1939.

3 *JRIBA* 12-1949, p. 54; *The Architects' Journal (AJ)* 4-9-1952, p. 293.

4 Association of Building Technicians, *Homes for the People*, 1946, p. 60.

5 *AJ* 28-4-1960, p. 653.

6 A. W. Cleeve Barr, *Public Authority Housing*, 1958, p. 110.

7 *AJ* 17-3-1955, p. 358

8 R. M. E. Diamant, *Industrialized Building: 50 International Methods*, 1964, pp. 49–51.

9 P. Dunleavy, *The Politics of Mass Housing in Britain*, 1981, pp. 184, 188; *Housing Review*, 7/8-1978, pp. 86–7.

21

SOLVING LATE-VICTORIAN LONDON'S TRAFFIC PROBLEMS

Sir John Wolfe Barry

Source: *Journal of the Society of Arts*, 47 (1898–9), pp. 12–16, 18–20

I propose now to consider the new streets and widenings of streets which have been effected in the last 50 years with a view of facilitating the constantly growing streams of traffic.

In comparison with the large expenditure by railway and tramway companies we cannot but be struck with the smallness of the mileage of streets made and of the capital expended.

The following are the most prominent of the street improvements executed within the last 40 years, exclusive of the Thames Embankments:

Date		Length (furlongs)	Cost £
1854	Cannon Street	$3\frac{1}{2}$	500,000
1864	Southwark Street	5	366,000
1870	Holborn Viaduct and streets connected with it	6	2,552,000
1871	Hamilton Place	1	111,000
1871	Queen Victoria Street	5	1,076,000
1876	Northumbertand Avenue	$1\frac{1}{2}$	711,000
1882	Tooley Street	6	405,000
1883	Hyde Park Corner	1	11,000
1884	Eastcheap	$1\frac{1}{4}$	600,000
1886	Shaftesbury Avenue	5	779,000
1887	Charing Cross Road	4	584,000
	Totals	4 m. $7\frac{1}{4}$ f.	£7,695,000

This total amount, small as it is for the metropolis during nearly 40 years, is really an over-statement, as it takes no account of the recoupment by sale of surplus property, improved rents, or the like. On the other hand, there are no doubt many street improvements which I have not noticed which are of less prominence. But in the aggregate, and when all are recognised, the amount of expenditure in so long a period cannot but appear very small.

In the matter of street improvements in London also, one cannot but notice an almost entire absence of grasp of a large subject. We can, no doubt, record some useful and even fine undertakings, but in the history of the past 40 years, we look in vain for any new arterial thoroughfares traversing Inner London from end to end, and proportioned in width to the demands upon them at different parts of their route. On the contrary, we find in the new streets, as in the old ones, that the nearer they are to the heaviest of the traffic, the narrower they are in absolute dimensions.

Cheapside, Fleet Street, Piccadilly, the Strand, Oxford Street, Marylebone and Euston Roads remain very much as they were 50 years ago, when the traffic was a mere fraction of what it is now, and there has been, practically speaking, no attempt at improvement in these most important thoroughfares. In the case of the Strand, no doubt some relief has been found from the Thames Embankment, promoted, though it was only in a minor degree, as a thoroughfare, but the monstrous condition of the Strand at crowded hours of the day and night shows that the embankment roadway, which has been opened 28 years, has failed altogether to afford the necessary relief. At such times, one can say, in the words of Hood, that there is

'No road, no street, no t'other side the way.'

I desire to speak with all respect of those who years ago conceived and executed the idea of Regent Street. I believe we have to thank Nash the architect and the Prince Regent for having made an improvement in the arterial line from north to south in a large-minded way and with some conception of the requirements of that time, and of provision for the future.

If we can imagine that in 1813, when Regent Street was designed, the four east and west routes by Pall Mall, Piccadilly, Oxford Street, and the New Road were adequate for the east and west traffic, we can see that Nash's wide Waterloo Place and Regent Street with its circuses at Piccadilly and Oxford Street, and joining the only 100 feet street in Mid-London, Portland Place, except Whitehall which is 125 feet wide, was a work conceived in a large-minded way, and was a real effort to deal with the requirements of the traffic north and south at the western end of London.

When, however, we come to consider more modern street improvements, most of them at least seem piecemeal and patchwork enterprises narrowed to the very least dimensions which would pass muster, and without any but the most meagre provision for the future, or, except in the case of Regent Street, the slightest attempt at systematic artistic treatment of the question. In fact there has been and is now a hitherto incurable *petitesse* in dealing with such matters in London, which is a great contrast to what we see in foreign cities of far less importance and of far less wealth than those of the English metropolis.

For example, when Queen Victoria Street was made - a highly useful and costly undertaking - not only was the inadequate width of 65 feet adopted for a street with shops and warehouses on both sides, which is in prolongation of the 110 feet of the Victoria Embankment, which has no buildings upon it at which vehicles stop to load and unload, but where Queen Victoria Street crosses Cannon Street, a street (New Earl Street) made in former years, which had only a width of 50 feet, was allowed to remain and form part of the new Queen Victoria Street.

Again, where the new street was to reach the Mansion House no provision was made by any widening of existing thoroughfares to carry forward the traffic, and as Queen Victoria Street in the nature of things was only a means to an end, viz., for traffic to reach the heart of the city, the congestion at the Mansion House could not be relieved by it, and we daily see this highly expensive street blocked by traffic at Cannon Street and at the Poultry, and the confusion at the Mansion House worse confounded.

In the instance too, of the important widenings of Eastcheap and Tower Street, made - only some 15 years ago - at the joint expense of the Metropolitan Board of Works, the City, and the Metropolitan and District Railway Companies, a width of only 60 feet was adopted, though I remember the late Colonel Haywood, the experienced engineer of the Commissioners of Sewers stating that even 80 feet would soon be found inadequate. His prophecy has been already fulfilled. The widening of Ludgate Hill, but very recently completed, only leaves this most important approach 59 feet wide, crowded with shops on both sides, and Fleet Street remains untouched with a width of 45 feet.

The highly useful and well-designed Holborn Viaduct improvement was a local and not an arterial improvement, for it left Newgate Street and Cheapside unaltered, and was the means of facilitating the bringing of traffic to another point of congestion.

In more modern improvements the width of 60 feet has been adopted for Shaftesbury Avenue, and the Charing Cross Road, and one can even now easily see that in a few years they will be found much too narrow, though the north and south traffic does not probably call for such widths as does that from west to east.

Even in the case of Northumberland Avenue one cannot but notice the same fatal parsimony. If ever there was a case for a fine street, as approaching the Thames Embankment, this surely was one, and at the time at which it was made, urgent remonstrances were made in favour of a width of 100 feet. But it was in vain, some not very important banking house would have been required, and to save this the line of the avenue was altered, and a width of 80 feet adopted as an approach to the embankment roadway which is 110 feet wide.

Now of course it must be admitted that a street 80 feet or 100 feet wide will cost more than one 60 feet wide, but it will not cost proportionally more, and this is specially the case when a street is widened and not cut transversely to existing streets.

If one assumes a street of 60 feet from house to house, about 12 feet on each side must be devoted to footways, leaving 36 feet for vehicular traffic. One must deduct about 9 feet on each side for drays and carts standing at the houses, and all that remains is 18 feet, which is only adequate to two or at most three lines of moving vehicles.

But when one criticises these dimensions, what is to be said of the Strand (50 feet wide constricted at one of the busiest parts to 40 feet) and Piccadilly from Sackville Street to Regent Street (55 feet wide), or Cheapside (50 feet wide), except that they are ludicrously inadequate to the demands of a great city flooded daily with an enormous influx from a closely populated area of 688 square miles.

What do we find in continental cities? In Paris, the old Boulevards have a width of 100 feet or over, the Rue de Rivoli (made 60 years ago) of 80 feet, the new Boulevards, a width of 80 to 130 feet. In Vienna, the Ringstrasse has a width of 175 feet, in Buda-Pesth the Andrassy Strasse must be 140 feet wide, and in New York the principal Avenues have widths of 80 to 150 feet.

If we take Paris alone by way of contrast with our metropolis, we find that in about 40 years the following important street improvements have been effected, besides many others:

Completion of	Length (miles)	Width (feet)
Rue de Rivoli	1.4	80
Boulevard Sebastopol	2.5	95
” Strasbourg	2.5	95
” Haussmann	1.7	95
” Malesherbes	1.7	125
” du Palais	0.2	130
” St. Michel	1.1	80
” St. Germain	2.0	125
” Magenta	1.2	90
” Voltaire	2.0	95
Avenue de l'Opera	0.5	95
Rue de quatre Septembre	0.6	80
Avenues leading to the Champs Elysées	3.5	100
Total	18.4 miles.	

As in my reference to the principal street improvements of London, so in the case of Paris, there must be a large number of less prominent works the aggregate of which must be very considerable, and more in proportion than the London improvements not included in my list.

It is to be carefully borne in mind that the want of accommodation for the traffic in the streets of London is not merely a question of grumbling of those who suffer from it, but involves many other considerations. In the first place there is daily and hourly loss of much money in the delays to men of business or in professions, and to the operative classes. We must add loss of time to vehicles and horses, the practical impossibility of introducing cheaper and more expeditious means of transit, such as by electric tramways, and great want of free circulation of air. [. . .]

In the consideration again of wide and narrow streets, we can recognise that, apart from the circulation of air, the ground on each side cannot be so well utilised by lofty buildings in the narrower thoroughfare. It is rightly a regulation in London that an angle of 45° at the pavement should be subtended by the opposite side of a new street. Thus a street 60 feet wide will permit of 40 feet less in height on each side of a street being devoted to additional stories as compared with a street 100 feet wide. Consequently, three stories on each side of the street are lost by the narrower street. In these days of rapidly moving lifts, which give value to the upper stories, such a consideration ought to be at least a great help towards the extra cost of the land required for the wider street.

Again, one broad continuous thoroughfare is, so far as the circulation of air is concerned, a vastly different thing to the same width cut up into two narrow streets, and would tend to improved health and sanitation.

Apart from time saved and the benefits to health the question of accidents to life and limb, which might be avoided by less crowding of vehicular traffic, should not be lost sight of. At present no less than 150 persons per year, or an average of nearly one person every second day, lose their life by street accidents in London, and the number of those injured amounts to 8,000 per year.

My plea is then that to meet the requirements of the traffic of London what is wanted is not so much additional railways, underground or overground, traversing the town and connected with the suburbs, but wide arterial improvements of the streets themselves. Railways which are strictly urban, or rather I should say trains which only traverse the town itself through carrying immense numbers of passengers, have not dealt with the question and will not produce the desired result in relieving the streets; on the contrary they tend to add to the congestion from the point of view of both urban movement and suburban influx.

It is an open secret that if the underground railways could only rely on strictly urban passengers, they would be nearly bankrupt. A reason for this is not far to seek. The urban journey is generally a short one, and when the little losses of time necessary to reach a station, catch a train, and to go from the station to one's destination are all reckoned, the omnibus which carries a man directly from one door to the other, at fares which are frequently cheaper than even a third-class ticket, is found the preferable mode of transport. The Metropolitan and the District Railways alone between them carry about 140,000,000 of passengers annually, but the bulk of this great traffic is not urban, but is of suburban origin, deposited in urban localities. Moreover we must take note of the fact that third-class traffic is in numbers .78 per cent of the whole traffic on the underground lines. This indicates that the great travelling public is one which scrutinises expenditure narrowly, and the fact implies that so long as omnibuses can carry people for penny fares, and still more if the halfpenny fares of Glasgow are introduced in the metropolis, the vehicular traffic of the streets will show more and more increase as compared with the journeys made from one part of the town to another by railways. But apart from these considerations, the extension of the underground railways into the suburbs more than counterbalances what relief they afford to the traffic of the streets by carrying strictly urban passengers.

I do not lose sight of the fact that the new Central London Railway, now nearly approaching completion, will be made under the existing thoroughfares of Oxford Street, Holborn, and Cheapside, but I do not think, from reasons already given, that any underground line can do very much to relieve the ever-growing traffic of the streets. No doubt the new railway will carry multitudes of people, but in the first place it will create a new urban and suburban traffic of its own, and, secondly, the experience of the Metropolitan and District lines is that surface transit for short distances continues to increase, in spite of the competition of the underground railways, even if it be not fostered by them. [. . .]

In my judgment, the question of street improvements in London should be considered as a whole and in a large-minded way, unless we are to be doomed to perpetual disappointment. We should endeavour to enlarge our views of present requirements, and provide for our successors. A scheme of new main thoroughfares of adequate width for present and future traffic should be laid down, and this should be realised as time and finance will permit. It should be a scheme worthy of London, and such as that which was laid down by Sir Christopher Wren and published in 1666 for the renovation of the City after the fire of London, but unhappily laid aside; or as was designed for the improvement of Westminster, presented to the public by my father, Sir Charles Barry, in 1857, and published in the memoir of his life.

There should be continuity of effort towards radical amelioration, by the construction of great main lines of through communication as distinguished from merely local improvements; and all local improvements should, as we have seen in Paris and other foreign cities, be so devised as to form parts of a harmonious whole. Thus I venture to think that the first thing to be done by our municipal rulers, is to realise what is wanted, and to employ the highest and most experienced talent of the age to lay down the best lines for arterial thoroughfares.

Much has to be done without delay, but it is not necessary to urge that the whole expenditure should be immediately or concurrently undertaken. The vital point is to endeavour to realise requirements and to make every improvement which can be put in hand subserve the purposes of a thorough conception of the problem of how best to deal with the traffic of London in the great streams from east to west and from north to south.

Apart from wide streets, a matter which has been too much lost sight of, is the provision of means for allowing the north and south traffic to cross the east and west traffic with the least possible confusion. With the exceptions of the north approach of London Bridge crossing Thames Street, of the famous Highgate Archway, of the Holborn Viaduct improvement undertaken by the Corporation of London, and of the arches which carry the south approaches of Waterloo, Blackfriars, and Southwark Bridges over a road which closely adjoins the river bank, there are no means of such crossings in London otherwise than on the level. The Holborn Viaduct, and the streets leading to it, form collectively a very fine and creditable work, and they have been an enormous benefit to Londoners. The value of the undertaking can only be appreciated by those who remember the steep slopes of Holborn Hill and the continual congestion of traffic, in consequence of the level crossing of Holborn and Farringdon Street.

A very easy and obvious improvement of this nature, on a small scale, could be readily effected by passing the north approach to Southwark Bridge over the crowded thoroughfare of Upper Thames Street.

If some means could be devised for the crossing of north and south traffic over or under the east and west traffic at such places as Hyde Park Corner, Piccadilly Circus. Ludgate Hill, the south end of Tottenham Court Road, and Wellington Street (Strand), the relief to the main thoroughfares at the spots in question would be enormous.

For example, a scheme is now being promoted for a new street from the Strand to Holborn, in a line, or nearly so, with Waterloo Bridge; but its utility, great as it may be in itself, will be to a great extent marred if no means are provided for dealing with the traffic from the south of the Thames by Waterloo Bridge, and crossing the Strand, or coming from the north, and crossing Holborn, otherwise than interjecting it athwart the east and west traffic in those crowded thoroughfares.

Of course, all such works would be very costly, for they involve not merely the crossing itself of leading thoroughfares by means of bridges and viaducts with approaches of about 1 in 40, but the connection of the streets on the level must, of course, be also maintained, so that traffic desiring to join the streets at right angles on the level might be able to do so. The results would, however, be well worth the expenditure, heavy as it might be, and I venture to contend that as in the case of the provision of wide arterial streets, the improvements of right-angled level crossings by means of sunken or raised roads and bridges ought not to be put aside as impracticable because they involve the same difficulty of cost. In this they would be as efficacious in their way of systematically meeting the wants of London as the other more obvious work of widened thoroughfares, while they possess this advantage, that they could be put in hand at once without waiting for the completion of the great through routes. There is no reason whatever why such works should be unsightly; on the contrary, they might be made highly artistic, and be architectural embellishments of London. When we contemplate what has been voluntarily spent by railway companies in getting rid of junctions on the level for similar advantages to their main lines of traffic, and when we remember the expenditure which is properly laid upon them by Parliament when they carry their lines across public streets, the outlay necessary for providing over or under crossings for the enormous vehicular and pedestrain traffic of London ought not in itself to be considered prohibitory. [. . .]

It appears to me that electric tramways on the surface of the ground are what will be wanted in the near future of London, on at least one great through route east and west, and on two, or perhaps three routes north and south. We have, I think rightly, in view of the narrowness of our streets, stopped the tramways south of the Thames and at the Euston Road, but it cannot be doubted that this necessity is most unfortunate for the welfare of the poorer classes of the metropolis. [. . .]

I fear that in the street improvements which I have sketched I shall be thought to have extravagant and utopian views, but when one contemplates the vast sums spent by railway and public companies to bring traffic to London, or realizes what has been done by continental nations to improve their means of transit in their capitals and to embellish them, I cannot see that Londoners should consider the cost of such measures in the metropolis of the kingdom and may I not say of the Greater Britain as prohibitory. I have above given an indication that a wise and liberal expenditure on street improvements brings with it an immediate return in public convenience, though it does not appear in the same direct way as do fares paid to public companies for the use of their means of locomotion.

One difference attending continental improvements, as compared with those of this country, is that they are largely paid for by indirect taxation of octroi duties, while our only system, since the short-sighted repeal of the coal and wine duties, is that of direct rating. The money equally comes out of the pockets of the inhabitants, but our present system undoubtedly tends to cramp the views of our administrators. No doubt it is right that those who pay should have a directly preponderating voice in the expenditure. This they have here, and no doubt the electors narrowly scrutinise the rates, but I should not despair of educating the public of London to an appreciation of the value of good arterial means of communication so that those who favour them would not run the risk, which our local representatives so much dread, of being accused of extravagance when the day of re-election comes. Expenditure may be wise or foolish, and it is foolish expenditure on ill-considered schemes and piecemeal projects which should really be deprecated. At any rate let the subject have its due attention and let the best minds devote themselves to the problem which is urgent now and becoming intolerable.

All roads were said to lead to Rome. How much more do they now lead to London from every part of the civilised world. London is *par excellence* the city of Europe, Asia, Africa, and America. England is the home of our fellow subjects from every colony, and I think I may say also of many of our American Cousins. London is the richest city ever known and the most populous, but when this is acknowledged we have to admit that its streets are getting day by day, in spite of the improved means of transit afforded by underground lines, more and more impassable, and to look more and more mean, while we must further remember that in 33 years we shall be face to face with a population of at least $8\frac{1}{2}$ millions in Greater London.

Is it not time then to approach the subject of its streets in a large-minded and systematic way, laying down the great principles to be carried out and working steadily towards a great result?

The expenditure involved, heavy as it no doubt would be, should not affright us. It would be repaid by the increased facilities both for trade and pleasure. London would be more attractive than it now can be. The stress of life would be lightened, and the saving of time to the millions using our streets would be enormous, though, as I have said, a money value to that saving cannot possibly be appraised. That saving would accrue to the poor even more than to the rich, for to the poor man who has nothing to sell but his labour, time is of cardinal importance. The crowding together of our labouring classes would be diminished as better facilities of easy and cheap locomotion enabled them to reside in the outer ring of London, increasing their means of inhabiting healthy homes and adding greatly to the happiness of themselves and their families. [. . .]

22

BUILDING SELFRIDGE'S: CONSTRUCTION ON AN AMERICAN SCALE IN A LONDON SETTING

Jeanne Catherine Lawrence

Source: J. C. Lawrence, 'Steel Frame Architecture versus the London Building Regulations: Selfridge's, the Ritz, and American Technology', *Construction History* 6 (1990), pp 23–46

Between 1906 and 1909, there unfolded an Anglo-American drama which had a significant impact upon retailing practice in London: an American businessman, H. Gordon Selfridge, arrived from Chicago in order to found a department store at the west end of Oxford Street. It would be, he declared, 'the best thing of its kind in the world'.[1] Selfridge's enterprise was described as the 'American Invasion of London'[2] by the daily and drapery trade presses, which accorded the venture extensive, and generally hostile, coverage as it evolved. However, the American methods of retailing thus ostensibly introduced to the British shopping and shopkeeping public were not the only trans-Atlantic innovations which Selfridge's 'gigantic building'[3] brought to public attention. The success of his scheme was dependent upon the size and appearance of the store itself. Selfridge envisioned a truly monumental retail emporium which would help him to achieve his ultimate goal, that of raising 'the business of a merchant to the Dignity of a Science'[4] (Figure 22.1).

The modern methods of steel-frame and reinforced-concrete construction being used in Chicago and elsewhere in the USA at the turn of the century were critical to Selfridge's vision of an enormous, technologically advanced department store. However, the London Building Regulations contained no provisions for structures of this kind, and therefore hindered the construction of buildings with the wide internal spaces and vast street-level windows which Selfridge desired, and with which his architects and engineers were familiar. These regulations were contained in the London Building Acts of 1894 and 1905. The 1894 Act incorporated all previous Acts from 1844 to 1893, and was aimed at the regulation of 'widths of streets, lines of frontages, open spaces to dwellings, heights of buildings and projections therefrom, ventilation and height of habitable rooms and the control and prevention of the spread of fire'[5] . The London Building Acts (Amendment) Act 1905 required new buildings to be equipped with means of escape from fire.

The 1894 and 1905 Acts impeded the construction of the Selfridge building through their regulations for (1) fire prevention (which entailed restrictions placed on cubic footage between party walls) and (2) structural stability (for which the 1894 Building Act prescribed the required thickness of external walls). Although reinforced-concrete flooring could be used to create larger (yet fire-resistant) spaces, and structural steelwork could be

Figure 22.1 Selfridge's as foreseen by the *Illustrated London News*, 25 July 1908

employed to support the loads and stresses of a building (thus making load-bearing external walls unnecessary), the building regulations served to inhibit the erection of large structures whose interior and exterior appearance fully benefitted from these advances in technology.

The Building Acts were finally reformed after years of agitation by engineers, architects, and businessmen for legislation to allow the construction of large open premises supported by structural steelwork. The LCC (General Powers) Act of 1908 allowed greater cubical extent, and dealt with the uniting of buildings by openings in internal and external walls. The LCC (General Powers) Act of 1909, popularly titled the Steel Frame Act, officially recognised steel-frame construction.

In the reform of the Building Acts to accommodate new construction methods, Selfridges department store played an important and instrumental role. The building was not solely responsible for legislative change. However, the highly publicised construction techniques employed by structural engineer Sven Bylander, first on the Ritz Hotel (1904–5) and then on Selfridge's daring commercial and architectural venture, were an important part of the process which led the LCC to take account of progressively more sophisticated methods of steel and reinforced-concrete construction. In effect, the Selfridge store was a transitional building, erected under the prevailing regulations, but with the knowledge that they were soon to change. H. Gordon Selfridge fully expected, and therefore anticipated, legislative reform; he consistently petitioned for waivers from the regulations, and, through his persistence, helped the building reforms come to pass. Selfridge's department store therefore became the first large building in London to fully exploit steel-frame and reinforced-concrete construction so that both the interior and exterior of the building revealed the use of these modern methods of structural engineering.

Complex foundations

> At the Corner of Oxford Street and Duke Street, Mr H. G. Selfridge, formerly of Chicago, is erecting a large department store . . . It is aggressively big in scale and entirely at odds with everything else in Oxford Street, a matter which is not altogether to be regretted because Oxford Street is one of the ugliest streets in the world, and everything that pertains to architecture has been until recently conspicuous by its absence.
>
> (Francis Swales, 'Notes from Europe', *The American Architect* XCIV 28 October 1908, p. 140)

Subsequent building programmes have altered and extended Selfridges premises, involving a number of architectural and building firms, and resulting in a tangled web of architectural history. Yet even as the first Selfridge premises (now the south-east wing of the building) opened to the public in 1909, the store was known to have had complicated origins. The entire enterprise was dependent upon a network of connections which linked together Selfridge, his business associates, and the architectural and engineering firms engaged to carry out the Selfridge store's construction.

Selfridge was no newcomer to the department store scene. He began his career as a sales assistant at Marshall Field's, Chicago's premier department store, in 1879, and become one of Field's junior partners in 1889. [. . .] In 1904 Selfridge left Field's to go into business on his own, buying Schlesinger and Mayer's department store, the steel-frame building (1899–1903) designed by Louis Sullivan, which was located just down the street from Marshall Field's. But Selfridge disliked competing with his old employer, and within a few months sold the store to Carson Pirie Scott & Co. He next turned his attention to London, which he believed was in need of a progressive and modern department store.

Across the Atlantic, Harrods reigned over the vast London retail field which included Whiteley's (the self-proclaimed 'Universal Provider'), D. H. Evans, John Lewis, John Barker of Kensington, and a number of other large drapery concerns. Most of these stores had evolved piecemeal from small shops, gradually adding departments and taking over neighbouring buildings; many had roots going back to the 1860s or even earlier. Purpose-built structures to house these retail establishments were, therefore, rare: in fact, despite its unified appearance, Harrods grandiose building of 1901–5 was actually a re-building, in stages, over the old existing structures.

Once in London, Selfridge secured English support for his 'American Invasion'. He entered into partnership with Samuel J. Waring, of Messrs Waring & Gillow, London's largest furniture and furnishings emporium, and highly successful interior decoration firm, with headquarters located at 175 Oxford Street. The businessmen formed a company, Selfridge and Waring Ltd, 'to purchase land and carry on the business of drapers, tailors, hosiers, . . .'[6] and Selfridge joined the board of directors of Waring & Gillow Ltd. The partnership of Selfridge and Waring was short-lived: the company was dissolved in 1909, and Selfridge's new company, Selfridge & Co. Ltd, bought out Waring's interest in the venture with the understanding that the store would not sell furniture. From 1906 to 1908, though, Selfridge and Waring Ltd had bought up a number of the property leases on the proposed Oxford Street/ Duke Street site (owned by the Portman

Estate), which was occupied by a 'medley of small shops and private houses'.[7]

Builders, engineers and architects

In addition to aiding Selfridge's enterprise, initially with both his capital and his knowledge of the London retail scene, Samuel J. Waring's interest in the venture extended to the building of the proposed store, for he also controlled a construction firm, the Waring White Building Co. Waring's partner in this business was James Gilbert White, an American engineer who had undertaken a number of large projects for English entrepreneurs in Australia. In 1900 White had founded an English branch of his firm, J. G. White and Co. Ltd, through which he supervised several power plant and electric railway works. Waring and Gillow had entered the construction business in order to build their own new eight-storey premises near Oxford Circus. In 1904 Samuel J. Waring and J. G. White merged their interests to take construction contracts over from Waring & Gillow Ltd: these included, notably, that for the Piccadilly hotel which would become the Ritz. Two years later, the company re-registered as Waring and White (1906) Ltd, and took several contracts over from the Waring White Building Co., including that of 13 November 1906 for Selfridge and Waring Ltd to erect 'Stores in Oxford Street'.[8]

Chief engineer for the Waring White Building Co. was Sven Bylander, a Swedish born structural engineer who had designed large steelwork buildings in Germany and America, prior to moving to London in 1902. Bylander designed the Ritz and Selfridge's steel frames, and also that of the Royal Automobile Club (1910–11), in accordance with precedents set by Chicago and New York commercial high-rise architecture in the late nineteenth century, using, in fact, the Carnegie Steel Company's Handbook, issued in 1897.

Selfridge also counted upon the expertise of Daniel Burnham's Chicago architectural firm. Burnham was the architect to whom Marshall Field consistently turned for his retail buildings and warehouse structures. Selfridge, an integral member of the Field organisation for over 20 years, was well aware of Burnham's mastery of the technical requirements of department store construction such as fire-proofing, elevator placement and electric lighting. [. . .]

Burnham's office supplied Selfridge with a complete set of drawings in 1906. During a trip to England in April 1907, Burnham visited Selfridge, perhaps to finalise plans for the store. However, Burnham's firm was but the first of a series of 'supernumerary cooks'[9] involved in the building's design. The Burnham elevation was soon altered by another American, Francis Swales. Swales modified the building's external appearance, including, as he explained, the introduction of triple windows in the friese, and 'the change in style of detail from the neo-Grec to that of Louis XVI'.[10] A shrewd self-publicist, Swales praised his own contributions to the building's design in the *Architectural Record.*[11]

Burnham's firm bowed out of the project altogether when the London building regulations, which required that commercial premises be split into cellular compartments of no more than 250,000 cubic ft each, became too difficult for them to deal with trans-Atlantically. This measurement of a building's 'cubical extent' meant 'the space contained within the external surfaces of its walls and roof, and the upper surface of the floor of its lowest storey' (irrespective of horizontal divisions created by floors).[12] The internal walls of a building (referred to in the Act as 'party walls') could contain openings of no more than 7 ft in width and 8 ft in height, and, taken together, these openings could not exceed one half the length of the party wall in which they occurred. Such openings were required to be fitted with wrought iron doors or shutters to prevent the spread of fire; otherwise the two connecting spaces could not, taken together, exceed

250,000 cubic ft. The 250,000 cubical extent limit could be waived, but the absolute maximum was 450,000 cubic ft. The D. H. Burnham & Co. plans submitted to the LCC in February 1907 showed divisions exceeding 450,000 cubic ft, and permission to erect the building was initially refused because 'no power is given to the Council under the London Building Act 1894, to consent to the erection of buildings of the warehouse class with divisions of a greater cubical extent than 450,000 cubic ft'.[13] In order to realise Selfridge's vision of a spacious store, a London-based architect was needed, first, to petition the LCC for permission to divide the building into compartments of 450,000 cubic ft each; and, secondly, to alter the Burnham plans (presumably by adding more internal walls) in order to bring the cubical extent within the divisions down to the 450,000 maximum.

R. Frank Atkinson, Waring & Gillow's architect, was contracted to carry out the project. Like Waring, Atkinson had moved to London from Liverpool; in his case, he had studied architecture there. The new store premises for Waring & Gillow which opened in June 1906 had been designed by Atkinson, who was, therefore, familiar with the process of erecting buildings on Oxford Street.

Bypassing the building regulations

Atkinson's main diplomatic chore in London throughout 1907 and 1908 was petitioning the LCC Building Act Committee on behalf of the Selfridge venture to exceed the regulations for structures of the 'warehouse class' as defined and laid out in the London Building Acts of 1894 and 1905. According to the definition used in those acts:

> The expression 'building of the warehouse class' means a warehouse, factory, manufactory, brewery, or distillery, and any other building exceeding in cubical extent one hundred and fifty thousand cubic feet, which is neither a public building nor a domestic building.[14]

Structures corresponding to this description were not to exceed 250,000 cubic ft without party walls; hence the existing London department stores (including Harrods) each consisted of a series of separate, but interconnecting, rooms, like individual stores side by side, rather than departments within a single building. It is worth noting that another large London drapery establishment, D. H. Evans, also petitioned the LCC for permission to exceed 250,000 cubic ft in divisions of their proposed Oxford Street extension throughout 1907. Pressure on the LCC therefore came from others in the drapery trade as well, quite possibly in response to the threat of increasing competition posed by Selfridge's much publicised grand scheme.

By 1907 the authorities were beginning to realise that adherence to this regulation was not always essential. *The Builder* noted in 1907 that during the 1905–6 construction year 12 businesses had petitioned the London County Council for permission to exceed 250,000 cubic ft; six of the requests were granted, and six were refused.[15] In August 1907 *The Builder* reprinted an LCC Building Act Committee report which urged that the current regulations be amended 'so as to remove all restrictions on the Council's power to allow increased cubical capacity for buildings of the warehouse class'.[16] The London and District Association of Engineering Employers had initiated the proposal, reasoning that 'these restrictions made it almost impossible for engineering firms to carry out their work in London in accordance with modern requirements'.[17] The proposed amendment would give the Council discretionary power to allow for horizontal separations within buildings; openings in party walls; fire-resisting doors of materials other than iron; and the uniting of buildings through wall openings. Much of this proposal was eventually passed in the LCC (General Powers) Act of 1908, but first it was defeated on the grounds that 'the erection in London of buildings of great cubical extent, not subdivided by party walls, cannot fail to expose London to the risk of conflagrations . . .'.[18]

Fear of fire

The risk of fire was of great concern to the Council's Building Act Committee. As noted in *The Builder,* of 112 fires involving questions of structural safety which occurred within the London County boundaries in the year 1906, 97 occurred in buildings coming under the Council's building regulations; in these fires 24 lives were lost and 136 were endangered.[19] The Council clearly felt responsible for the structural safety of buildings coming under its jurisdiction, and was reluctant to pass any amendment which might result in further tragedies. Theatres, with their large crowd capacity, were notorious fire hazards. However, drapery houses too posed an especial threat, as fabrics, clothes and other dry goods were highly flammable. The danger was further heightened by the fact that in London a large proportion of drapery shop assistants lived-in, residing in crowded company-owned accommodation either on or very near the business premises. They were therefore spending 24 hours a day in a high-risk environment.

Widespread recognition of the need for fire legislation had resulted in the London Building Act (Amendment) Act of 1905. This Act required new buildings to be provided with so-called 'reasonable' means of escape in case of fire, and stipulated that plans and particulars be deposited with the Council before building work began. Still, these new regulations lagged behind technological innovations in fire-proofing and building construction which enabled structures of greater cubical extent to be essentially safe from fire. Such advances included concrete flooring, encased steel framing, and rolling steel shutters in place of iron doors. These were understood to be particularly relevant to the construction of large commercial premises, a connection clearly made in 1907 when a spokesman for the LCC Building Act Committee stated that 'the Building Act of London was obsolete', with the result that 'restrictions placed on trade in London were too great'.[20]

In addition to his repeated requests for greater cubic footage allowances, R. Frank Atkinson petitioned the Council for the use of rolling iron shutters to be used in place of iron doors for fire prevention, and also for wider and more numerous wall openings than the regulation allowance of 7 × 8 ft for interior walls. The interior openings Atkinson requested were 12 × 12 ft and the Building Act Committee eventually granted permission for them. Atkinson's petition for exterior window openings equalling more than half the area of the external walls was also granted. However, the architect was not so lucky in his requests for greater building height. The maximum allowed under the 1894 act was 80 ft, and Atkinson therefore had to subtract two stories from the proposed building, leaving a total height of five floors above ground, with three basement floors. Atkinson submitted the plans for the Selfridge store to the Council in early 1907. The Building Act Committee granted permission for exterior windows to exceed half the external wall area in June 1907; for divisions of the store to exceed 250,000 (but not 450,000) cubic ft in July 1907; and for internal openings to exceed the regulation size, along with the use of rolling iron shutters for fire prevention, in October 1908. Also at this time, and clearly related to Selfridge's and Atkinson's perseverance in petitioning for waivers from legislation, the LCC passed the London County Council (General Powers) Act 1908. Part III of this Act amended the 1894 Building Act, and allowed for horizontal divisions in buildings of the warehouse class, for cubic footage to exceed

250,000 (but not 450,000) cubic ft, and for the uniting of buildings by openings in party or external walls. These changes had the effect of vastly increasing the legal limits of internal spaces bounded by walls. But there were still no regulations for reinforced-concrete and steel-frame construction.

The steel frame

Although still a new departure from traditional building methods, the internal steel frame was becoming more and more common in Chicago and New York by the turn of the century, but Britain lagged behind in its adoption. Many architects resisted the use of steel in building construction because they dreaded the necessary study, or were reluctant to collaborate so closely with engineers. The RIBA held up the process of legislation for new building methods by insisting that very definite rules, arrived at after much discussion and research, were needed in order to control builders. Finally, it has been argued that the nature of Britain's design market, and the bifurcation of the architectural and engineering professions along the lines of building types and clients, hindered the acceptance of new building methods by providing no impetus (such as competition) for creativity or innovation.[21]

Yet steelwork was definitely being employed, albeit in idiosyncratic ways. As Sven Bylander later noted, when he arrived in London in 1902 it was usual practice to 'employ some steelwork in the internal part of the building only, or to carry the external wall at the first floor level on steelwork to permit large shop windows, and sometimes steel pillars were used to strengthen external walls', while little precaution was taken for the stability or fire protection of individual steel members.[22] One of Bylander's engineering colleagues affirmed that at that time 'builders, in using steelwork in building simply piled one piece on top of another, stuck a few bolts in and called it constructional steelwork', a practice he described as 'ironmongery'.[23] Because of the haphazard ways in which steelwork was employed, and the lack of standardisation in either methods or materials, it is virtually impossible to pinpoint the 'first' steel-framed building in Britain - although claims have been made for, among others, Robinson's Emporium in West Hartlepool.

The issue is further complicated by the fact that no standard *definition* of 'steel-frame construction' existed at the time - the term 'steel frame' was often used to describe any structure that employed some steelwork, in some way. The discrepancies in use of the term, and the ensuing difficulties in understanding just how various buildings had been constructed, were brought up in discussion at a meeting of structural engineers as late as 1913. There a speaker noted that a number of terms were used quite loosely in reference to steel construction, namely: 'steel-cage construction', 'interior skeleton', 'steel skeleton', 'skeleton construction', and 'cage construction', and this lack of firm definition prevented one's understanding of exactly how and where the loads and stresses were being carried in buildings utilising structural steelwork.[24] At stake was the question of whether the steel frame supported solely (or primarily) the floor loads of a building which had self-supporting external walls, or whether the frame actually supported *all* loads and stresses, including those of the walls.

In early twentieth century London there was no real incentive to erect a steel-frame building with non-loadbearing external walls. This was because Part II of the 1894 London Building Act defined the necessary thickness of walls for large buildings: depending upon the height and length of the walls. Their base measurement was to range, for example, from 13 in. (for a wall of 25 ft or less in height) to 31 in. (for walls of between 100 and 120 ft in height and 45 ft in length); a taller or longer building would require even more substantial walls.

Further, no wall was to be less in thickness than one fourteenth part of the height of the storey.

The Ritz and American technology

The Ritz Hotel, the first London steel-framed building of 'importance'[25] was designed with a 'complete steel frame'[26] which carried all loads on steelwork, including the reinforced concrete fire-proof floor system. Still, the structure had also to conform to the LCC 1894 and 1905 Building Acts, and therefore the Ritz walls measured 39 in. in thickness at street level and 14 in. at sixth floor level. The hotel, designed by Mewes and Davis with Sven Bylander as structural engineer, and constructed by Waring White Building Co., was built in 1904–5, amidst great excitement in the architectural community. The most minute details of the building's steel framework, and every 'Americanism' inherent in its construction, were fully recorded in the architectural press. *The Builder's Journal and Architectural Record* ran a lengthy series on the engineering side of the Ritz construction for a full year, from 28 September 1904 through 13 September 1905. Readers were assured that they would 'be introduced to various methods employed in modern American contracting practice new to this country,' because the hotel was 'being erected under the management and supervision of men from the United States who have had wide experience of large building works there'[27] (i.e. J. G. White and Sven Bylander). Full details were provided of such innovations as the 'American cranes' used in raising the steelwork, including 'a derrick of American pattern specially constructed for the builders' with a 360° arc and, importantly, Bylander's standardised drafting procedures.[28]

Bylander credited his method of preparing drawings to his experience in America, where 'every office in good standing has a set of standard tables' which 'are used throughout the office by each member, and this produces uniformity in methods and design'.[29] Numerous reproductions of the engineer's drawings and extensive photographic coverage of the Ritz frame as it went up testified to the interest in, and importance of, the new form of construction (Figure 22.2). Two of the *Builder's Journal* articles included 'Notes on the Steelwork by S. Bylander,' in which the structural engineer explained how to read the accompanying framing plans, described the use of standardised parts which eliminated the need for on-site templates, outlined the numbering system used to distinguish each piece of steelwork, provided factual information on loads and stresses, and, importantly, reassured the public that 'the construction practically conforms to the latest standards for steel-framed office buildings in America'.[30]

Figure 22.2 The Ritz Hotel under construction (from *The Builder's Journal*, 12 April 1905)

Reinforced concrete construction

After the Ritz was completed and work on Selfridges had begun, a new professional organisation, the Concrete Institute, was founded to study and promote the use of reinforced concrete in construction. The birth of the Concrete Institute in 1908 (renamed the Institution of Structural Engineers in 1922) coincided with the creation, by the Institution of Civil Engineers, of a special committee to report on reinforced-concrete construction in response to prevailing doubts about its safety. Just as the LCC building regulations hindered the adoption of steel-frame construction, so too did they inhibit use of reinforced concrete. Although several systems of this type of construction were well known in Britain, and the Hennebique system in particular was commonly used outside London, the LCC regulations contained no provisions for the use of concrete in building. New methods or materials could not be used unless a waiver was obtained, and the LCC had no power to grant waivers for the use of either reinforced concrete or structural steelwork.

In the absence of any regulations for concrete construction, the 1894 Building Act required that, as the Concrete Institute put it, 'every building must practically be enclosed with brick or stone or concrete walls of an unnecessary thickness', whereas 'by the use of reinforced concrete, as by the use of steel skeleton construction, this unnecessary expense may be saved'.[31] The Ritz Hotel was a victim of this legislation; although the building was constructed with a load-bearing steel frame, it was clad with masonry of loadbearing thickness, as if the steel frame were not there. Sven Bylander was a member of the Concrete Institute, which pushed for Council authority to waive the existing Building Act rules where steel framing and/or reinforced concrete were involved, and agitated for the official recognition of these building methods in the LCC regulations.

Selfridge's

In this atmosphere of increasing pressure for the repeal of antiquated legislation, and the enactment of new regulations appropriate to the erection of steel-framed buildings, the Selfridges department store was constructed along the lines of American high-rise technology. It incorporated a steel frame together with staircases, flooring, and one retaining wall of reinforced concrete. In light of the Selfridge store's incubation as a Burnham design, the internal steel framework is entirely understandable. So, too, is Bylander's involvement in the project: not only was he chief engineer for Waring & White (1906) Ltd, the construction firm headed by Selfridge's short-lived partner, Samuel J. Waring, but his approach and methods had received wide publicity through the Ritz Hotel project. *The Builder's Journal* had promoted Bylander's 'considerable experience in the design of steel-framed or skeleton buildings in the United States'.[32]

Work on the Selfridge building progressed at exceptional speed: the structure was completed in twelve months, 'the erection of the steelwork, amounting to 3000 tons, occupying less than half this time'.[33] Much of the facility with which Selfridge's was erected was due to Bylander's organised system of preparing the engineering drawings and specifications, which enabled the steelwork to be cut (and in some instances shaped or riveted) in the shop (Figure 22.3). For Selfridge's, Bylander prepared 12,000 blueprints, and construction was carried on at the rate of about 125 tons per week; as Bylander noted, 'The shop details prepared per week was equivalent to 100 tons of steel'.[34] (Figures 22.4 and 22.5 show interior and exterior views of the constructional steelwork).

The internal steel frame which Bylander

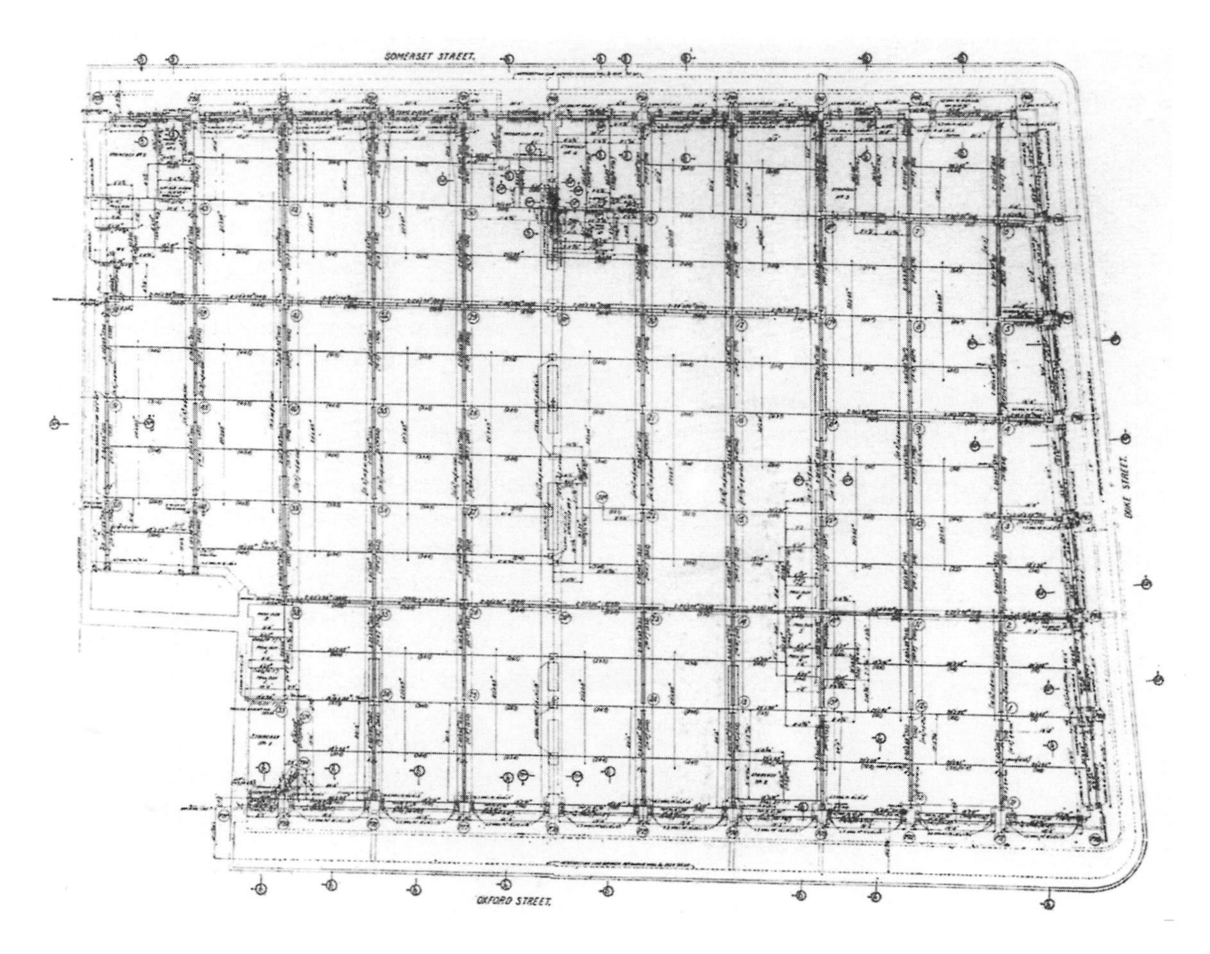

Figure 22.3 Selfridge's: the ground-floor framing plan and the completed layout
Source: Concrete and Construction Engineering, March 1909

designed for Selfridge's corresponded to the building's exterior, as well as interior, appearance. The LCC regulation wall thicknesses for buildings of the warehouse class were bypassed, allowing not only much thinner walls, but far greater window area. Traditionally, the width of window openings had been determined by the safe span for a stone lintel. At Selfridge's the steel frame, combined with the use of cast iron window surrounds and entrances, allowed a much larger proportion of the facade to be taken up by windows. Very large plate glass windows were installed, some as large as 19 ft 4 in. long by 12 ft high. In fact, the window area was greater than half the area of the external walls on both the Oxford Street and Duke Street frontages, and permission for this had been granted by the LCC in 1907.

The steel frame carried the weight of the interior walls and the reinforced concrete floors; the ground floor piers were built 'sufficiently large in blue brick to carry the external wall as well as the load from the floors'.[35] The 'external wall' actually amounted to masonry strips supporting the pillars on the building's

Figure 22.4 The constructional steelwork of Selfridge's
Source: Concrete and Constructional Engineering, March 1909

facade. (Figures 22.3 and 22.6 show the correlation between the engineering and architectural plans of the building). Bylander contributed a 13 page, fully illustrated account of the Selfridge store's construction to the March 1909 issue of *Concrete and Constructional Engineering* in which he explained that:

> All the interior walls, except the west party wall, are carried on steel framing, and the floors are built independent of the walls. The exterior wall to Oxford Street and Duke Street is faced with Portland Stone and the backing is blue brick for piers. The frontage to Somerset Street is brick. One of the most noticeable features of the building is the great distance between the columns, and the omission of brick or stone mullions. The window area, therefore, is very large, and good lighting has thus been obtained, also the weight of the exterior walls has been materially reduced. The window frames and mullions are of cast iron.[36]

Importantly, the internal division walls were carried on steel and not self-supporting, so they could be removed at a later date if and when the LCC would permit more than 450,000 cubic ft in each section of the building. Placed at approximately 40 ft intervals, many of these walls were taken down 20 years later, when

Figure 22.5 Selfridge's: the steelwork erected to the third floor except on the last bay
Source: Concrete and Constructional Engineering, March 1909

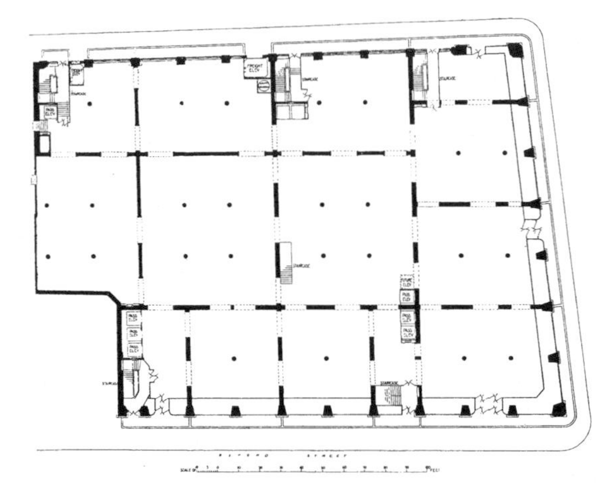

Figure 22.6 Selfridge's: ground-floor plan
Source: *The Architectural Review*, June 1909

legislation allowed a greater cube for divisions within buildings.

The open interior spaces, combined with the great degree of window area, created 'an impression of lightness and brightness' throughout the floors of the store.[37] Shop fixtures such as counters were purposely built lower than the usual height so that, Bylander claimed, one could 'see from end to end of the building'.[38] (Figure 22.7 affords an interior view of the newly-opened premises). These wide internal spaces were allowed under the 1908

Figure 22.7 A showroom in the completed Selfridge's
Source: *The Architectural Review*, June 1909

LCC (General Powers) Act which, in effect, the Selfridge venture had helped to enact. As noted in the editor's introduction to the article on Selfridge's which Bylander contributed to *Concrete and Constructional Engineering:*

> . . . The building is one of the first, if not the first, in the Metropolis to which the recent amendments to the London Building Act have been applied, and which thus comprises a number of compartments of 450,000 cubic feet each, separated from one another by divisional walls, in which the door openings are also of larger area than was allowable before the passing of the new Act - 12 ft by 12 ft.[39]

The journal's editor also praised Selfridge, who 'by his perseverance did much, not only to obtain a building of very high qualities, but also to improve the legislative conditions under which it was executed'.[40]

The Steel Frame Act

The efforts of H. G. Selfridge, his architect R. Frank Atkinson, his engineer Sven Bylander, and the Concrete Institute were instrumental, also, in the enactment of the LCC (General Powers) Act of 1909. Known as the Steel Frame Act, this was the legislation which finally gave the Council the power to regulate the construction of reinforced-concrete structures, and decreed that:

> . . . it shall be lawful to erect subject to the provisions of this Section buildings wherein the loads and stresses are transmitted through each storey to the foundations by a skeleton framework of metal, or partly by a skeleton framework of metal, and partly by a party wall or party walls . . . [41]

The Act provided guidelines for the encasement of steelwork with fire-resistant material, and laid out required wall thicknesses of 8.5 in. for the topmost 20 ft of a building and 13 in. for the remainder of its height, but allowed for this regulation to be modified or waived.

Yet in some ways the passage of the Steel Frame Act complicated, rather than simplified, the problems faced by structural engineers. There was no consensus on how to interpret the legislation, for it contained insufficient details. 'When the Act first came out, engineers had to consult the district surveyor in each particular district in order to find out what was his reading of it, because it would probably be opposite from that of his neighbour'.[42] The new Act did not apply to buildings designed under the old regulations. And difficulties also stemmed from the fact that the legislation was in the form of an Act of Parliament rather than a local bye-law. As evidenced by the time taken to pass the General Powers Acts of 1908 and 1909, innovations in building technology could not quickly be introduced, and the process of amendment was slow. Finally, 450,000 cubic ft remained the maximum space allowed for divisions within commercial buildings.

Conclusion

Selfridge's department store opened its doors to the public on 15 March 1909. The building occupied the whole of its 250 by 175 ft site. There were nine passenger lifts, two service lifts, and six staircases. The store's eight floors (five above ground and three below) averaged 15 ft in height, and housed over 100 departments as well as a vast range of amenities, including an Information Bureau; First Aid Room with trained nurse; French, German, American and Colonial retiring-rooms 'typically furnished'; restaurant, luncheon-hall, tea-room and roof garden. Some 1400 employees had been hired to ensure the smooth running of Oxford Street's new commercial palace.

In *The British Building Industry*, Marian Bowley suggests that 'the need for new buildings for new purposes' may stimulate innovation in construction methods.[43] Selfridge's provides an example of such an enterprise and demonstrates, indeed, that modern construction techniques were critical to the success of the venture. However, in order to construct the store H. G. Selfridge envisioned, the obstacle of the London building regulations had to be overcome. The highly publicised raising of the Ritz steel frame in 1904–5, followed by the much promoted Selfridge undertaking in 1906–9, were important events in the transformation of the building regulations to permit recognition of steel-frame and reinforced-concrete methods of construction. Both projects drew attention to modern American methods of structural engineering, and introduced British engineers to standardisation techniques in the production of drawings and specifications, as well as in the actual steelwork. The Ritz, constructed under the old Building Act of 1894, was covered in masonry of the required thickness, and its steel-frame construction was therefore not apparent. But Selfridge's, with its wide plate glass windows and near-absence of external walls, was clearly a different sort of building, achieved only after much negotiation with the LCC. The Building Act reforms of 1908 dealing with cubic footage, and the steel-frame and reinforced-concrete sections in the 1909 (General Powers) Act were, at least in part, due to H. G. Selfridge's determination to build a Chicago-style department store in London.

Selfridge's novel approach to building, that of planning in anticipation of legislative changes, and then pushing for the necessary reforms, surfaced again in 1919. At that time he determined to construct a massive 300 ft tower on top his emporium, as part of the western extension designed by Sir John Burnet and Thomas Tait in association with Burnham & Co.'s Chicago successor firm, Graham, Anderson, Probst & White. Building work, including the laying of foundations for the tower, was begun in 1919 and completed in 1924. In this instance, however, the LCC Building Act Committee could not be persuaded to waive their regulations. The Committee remained unconvinced that Selfridge's should be allowed a monumental tower which would vie with the dome of St Paul's; permission to exceed the 80 ft height regulation was not granted and the tower was never constructed.

In effect, the construction techniques introduced to London through the 1909 Selfridges building had a significant impact upon the urban landscape. The steel frame allowed wide interior spaces and permitted the installation of very large plate glass windows which provided the store with considerable natural light, whilst also creating grander window shopping possibilities for passers-by. As stated in *The Architects' Journal* in 1920, 'The building gave a new scale to Oxford Street and has exercised a strong influence over the design of many big structures that have since been erected in the metropolis'.[44] These features were already becoming standard architectural design for retail premises elsewhere; to the extent allowed by the revised but still restrictive Building Acts, H. G. Selfridge's monument to commerce brought modern American department store design to London.

Notes

1 *Draper*, 13 March 1909, p. 256.

2 *Drapery Times*, 20 Feb. 1909, pp. 359–60.

3 *Draper*, 13 March 1909, p. 256.

4 *Drapery Times*, 13 March 1909, p. 523.

5 Quoted in C. C. Knowles and P. H. Pitt, *The History of Building Regulations in London 1189–1972* (1972), p. 93.

6 PRO BT31/11562/89184.

7 Lease Catalogue and Company Books, Selfridges' Archives; *Draper's Record*, 10 April 1909, p. 105.

8 PRO BT31/10695/81048 & BT31/11783/91435.

9 *American Architect*, XCIV (28 Oct. 1908), p. 140.

10 Francis Swales, 'The Influence of the Ecole Des Beaux-Arts Upon Recent Architecture in England', *Architectural Record* 26 (Dec. 1909), pp. 422–3.

11 *Ibid.*, pp. 421–3.

12 W. R. Griffiths and F. W. Pember, *London Building Act, 1894 (1895)*, p. 12.

13 *LCC Minutes* (April–June 1907), Appendix A, 'Applications under Building Acts', no. 1023.

14 Griffiths and Pember, *The London Building Act, 1894*, p. 13.

15 'A Year's Work Under the London Building Act, 1894,' *Builder*, 12 Oct. 1907, p. 393–4.

16 *Builder*, 3 Aug. 1907, p. 145.

17 *Ibid.*

18 *Builder*, 19 Oct. 1907, p. 417.

19 *Builder*, 26 Jan. 1907, p. 92.

20 Capt. Hemphill of the LCC Building Act Committee, quoted in *Builder*, 18 May 1907, p. 608.

21 Bowley, *British Building Industry*, pp. 33–4

22 *Concrete Inst. Trans. and Notes* V (Oct. 1913), p. 57.

23 W. G. Perkins, discussion of Bylander's paper, ibid, p. 107.

24 E. Fiander Etchells, discussion of Bylander's paper, *Concrete Inst. Trans. and Notes* (Oct. 1913), pp. 109–10.

25 S. Bylander, 'Steelwork in Buildings – Thirty Years' Progress,' *Structural Engineer*, Jan. 1937, p. 2. Because of confusion surrounding the term 'steel frame', Bylander himself was not sure that the Ritz was the first such building in London.

26 S. Bylander, *Steel Frame Buildings in London*, p. 71.

27 'The Ritz Hotel,' *Builder's Journ.*, 28 Sept. 1904, p. 165.

28 'The Ritz Hotel,' *Builder's Journ.*, 2 Nov. 1904, pp. 235–6.

29 S. Bylander, *Concrete Inst. Trans. and Notes* IV (July 1912), p. 88.

30 'The Ritz Hotel,' *Builder's Journ.*, 22 March 1905, p. 148–56; 13 Sept. 1905, p. 146–8.

31 *Concrete Inst. Trans. and Notes* (Feb. 1909–Dec. 1910), p. xi.

32 'The Ritz Hotel,' *Builder's Journ.*, 22 March 1905, p. 148.

33 S. Bylander, 'Steel and Concrete at the Selfridge Stores, London', *Concrete and Constructional Engineering*, March 1909, p. 26.

34 Bylander, *Steel Frame Buildings in London*, p. 70.

35 *Ibid.*, p. 71.

36 Bylander, *Steel and Concrete at the Selfridge Stores, London*, p. 22.

37 S. Bylander, 'Concrete and Steel Construction at the Selfridge Stores', *Builder's Journ.*, 31 March 1909, p. 280.

38 *Ibid.*

39 Editor's preface to Bylander, *Steel and Concrete at the Selfridge Stores, London*, p. 9.

40 *Ibid.*, p. 9.

41 'Steel Frame Act,' H. D. Searles-Wood & Henry Adams, *Modern Building* (1921), pp. 157–62.

42 E. Lawrence Hall, discussion of Bylander's 'Steel Specifications' paper, *Structural Engineer* IV (March 1926), p. 120.

43 Bowley, *British Building Industry*, p. 35.

44 'Retaining Walls and Foundation for Selfridge's New Building,' *Architects' Journ.*, 18 Feb. 1920, p. 222.

23

PARIS AND THE AUTOMOBILE

Norma Evenson

Source: N. Evenson, *Paris: a Century of Change, 1878-1978*, New Haven, Yale University Press, 1979, pp. 49-63, 65, 71-2, 75

When Edmondo de Amicis left the Gare de Lyon in 1878, he found himself suddenly immersed in Parisian traffic which, judging from his description, was fully as congested, chaotic, and dangerous as in our own time. In spite of the width of the street, circulation was clogged, and, he noted: 'Our carriage is obliged to stop every moment to wait until the long line which precedes it is in motion. The omnibuses, of every shape, which seem like perambulating houses, pursue each other madly. The people cross each other, running in every direction, as if playing ball across the street, and on the sidewalks, they pass in two unbroken files.'[1] Describing his first exposure to dinner-hour traffic, de Amicis reported: 'The commotion is simply indescribable. Carriages pass six in a row, fifty in a line, in great groups, or thick masses, . . . making a dull, monotonous sound, resembling that of an enormous unending railway train which is passing by.'[2]

In addition to the noise and confusion, Parisian traffic presented a 'hell for horses,' and to sensitive observers the street was a theater of cruelty. In 1906 it was noted: 'Our boulevards are still dishonored by the pitiable nags who, with dragging backs, pull the carriages.' Describing the scene near a freight station, a writer observed that 'there the unfortunate percherons almost strangle themselves straining their muscles, when fatigue halts them. Then they are made to move with blows of whips, kicks, and clubbing. It's a tumult of curses and whip cracks. The poor beasts foam, fall: there are scenes of indescribable barbarism.'[3]

In addition to providing ample scope for brutality to animals, the age of horse-drawn traffic made the boulevards a continuous repository for manure, and the beauty of many streets was accompanied, according to contemporary observations, by some rather persistent smells.

In a book entitled *Les Odeurs de Paris,* published in 1881, it was noted that 'the public street is the great receptacle for a mass of debris and decomposing matter: it is the principal ground where offensive and toxic smells originate and propagate; it is the common reservoir, par excellence, where one gets the majority of diseases engendered by an unwholesome locale.'[4]

It was the horses, of course, who, 'by their always increasing numbers, and the quantity of their excrement, are the principal contributors to the infection of the public street. Who has not observed, near the stands of omnibuses and carriages, the repugnant dung heaps which persist in a permanent state?'[5] The author had no suggestions to make, however, except to urge more frequent street cleaning. Looking ahead, in 1885, an engineer expressed the hope, 'but perhaps it is only a dream, that the horses will be freed one day, and that they will be replaced in large part, in the cities, by mechanical devices, stronger, more docile, more agreeable, and more economical.'[6]

Such mechanical devices, needless to say, were not long in coming, giving rise almost at once to complaints about their speed, noise, smell, and general destructiveness. It was observed in 1909 that 'the inconveniences, the dangers which circulation presents, are aggravated. . . . It is because, since the beginning of the twentieth century, a veritable revolution has taken place through the entry on the scene of more and more automobiles. . . . It was one thing, previously, to protect oneself from a carriage when the horse had a consistent and relatively slow speed; it's another thing now to shelter yourself from these dreadful vehicles which suddenly appear around the turn of a street, describing curves and zigzags in passing one another, and really terrifying people who thought themselves possessed of sangfroid.'[7]

Another writer commented gloomily: 'It's finished, the tranquility of our streets, and the charm of promenading either on foot or in a carriage . . . Paris belongs to the machines.' Added to the alarming speed of the new automobiles, was the frightening bulk of the motor bus, 'deformed, enormous, a masterpiece of ugliness. You can't imagine anything worse, and yet, one sees in the streets of Paris a machine more horrible yet: it is the truck. One can recognize it from a distance, from its formidable gasping, from its black plume of smoke, from the trembling of the ground and the shaking of the buildings. It's a terrifying mass, which poisons the atmosphere and troubles pedestrians, building occupants, and vehicles within a three-hundred-meter circumference.'[8]

Although horse-drawn traffic had not been without problems, efforts to ameliorate conditions had been limited to the creation of new streets. Vehicles of different speeds employed the same roadway, with circulation hampered through frequent intersections and confused through a lack of traffic regulations. Pedestrians were generally unassisted in their efforts to thread their own patterns of movement among the vehicles.

As the twentieth century advanced, however, and mechanized transport became dominant, theorists of the modern movement began to envision a total reordering of the urban fabric. The speed and power of the automobile seemed to demand a pattern of uninterrupted movement, separated from pedestrians and building lines. In the thinking of the modernists, the heterogeneous mixture of activities and structures which had dominated most existing cities was incompatible with efficiency, and a new urban form was conceived in which civic elements would be sorted out and separated. In France, the most influential of these visionaries was Le Corbusier, who in 1922 had produced his well-publicized City for Three Million People. This exhibition project presented an imaginary new city in which the traditional street had been replaced by the limited-access expressway. Building lines were freed from traffic lanes, and pedestrian circulation developed within ample park areas. As Le Corbusier continued to rework his concept of a renovated urban form, he frequently juxtaposed his new vision against the existing fabric of Paris. The dense pattern of building, the constricted block sizes, the shadowed courtyards, the clogged 'corridor streets' were continually contrasted with an open texture of superblocks, widely spaced high-rise building, and broad motor freeways.

Observing the street pattern of Paris, Le Corbusier noted that: 'It is into this tight network, locked in, infinitely fragmented, that modern speeds, twenty and thirty times increased, are thrust. It's useless to describe the crisis, the disorder: you can't move, you waste time, gasoline, and vehicles. You mark time in place.'[9]

In 1925 Le Corbusier exhibited the first specific application of his ideas to Paris in a project called the Voisin Plan. His proposal included a complete reconstruction of the center of the Right Bank, involving the demolition of all existing structures in an area of 240 hectares extending from the Rue du Louvre to the Place de la République, and from the Gare de l'Est to the Rue de Rivoli. A new district of geometrically spaced skyscrapers would be created along the axis of the Boulevard Sébastopol, with an area of redevelopment extending westward along the Champs Élysées axis. Essential to the proposal was the creation of a new east-west artery in the form of a limited-access expressway paralleling the existing Rue de Rivoli-Champs Élysées route (Figures 23.1 and 23.2). He was to refer to this motorway as the 'east–west backbone of Paris, crossing the entire city: opening up, making way. . . . In this way the 'Voie Triomphale' would be rescued from compromise, ambiguity, absurdity, and all of the traffic hastily thrust into the *cul-de-sac* of the Place de la Concorde would be reabsorbed.'[10] This proposal was to be redeveloped by the architect for many years. A version presented in 1939 to the sixth meeting of the Congrès Internationaux d'Architecture Moderne (CIAM) included the addition of a north–south expressway system intersecting the east-west route in the center of Paris. Beyond the city, these routes would connect with the regional highway system.

Because of its vast scale and the massive urban revovations that would accompany it, the circulation scheme of Le Corbusier was among the most destructive proposed for Paris. Like most advocates of urban renewal, he claimed that he was revitalizing a slum, pointing out: 'All the ancient buildings are preserved. The historic past of Paris (from the Étoile to the Hôtel de Ville) is outside of the plan.'[11] In examining his proposal one may see that such monuments as the Palais Royal, the Place Vendôme, and the Madeleine are retained, but in surroundings altered out of recognition. Although a few existing arteries such as the Rue Royale, the Rue Castiglione, and the Rue de la Paix were to be retained, the entire street and block pattern of the redevelopment area would be changed into a large grid. The open spaces of the Palais Royal and Place Vendôme would be surrounded by open spaces, while the Opéra would sit, disengaged, on the north side of the giant expressway.

Even people who respected Le Corbusier's talent and admired the imagination of his visionary urban designs might well have been appalled to see the center of Paris rebuilt according to his principles. Le Corbusier, however, seemed unable to understand why people didn't like his plan, and he continued to elaborate it throughout his career. Toward the end of his life, he reported: 'Since 1922 (for the past 42 years) I have continued to work, in general and in detail, on the problem of Paris. Everything has been made public. The City council has never contacted me. It calls me "Barbarian"!'[12]

When Le Corbusier produced his Voisin Plan, the number of automobiles in the Paris region was approximately 150,000. By 1930 the number had increased to 300,000, and on the eve of the Second World War it had reached 500,000. Following the interruption of the war and occupation, the volume of automobile traffic began to climb dramatically (Figures 23.3 and 23.4). There were over 1 million automobiles in Paris by 1960 and over 2 million by 1965. In 1970 the figure had reached 2.5 million. Meanwhile, since the beginning of the century, the street surface of Paris had increased by only 10 per cent.[13]

During the 1950s and 1960s, persistent efforts were made to expand traffic lanes at the expense of the adjoining sidewalks. Included in this program was the widening of the Boulevard Montparnasse, in which traffic lanes were increased from 13.50 meters to 21 meters, the Avenue des Ternes (16.50 meters to 22 meters), the Boulevard Malesherbes (14 meters to 22 meters), part of the Boulevard Haussmann (14 meters to 22 meters), and the Boulevard de Magenta (15 to 20 meters). The reduction of sidewalks, in addition to removing

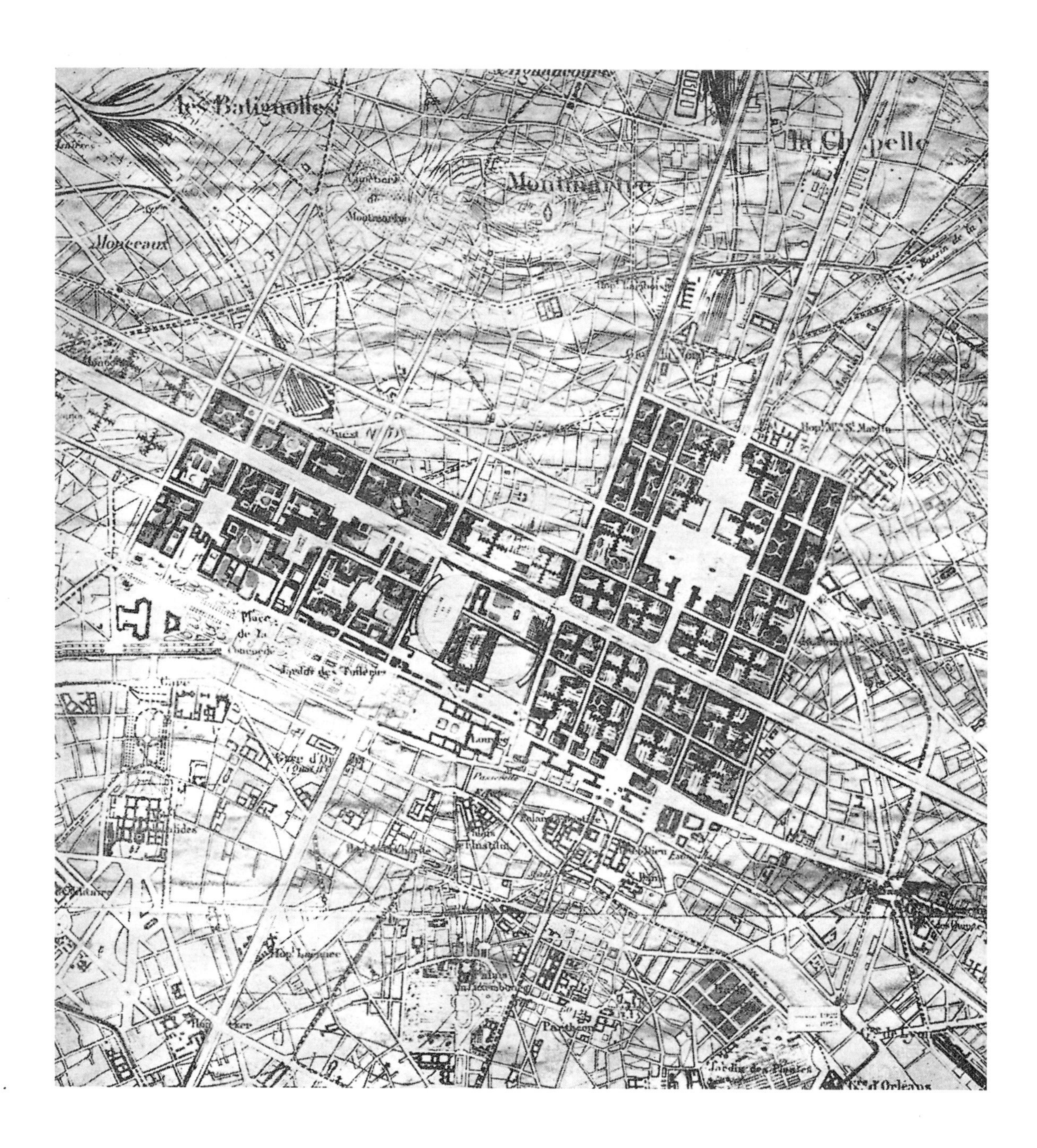

Figure 23.1 The Voisin Plan

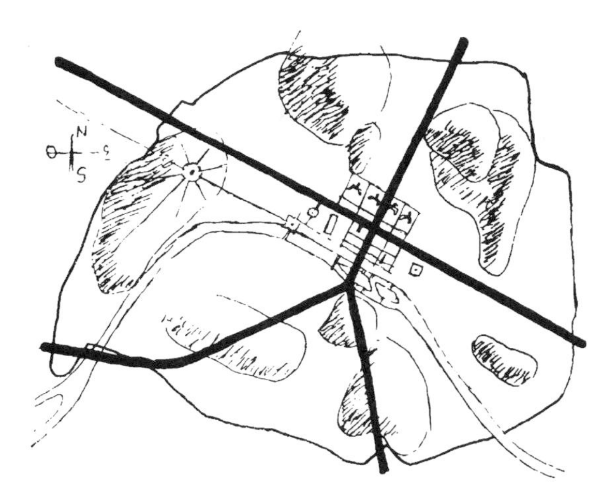

Figure 23.2 Le Corbusier's *grande croisée*

circulation space for pedestrians, notably diminished the number of trees in Paris (Figures 23.5 and 23.6). Sidewalks that had previously carried a double row were reduced to one, and other sidewalks lost all their trees. Along with the greenery, many of the old sidewalk structures disappeared. For years there had been complaints that the sidewalks were excessively cluttered. Now there were few sidewalks, in the old sense, left in Paris. The outdoor living room, the linear park-bazaar-cafe-circus-promenade, was giving way to a minimal utilitarian strip of raised pavement.

The modern world presumably supplied compensations for the decline of the street. What did they expect of the average Parisian sidewalk, the little people of the Second Empire, for whom, above

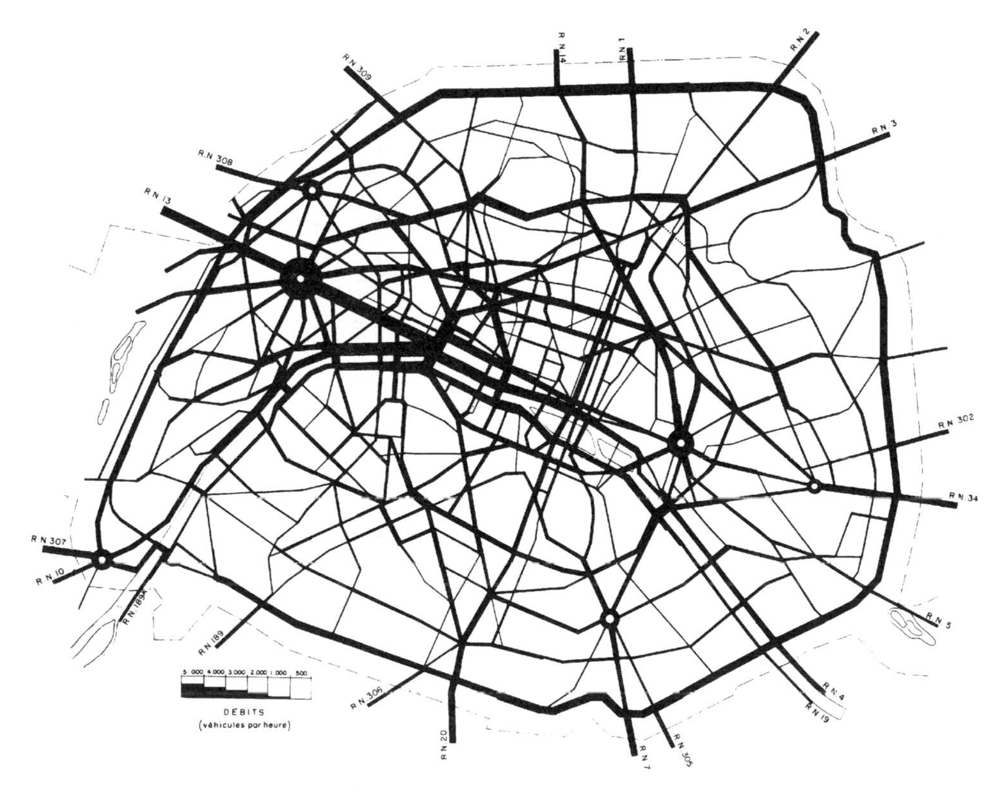

Figure 23.3 Vehicular circulation in Paris in 1957. The map shows areas of greatest traffic volume recorded during weekday afternoons

Figure 23.4 Traffic in the Place de la Concorde

Figure 23.5 Looking east from the Boulevard Saint Denis toward the Boulevard Saint Martin at the turn of the century. The Porte Saint Martin is at the left, largely obscured by trees

> all, it was developed? They wished to find there the essential things which they lacked in their own sad lodgings . . .. Today the great majority of Parisians enjoy sufficient comfort to find at home the varied services which the street used to furnish . . . . As to the slight incidents of the street, they seem pale next to the televised news. What does the Parisian of today wish to find in the street? In a city where private gardens disappear at great speed, where the meters of habitable surface are each day more restricted for each occupant, would it be a little fresh air, a foretaste of the country, trees and flowers which evoke the garden of their dreams around a charming pavilion? Vain utopia! The noise and the gasoline

Figure 23.6 Looking west from the Boulevard Saint Martin toward the Boulevard Saint Denis in 1973. The Porte Saint Martin is at right

> fumes will have discouraged the stroller who would like to relax and breathe on a bench! Perhaps the pedestrian aspires more deeply to recover a little of that human warmth which he found previously on the sidewalks of Haussman. The timid attempts at 'pedestrian streets,' will they be enough to satisfy this?[14]

Not only was an increasing amount of street surface, in the postwar years, given over to motor traffic, but under the tolerant eyes of the police it became commonplace for the already reduced sidewalks to be invaded by parked cars (Figure 23.7).

Meanwhile, many officials were convinced

Figure 23.7 Automobiles parked on the sidewalk of the Avenue Montaigne during the 1950s

that mere street widenings were inadequate to cope with vehicular traffic, and that more radical measures would be required to provide for the apparently insatiable needs of motor cars. Although Parisian urbanists had previously concluded that the dense fabric of the city did not lend itself to large-scale street alterations, the postwar boom inspired some urbanists to contemplate a rapid program of renovation. Paris was to be adapted to the future, and just as the tall building became, to some eyes, a symbol of economic power, so the automobile was seen as an inevitable accompaniment to prosperity and progress. According to the Prefect of the Seine, Paul Delouvrier, 'If Paris wants to espouse her century, it is high time that urbanists espouse the automobile.'[15] There was no stronger enthusiast for this concept than the president of France, Georges Pompidou, who insisted: 'Paris must adapt itself to the automobile. We must renounce an outmoded æsthetic.'[16]

As city officials began the long series of studies and proposals leading toward a master plan for Paris, the need to improve traffic circulation maintained a high priority. A proposal submitted to the Paris Municipal Council in 1951 attempted to project a traffic system for the succeeding hundred years. The old problem of circulation across the center of Paris was taken up again, and it was suggested that three north–south arteries be created, one of which would involve widening the Rue Saint Denis to fifty or sixty meters, and building a tunnel to connect the Rue Rambuteau to the Boulevard Saint Germain. This plan was considered too costly and destructive to be undertaken, however, and studies continued.

A somewhat different solution was proposed by the president of the Municipal Council, Bernard Lafay, in 1954. He acknowledged the problem of circulation through central Paris, but he did not believe that massive demolitions or street widenings should take place in this part of the city. Rather he sought to ease the flow of traffic by having vehicles bypass the center on a new motor expressway to be built in an oval loop around the historic core. This inner motorway would be connected by radial motorways to an outer loop which was to ring the city at its edge. Lafay believed that such a system would disencumber the center of many vehicles by absorbing a large volume of north–south and east–west traffic. Although the new motor route would have involved extensive demolitions in the districts through which it passed, Lafay was content that he had left the center of the city untouched.

As planning continued, Lafay's ideas exerted an influence. The concept of a peripheral motorway to be built around the city boundary was easily adopted, as it could be constructed on unbuilt land which had previously been part of the system of fortifications. The development of the traffic system was complicated by the fact that planning for the city of Paris coincided with the working out of a plan for the Paris region. Obviously the regional traffic pattern would need to be coordinated with that of the city. In 1956 the Plan d'Aménagement de la Région incorporated the principle of Lafay's radial system, in which regional highways connected with a peripheral loop. An internal loop, or *rocade,* was included, which would more or less follow the route of the exterior boulevards, and the radial routes were to extend into Paris as far as this circuit.

Various proposals were made for the form of this internal expressway. Some suggested that it be designed as a double-level structure, with underground sections in the Sixteenth Arrondissement. Others envisioned it as a single-level expressway eight lanes wide. Consideration of the motor route revealed a number of difficulties, in addition to the expense, technical problems, and physical destructiveness it presented. How, for example, could a large number of vehicles be ejected from this rapid motorway into the adjacent streets? Studies for

Figure 23.8 The Boulevard Périphérique looking north at the Bagnolet interchange. To the left may be seen the site of the old fortifications. The outer edge, previously part of a zone in which building was prohibited, was developed following the Second World War and includes widely spaced high-rise housing. The inner band represents the emplacement of the fort itself and was urbanized during the 1920s and 1930s as a district of relatively dense apartment housing

Figure 23.9 The Boulevard Périphérique looking south near the Porte Maillot. The buildings bordering the motorway on the left were constructed on the site of the fortifications. The high-rise hotel at left marks the site of the Porte Maillot. The Bois de Boulogne is in the distance

linking the expressway to the existing urban fabric provided no satisfactory solution.

The idea of the interior *rocade* was rejected by the Municipal Council in 1959. It was decided instead to facilitate traffic movement through a renovation of the exterior boulevards. However, studies of a north–south transversal at the Place de la Bastille continued, together with consideration of the question of radial streets. Included in the conception of radials was the Boulevard Vercingetorix, designed to lead southward from Montparnasse. Construction of this street began in 1976.

While discussions of the *rocade* were underway, construction began on the expressway surrounding Paris. [. . .] In 1956, however, when the new road was incorporated into the Paris regional plan, it took the form of a multilane, limited-access expressway (Figures 23.8 and 23.9). Thirty-five and a half kilometers in length, the road was designed to have thirty-three exit points. The first section, approximately five kilometers long, extended from the Porte de la Plaine to the Porte d'Italie, and it was linked to the main highway leading south. The final section, connecting the Porte d'Asnières and the Porte Dauphine, was put into service in 1973.

By the time this new artery, named the Boulevard Périphérique, was completed, Parisians had had ample time to experience the blessings of the automotive age. Although a government journal proudly referred to it as 'a gigantic operation of road building, which has formed the principal and the most spectacular renovation realized in Paris in several decades,'[17] a newspaper, in 1972, described the Boulevard Périphérique, then carrying over 170,000 vehicles daily, as an 'inferno' and a 'ring of death,'[18] producing an accident per kilometer per day.

While the Boulevard Périphérique serves to

facilitate traffic movement around the edge of Paris, it does not help internal circulation. Vehicles leaving the expressway are discharged directly into the heavily congested street system, and during rush hours the areas adjacent to the boulevard contain masses of stalled traffic.

At the time that the Boulevard Périphérique was begun, city officials were seeking a way of developing an east–west route through the city, and they decided to transform the river quais into motor expressways. The first segment of this system was begun on the Left Bank in 1956, where a river-bank expressway extending from the Pont de l'Alma to the Pont Royal was completed in 1960. Further work on the Left Bank was then suspended, and a scheme was conceived for an extensive motorway along the quais of the Right Bank. This motor route, approved by the city in 1964, was intended to permit drivers to traverse Paris in fifteen minutes. It was projected from Auteuil on the west to the Pont National on the east, a distance of thirteen kilometers. Connections to the Boulevard Périphérique were provided at each end. There had been some objections to the new expressway from the Departmental Commission on Sites, and some members of the Municipal Council wanted part of the road to be placed underground - the section running from the Place de la Concorde on the east to the Pont Sully, marking the western tip of the Île Saint Louis. In this way the expressway would have been invisible as it passed through the historic heart of the city. As it was eventually built, however, the Right Bank expressway dips into a tunnel only as it approaches the Louvre and emerges above ground as it reaches the Pont Neuf.

Although the Right Bank expressway, completed in 1967, provided Paris with a new east–west traffic artery without the need of building demolitions, it was soon evident that the city had lost one of its most cherished promenades (Figure 23.10). The calm, tree-lined quais, the photogenic refuge of fishermen and lovers, had been sacrificed for the rapid movement of automobiles.

Figure 23.10 The Voie Georges Pompidou passes the Quai de la Mégisserie on the Right Bank opposite the Île de la Cité. In the foreground a lone fisherman defies the traffic

The president of the Union Routière de France, not surprisingly, had only praise for the waterfront roadway. He admitted in a speech in 1968 that 'when the construction of the route along the river bank was decided, a veritable chorus of protest was heard.' He insisted, however, that driving along the completed roadway 'is a veritable joy. One travels at a moderate speed, advancing rapidly enough, but above all you see, in a few minutes of travel, one of the most moving scenic compositions in the world. The beauty of Paris is thus made available for the benefit of a far greater number of Parisians than before. Is this regression or progress? I respond without hesitation, it is progress.'[19]

'Progress,' however, was not without its critics, and as Parisians contemplated what had previously been one of the most beautiful river fronts in the world given over to noise, speed, and exhaust fumes, the Municipal Council began to consider the completion of the Left Bank expressway. The principal problem in extending the route lay with the section adjacent to the Latin Quarter. This was one of the most famous strolling grounds of Paris. At street level were the booksellers, carrying on an activity traditional to the quarter since the sixteenth century, while the river quais provided a heavily used recreation area. On the Île de la Cité, close to the shoreline, was the most beloved architectural monument in Paris, Notre Dame Cathedral. In an atmosphere of increasing public criticism, the council attempted to develop an acceptable proposal for extending the expressway.

In 1972 the Paris Prefecture exhibited three proposals for the development of the Left Bank in the Latin Quarter. The schemes were painstakingly presented and involved complicated designs. All made an attempt to conceal the expressway where it lay opposite Notre Dame Cathedral, by using tunnels or camouflaged roadways. Although some concessions were made to pedestrians, and some access to the water included, all the proposals involved a radical renovation of the existing riverfront. Few people looking at the exhibition models were convinced that the new Left Bank would be as agreeable as the old. The Municipal Council continued to study the project, but was unable to come to a decision.

Meanwhile a presidential election was held, and in 1974 Valéry Giscard d'Estaing came into office. In contrast to his predecessor, Georges Pompidou, who had enthusiastically promoted the modernization of Paris in the form of high-rise building, urban renewal, and motor routes, Giscard d'Estaing advocated a more conservative approach to the physical development of the city. Included in his program was the suspension of the Left Bank freeway, which was accomplished by withholding national government funds from the project.

While one could hardly maintain that in Paris the irresistible force of the automobile met the immovable object of the city and gave way, there was evidence by the 1970s that some degree of balance might be reached between the demands of circulation and the physical character of the city. The Prefect of Police in 1968 stated, 'I proceed on the assumption that we do not want to . . . or that we cannot . . . remake Paris from top to bottom to adapt it to the priority of automobile circulation.'[20] Such an announcement was doubtless welcome news to the many Parisians who apprehensively viewed the increasing inroads by motor traffic on the French capital.

The assumption that the needs of street circulation took precedence over almost all other urban considerations had long been dominant in Parisian planning. It had characterized Haussmann's thinking and had strongly influenced the work of his successors. [. . .]

By the 1970s the pedestrian, long ignored by Parisian planners, was achieving a few modest concessions. The construction of underground parking garages at the Place Vendôme in 1972, and in front of Notre Dame Cathedral in 1973, was accompanied by the creation of pedestrian areas at surface level. On the Left Bank, vehicular access to some of the narrow streets near the river was restricted, and the Place Saint André des Arts expanded to provide café space. With the long-contemplated extension of the Left Bank expressway definitely abandoned, the river quais extending from the Pont de l'Archevêché to the Pont d'Austerlitz were redeveloped as a waterfront park (completed in 1977). On the Right Bank, the creation of pedestrian streets accompanied the renovation of the Plateau Beaubourg.

Meanwhile, of course, traffic conditions in Paris have continued to worsen, and there is really no remedy. In peak hours the streets are gorged with cars that barely move. One of the most frequent complaints of city residents concerns traffic noise, and the volume of sound at the Place de l'Opéra has been found to be comparable to that of Niagara Falls. The percussion of iron-shod hoofs on paving stones, which de Amicis likened to the noise of an unending railway train, has been replaced by a maddening whine and roar. A small victory for noise abatement occurred in 1954, however, when a law was effected prohibiting the use of horns in Paris. To the astonishment of the world and, perhaps even more, to the amazement of the Parisians, the law has been more or less obeyed. [. . .]

In 1976, the catalog of an exhibition of Parisian street furniture began by stating, 'The traditional Parisian street is dead, or almost.'[21] The announcement was perhaps exaggerated, for in spite of erosion by the demands of motor traffic, in spite of noise and fumes and the loss of trees, the Parisian street scene has not been totally deprived of its richness. One may hope that in the face of continuing proposals for street renovation and textbook schemes for expressways and pedestrian malls, the old-fashioned city street, with its incompatible mixture of housing, commerce, pedestrians, and vehicles, may yet survive.

Notes

1 Edmondo de Amicis, *Studies of Paris*, New York: Putnam, 1882, p. 6.

2 *Ibid.*, p. 25.

3 Jacques Lux, 'Les Laideurs de Paris,' *La Chronique*, 17 August 1906, p. 224.

4 Jean Chrétien, *Les Odeurs de Paris*, Paris: Baudry, p. 9.

5 *Ibid.*, p. 11.

6 Jules Garnier, *Projet comparé d'un chemin de fer aérien*, Paris: Capiomont et Renault, 1885, p. 58.

7 Fernand Bournon, *La Voie publique et son décor*, Paris: H. Laurens, 1909, p. 71.

8 Lux, 'Les Laideurs de Paris,' pp. 223–4.

9 Le Corbusier, *Destin de Paris*, Paris: Editions Fernand Sorlot, 1941, pp. 28–9.

10 Le Corbusier, *The Radiant City*, New York: Grossman, Orion Press, 1967, p. 213.

11 Le Corbusier, *Oeuvre complète* 1910–1929, Zurich, Girsberger, 1929, p. 111. Reprinted, George Wittenborn, New York, 1964.

12 Le Corbusier, *The Radiant City* (1967 edn), p. 207.

13 Statistics from the *Bulletin d'information de la région parisienne*, No. 2, p. 30 (published by the Institut d'Aménagement et d'Urbanisme de la Région Parisienne).

14 Marie de Thézy (Librarian of the Bibliothèque Historique de la Ville de Paris), *Paris, la rue*, p. 79.

15 Premier Ministre. Délégation Générale au District de la Région de Paris. *Avant-Projet de programme duodécennal pour la région de Paris*, Paris, Imprimerie Municipale, 1963, p. 90.

16 Quoted in Pierre Lavedan, *Nouvelle histoire de Paris: histoire de l'urbanisme à Paris*, Paris, Hachette, 1975, p. 536.

17 Atelier Parisien d'Urbanisme (APUR), *Paris projet*, no. 10–11, p. 74.

18 *France Soir*, 22 November 1972, pp. 1–2.

19 Georges Gallienne, *Paris 2000 ou Paris 1900?* Conférence des Ambassadeurs, October 1968, pp. 15–16.

20 Maurice Grimaud, *La Circulation à Paris*, Conférence des Ambassadeurs, March 1968, p. 20.

21 *Paris, la rue* (exhibition catalog) Paris, Société des Amis de la Bibliothèque Historique, 1976, p. 7.

24

THE PARIS MÉTRO

Norma Evenson

Source: N. Evenson, *Paris: a century of change, 1878-1978,* New Haven, Yale University Press, 1979, pp. 91-5, 98-9, 103-19, 122

The Métro: the long debate

Although nineteenth-century Paris was served by a variety of transportation systems, facilities always seemed strained to the limit, and as the population grew conventional surface transport appeared increasingly inadequate to meet the needs of a modern metropolis. The idea of rapid rail transport was achieving wide acceptance as the solution to the problem of mass movement in cities. [. . .] In Paris proposals had been made as early as 1845 for an underground railway serving to transport goods, and by the 1870s city officials were giving serious consideration to such a system for public transportation. The term *réseau métropolitain* - metropolitan network - was employed by the municipal council in 1871 to describe the proposed rail system, which soon became popularly known as the *Métro.* After studying several proposals, the council decided in 1872 on the routing of an underground *chemin de fer métropolitain de Paris.* The line was to follow the *grande croisée* of Paris, with an east-west axis extending from the Place de la Bastille to the Bois de Boulogne via the Étoile, and a north-south route along the Sébastopol-Saint Michel axis. No request for a concession was produced, however, and the project remained dormant.

In 1875 the Prefect of the Seine presented a proposal of a totally different nature. He suggested the creation of a central railroad station underneath the Palais Royal, which would be linked by tunnels to the five major railroad stations. The concession for operating the system was to be given to the railroad companies. Although the Prefect's plan would have improved the service of the national railroads by providing access to the center of the city, the Municipal Council objected to it strenuously on the grounds that it was not designed to meet local transportation needs.

These two projects exemplified the issues in what was to become a twenty-year dispute between the city of Paris and the national government. The minister of public works and the Prefect of the Seine maintained that the proposed transportation system was a project of national interest and thus subject to national control. The Municipal Council, however, insisted on a local system serving the needs of the city and opposed the domination of Parisian affairs by the large railroad companies.[1]

In 1876 the Municipal Council sent a group of its members to England to study the London underground system. They returned convinced that Paris could develop a similar system independent of any national subsidy, and by 1883 a scheme employing six basic lines was proposed. The minister of public works, however, continued his opposition to local control of the system, and in 1886 he presented to the Chamber of Deputies a repetition of the proposal for linking the major railroads. Although this plan was rejected by the Chamber, the minister continued his campaign to prevent implementation of the Municipal Council scheme.

The prospect of a new transportation system seems to have had a stimulating effect on French engineers and, as the debate continued, a remarkable variety of audacious schemes were proposed. Some designers favored the government plan to link railroad terminals, while others sought to reinforce established patterns of local circulation. Many saw the value of improved transport in making the suburban areas easily accessible, enabling the Parisian working classes to obtain more salubrious and economical housing than was available in the center of the city. Others warned that the city was thus in danger of being depopulated, that commerce in the center would be ruined, and that 'grass would grow in certain quarters'.[2]

As the realization of the system approached, in 1895 a municipal councillor pessimistically predicted that, 'with our Métropolitain, all the life of the boulevards, the greater arteries, will disappear. The merchants, the manufacturers, the workers coming out of their offices and workshops will have but one objective: to run to catch the train. . . . There won't be any more intelligent beings. There will be only animals. In sum, with the face of Paris destroyed, the stores ruined, the small shopkeepers closing their boutiques, intellectual life no longer existing, . . . there will no longer be a Paris.'[3] [. . .]

Although the city, from the beginning, had been giving its most serious consideration to a system of underground transport on the London model, many argued that an elevated system based on the New York prototype was preferable to a 'sewer train.'[4] The opponents of the underground system maintained that construction costs would be prohibitive in a city as densely built as Paris. The work would disrupt the city and would involve damage to property, expropriations, and costly indemnities. It was pointed out that the subsurface of Paris was already encumbered with sewers, water mains, gas pipes, and cellars, and that the routing would thus be difficult to establish. Underground construction would be dangerous, involving accidents and illness for workers. The work site would frequently be flooded. There would be danger, moreover, in attempting to uncover the pestilential Parisian soil. [. . .] It was predicted that even after construction the underground system would continue to cause damage, as buildings would be affected by the vibration of the passing trains.

The most frequently repeated argument against an underground transport system was the presumed antipathy of the public. [. . .]

The term 'Nécropolitain' was suggested by one opponent 'to describe a railroad requiring the public to descend . . . into veritable catacombs.'[5] It was observed, moreover, that while travelers might be obliged to spend only a brief time underground, there would be a large number of employees doomed to spend their lives in this purgatory.

A French engineer who had tried the London underground reported: 'I came out of there absolutely oppressed, and in such a state of fatigue that I had to breathe hard the moment I was able to gain the surface.'[6] To some Frenchmen, the apparent success of such a dismal form of transport could only be attributed to the dismal nature of London itself. Another engineer commented in 1884: 'Indeed, what difference does it make to an inhabitant of London to be underground surrounded by vapor, smoke, and darkness; he is in the same condition aboveground. But take the Parisian who loves the day, the sun, gaiety and color around him, and propose that he alter his route to seek, in darkness, a means of transport which will be a foretaste of the tomb, and he will refuse, preferring the *impériale* of an omnibus.'[7]

The advocates of elevated systems maintained that the necessary structures could be built more cheaply than a series of tunnels, and without disrupting the city. Above all, they emphasized the joys of traveling in sunlight and breathing pure air, protected from the dust and smells of the street. The view of Paris obtainable from such a system might make it an outstanding tourist attraction, for, it was observed: 'It isn't everything to travel quickly, one wants also to travel agreeably. The proof is that as soon as the weather is good the *impériales* of omnibuses are absolutely invaded.'[8] [. . .]

In most proposals elevated systems were designed to follow major streets. The desire to avoid disfiguring the Parisian boulevards, however, led some to seek alternate routes, and the river quais were considered a potential site for rail lines. In 1885 a remarkable scheme published by Charles Tellier suggested using the river itself for the new Métropolitain (Figure 24.1). His project involved the construction of a four-track rail line to run on a viaduct elevated six meters above the surface of the Seine, following the river from the suburb of Alfortville on the east to Billancourt on the west. The system, it was noted, could be constructed without disturbing streets, buildings, or river navigation, and the structure would be sufficiently high to pass over existing river bridges.

Tellier anticipated some possible objections on æsthetic grounds. It was true that the proposed viaduct would block the view across the river. He pointed out, however, that as many of the river quais carried rows of trees, the vista was already somewhat veiled by foliage. Moreover, he maintained, if people in the past had worried excessively about change, Paris would never have developed the requisites of a modern city. [. . .]

Although there were sufficient proponents of elevated transport in 1887 to form a 'Ligue Parisienne du Métropolitain Aérien,' there was also sufficient opposition to form a 'Société des

Figure 24.1 Tellier scheme, 1885

Amis des Monuments Parisiens,' which provided an energetic and effective campaign to prevent the 'destruction of France by France.'[9] Created in 1885 by Charles Normand, this organization had the designer of the Paris Opéra, Charles Garnier, as president and Victor Hugo as honorary president. The Société, which included a large number of art historians and archæologists among its members, not only opposed the creation of elevated rail systems, but also sought to insure that an underground system would not endanger monuments and sites.

Just as the elevated concept attracted a number of farfetched schemes, the idea of a subterranean system inspired some rather bizarre proposals. In 1880 J. Mareschal suggested an underground railway where the vehicles would run down inclined tracks, powered by gravity, to be lifted up at each station by elevators. Another scheme included a system of tunnels to be used for passenger trains by day and as sewers by night. A system of ventilators would remove the smell each morning.

One of the most thoroughly developed underground schemes was produced by the civil engineer J. B. Berlier, who began outlining his proposals in 1887 (Figure 24.2). He envisaged an underground system of metal tubes carrying electrically powered trains, and the method of tunneling he proposed would presumably permit construction without disrupting the surface. Reviewing the arguments concerning underground and elevated systems, he deemed it 'inadmissible to install on the Rue de Rivoli and on the *grands boulevards* an elevated line which would be odious from the aesthetic point of view and would never be supported by the owners and tenants of the adjoining buildings.'[10] Berlier maintained that a tunnel, 'abundantly ventilated and luxuriously illuminated,' would be more agreeable than riding in an omnibus, and that Parisian workers would appreciate a means of rapid travel, even if it had to be below the surface.

Although the long period of controversy over the Métro had inspired many ingenious and imaginative proposals for elevated railways, city officials continued to give their most serious consideration to an underground system. The Municipal Council was particularly impressed by Berlier's scheme, and in 1892 they agreed to grant his request for a concession to build. He was unable to raise sufficient funds, however, and the enterprise lapsed.

The Métro: realization

The factor that finally brought about the construction of the much-debated Métro in Paris was the approach of the 1900 Exposition. The prestige of France was embodied in the image of the French capital as a modern city reflecting the advanced standards of French science and technology. At the time of the 1889 Exposition, existing transportation had proved barely adequate to handle the large crowds of visitors, and it seemed clear that to accommodate the even greater influx expected in 1900 Paris would need a new system of rapid transport. Rather than prolong the dispute between city and state, the minister of public works, Louis Barthou, yielded power to the city of Paris in 1895 to develop and control the new Métropolitain.

While planning the system, however, city officials still feared an eventual encroachment by the major railroads. To forestall any possible use of the track network by national lines, the city proposed using a track gauge of 1.30 meters instead of the standard gauge of 1.44 meters. As finally approved, however, the track width conformed to existing standards, and the exclusion of national rolling stock was effected by constructing smaller tunnels than those normally used for the railroads. (The Métro system was to employ a car 2.40 meters in width, as compared to 3.20 meters for the national railroads.)

Although the city would construct the infrastructure of the system, it was decided that the

Figure 24.2 Underground railway proposed by J. B. Berlier in 1887

operation would be directed by a private concessionaire who would be responsible for building the necessary superstructures. The concession was to last twenty-five years, during which time the city would receive a portion of the fares collected. At the expiration of the concession, the entire system was to revert to the city. From among several applicants for the concession, the city selected the Compagnie Générale de Traction, which subsequently formed the Compagnie du Métropolitain de Paris (CMP).

The creation of the Métro, described as 'the most important work Paris has experienced since its foundation,'[11] was directed by Fulgence Bienvenüe, a government engineer. Included in his previous accomplishments were the construction of the Avenue de la République, the creation of the Belleville cable railway, and the development of the park of Buttes-Chaumont. His involvement with the Métro became a lifetime vocation, and he remained in charge of its development from 1895 until his retirement in 1932. Fond of the classics, he once observed: 'By the enchanted lightning of Jupiter, the race of Prometheus is transported to the depths.'[12]

In contrast to the London Underground, which was constructed in tubes lying far below the surface, the Paris Métro system was built close to ground level. The top of the elliptical vault housing the tracks generally lies about a meter below the street, and the depth of the tunnel averages about eight meters. This eight-meter level is the depth at which most underground constructions - sewers, water pipes, electric conduits, etc. - are located. For passengers, entrance into the Métro requires a descent of about six meters, with a ticket level located halfway.

The general method of construction involved an excavation of the upper portion of the tunnel, which was then covered with a masonry vault. The rest of the tunnel would then be excavated, and the side walls and floor completed in concrete. In order to avoid damage to private property and the payment of indemnities, all lines of the Métropolitain system were projected to follow the path of wide streets. In this way, the system reflected already established patterns of movement through the city.

Because of the need to have the system in operation for the 1900 Exposition, construction activity, employing two thousand workers, continued day and night after its beginning on October 19, 1898. Needless to say, the city was greatly disrupted, and Jules Romains reported: 'You heard everyone complain that Paris was odiously encumbered. The *chantiers* of the Métro, which rose up everywhere like fortresses of clay and planks armed with an artillery of derricks, succeeded in strangling the streets and blocking the intersections.'[13]

The first segment of the Métro to be constructed followed the east–west axis of the *grande croisée,* extending between the Porte de Vincennes and the Porte Maillot, a distance of eleven kilometers. Partially completed, the line was inaugurated on July 19, 1900. In the same year two short sections were added, one running from the Étoile to Trocadéro, and the other from the Étoile to the Porte Dauphine. In spite of predictions that Parisians would never consent to ride underground, the Métro seems to have been accepted immediately. It was reported in *Le Galois,* following the opening, that 'the success of this new means of transport has been very big. It is due primarily to its rapidity. . . . Travelers yesterday were enthusiastic.'[14] The 1900 Bædeker guide stated that the Métro 'now takes precedence of all other modes of locomotion in the interior of the city.' Readers were informed, however, that as in London the stations were below the level of the street, 'and the atmosphere is similarly oppressive to susceptible people.'[15]

The initial concession for the Métro had involved the construction of six lines (Figures 24.3 and 24.4). Following the completion of

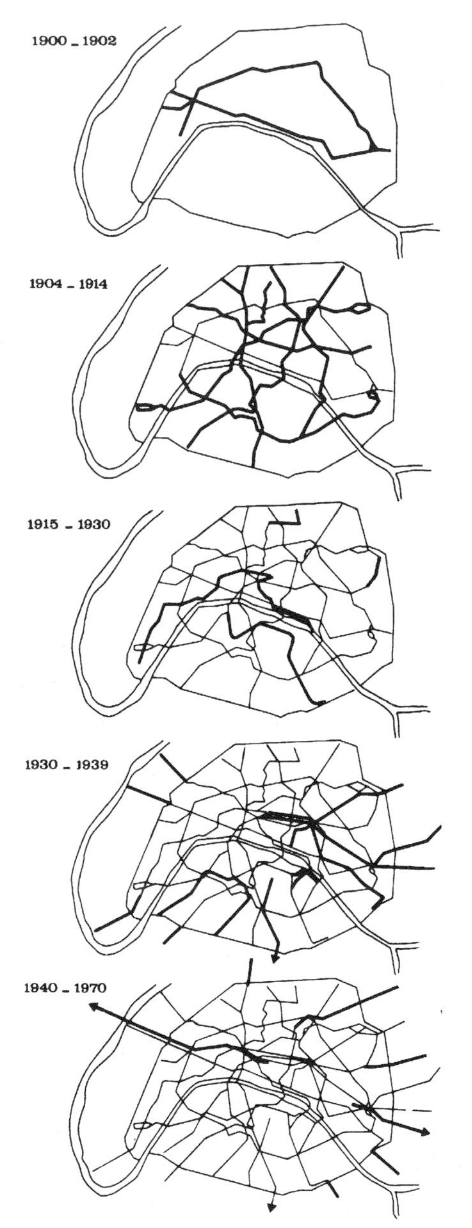

Figure 24.3 The evolution of the Paris Métro from 1900 to 1970

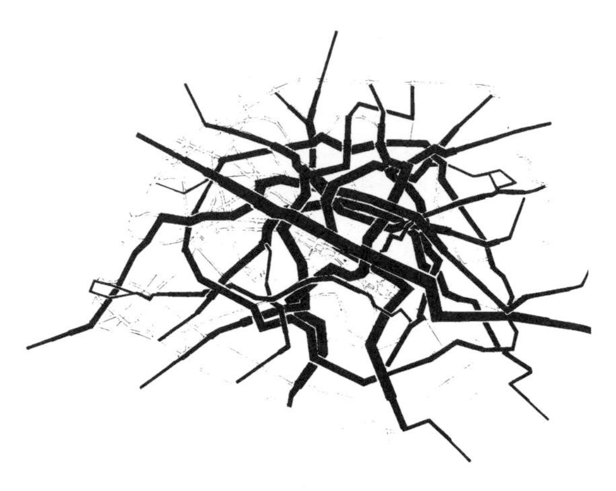

Figure 24.4 Diagram showing passenger volumes on the Métro in 1960. Line One continues to be the most heavily traveled route

Line One, the second line of the system, extending from the Porte Dauphine to the Place de la Nation via the Étoile and the exterior boulevards, was built between 1901 and 1903. This line was carried on viaducts for about one third of its length. Between 1904 and 1905 a third line, extending from the Porte de Champerret to the Place Gambetta, via Opéra and the Rue Réaumur, was opened, to be followed in 1906 by Line Five, extending from the Étoile to the Place d'Italie by way of the southern exterior boulevards. The circuit begun by this line was completed in 1909 when Line Six was opened between the Place de la Nation and the Place d'Italie. Line Four, a north–south line extending from the Porte de Clignancourt to the Porte d'Orléans, was completed between 1908 and 1910, and it marked the first use by the Métro of a tunnel passing below the bed of the Seine. A 'complementary network' was added to the system before 1914. This included two additional lines: Line Seven, extending from the Porte de la Villette to the Opéra, and Line Eight, running from the Place Balard to the Porte de Charenton.

It may be recalled that an engineer, J. B. Berlier, had been granted a concession for the Métro in 1892 but had been unable to raise sufficient funds. He was eventually awarded compensation of five hundred thousand francs by the city. In 1905, supported by a company called the Société du Chemin de Fer Électrique Nord-Sud de Paris, he obtained a concession for a separate underground line running from the Gare Montparnasse to Montmartre. Although his previous proposals had involved metal tubular tunnels, the construction of this line followed prevailing methods. The 'Nord-Sud' line was merged with the CMP in 1930.

The length of the Métro system had been expanded from ten kilometers in 1900 to eighty in 1914, the increase in track being accompanied by a corresponding increase in the number of passengers. In its opening year, between July and December 1900, the Métro carried 17,660,286 people. The following year the figure reached 55,882,027, and had grown to 149 million in 1905. By 1914, 400 million passengers were carried annually.

Although some Parisian officials had hoped that the construction of the underground system would relieve traffic pressure on the streets, the opposite seemed to be the case. A government report in 1910 suggested that the availability of rapid, cheap transport seemed to set off a demand for more transport, and that since the construction of the Métro Parisians had become accustomed to moving about with much more frequency. The construction of underground transport, it was believed, had not replaced surface movement but instead encouraged it.

In addition to the network of underground constructions, the Métro system also provided for the creation of a series of stations and entrances which, in their ingenious exploitation of Art Nouveau, created virtually a 'Métro style.' The Société des Amis des Monuments Parisiens, in the course of its campaign to prevent the establishment of an elevated transport system, had expressed the wish that no Métro station be allowed to extend above the sidewalk, and that they should preferably be installed in shops. Others, however, anticipated that the new transport system could be accompanied by distinguished artistic efforts. In 1886 Charles Garnier had written to the minister of public works: 'The Métro, in the view of most Parisians, should reject absolutely all industrial character to become completely a work of art. Paris must not transform itself into a factory; it must remain a museum. Don't be afraid to abandon lattice girders and thin metal framework; gather to you stone and marble, summon bronze sculptures and triumphal columns.'[16]

The first Métro stations embodied designs by the architect Hector Guimard. His apartment house, the Castel Béranger, which reflected an imaginative application of the currently fashionable Art Nouveau style, had just won a prize for facade composition. As the Métro itself represented a new form of transport, it may have seemed appropriate that it be accompanied by innovative design. The contribution of Guimard was in many cases limited to the embellishment of stairways leading from the sidewalks to the subterranean stations (Figure 24.5). Metal railing employing the sinuous foliate forms typical of Art Nouveau would

Figure 24.5 Métro station entrance by Guimard, Place des Abbesses

surround the opening, with tall, fancifully curving lampposts illuminating the entrance. The elevated signboard, in harmony with the ensemble, embodied decorative Art Nouveau lettering. In some instances, the staircase would be covered with a glass roof.

At some of the more important Métro stations, such as the Portes of Vincennes, Nation, Maillot, and Dauphine, glass pavilions marked the entrances, with perhaps the finest of these structures appearing at the Place de la Bastille (Figure 24.6). Lightly framed in metal, with projecting glass canopies, these stations were likened to 'dragonflies spreading their wings' or 'fragments of the skeleton of an ichthyosaur.'[17]

Because of their novelty the Guimard stations were somewhat controversial, and as Art Nouveau began to decline in popularity, strong opposition to them developed. One of the focal points of debate was the station entrance opposite the Opéra. [. . .]

The designer selected for the Opéra station was Joseph-Marie Cassien-Bernard, the architect of the Pont Alexandre III, who provided the entrance with a classical stone balustrade. [. . .]

From this time on, classical stone balustrades were the only type of Métro entrance considered suitable for monumental sites, and they were employed at the Place de la République, the Gare Montparnasse and Gare de l'Est, the Madeleine, and the Place de la Concorde. Such entrances eventually replaced the Guimard pavilions at the Étoile. Although metal would continue to be used for many Métro installations, directness and simplicity would govern the design.

During the period following the Second World War, many of the Guimard Métro pavilions were destroyed as stations were modernized, and in 1962 the elegant and fanciful 'pagoda' at the Place de la Bastille was razed. By this time the Guimard structures, no longer an eccentric novelty, were viewed by many as a valuable part of the artistic heritage of Paris, and in 1965 several of the Métro station entrances were classified as protected monuments. [. . .]

When the Métro was first constructed, no line extended outside the Paris boundary,

Figure 24.6 Métro station, Place de la Bastille, now demolished

thus reinforcing the conception of a purely local system. Between 1931 and 1939, however, a few modest penetrations were made into the neighboring suburbs. At the same time, the constant growth of the suburban region prompted consideration of an additional Métro system designed to join with suburban transport lines. As a first step in the establishment of such a network, the Ligne de Sceaux, a suburban rail line extending south of Paris, was made part of the Métro system in 1938. This line had been electrified the previous year.

By 1937 the Métro system included 150 kilometers of track, and in 1938 it carried 761 million people. The depression seems to have reduced passenger use somewhat, as this figure had reached 888 million in 1930. Because of the relative density of Paris, the number of passengers per kilometer of line was high, in 1935 reaching 5,800,000, as compared with 1,680,000 in London, 4,960,000 in New York, and 2,740,000 in Berlin.

The greatest pressure on the Métro system, however, was to come with the German occupation of Paris during the Second World War and in the period immediately following. During the war years the absence of gasoline rendered almost all motor vehicles inoperable, and virtually the entire burden of public transport was handled by the Métro. Crowded into increasingly antiquated and badly maintained facilities, travelers in the Métro reached over 1 billion in 1941, 1 billion 230 million in 1942, and 1 billion 320 million in 1943. The highest number of travelers was recorded during 1946, when passenger traffic reached 1 billion 598 million. Although the Métro had been crowded during the war years, it was one of the few places in Paris that was warm, and a French official reminiscing about this time observed that, 'having all become poor and deprived of heat, we dreamed of prolonging our trips to avoid shivering in our offices and homes.'[18]

During this period a reorganization of the transport system was begun. In 1942 it was decided to unify surface and underground transport, using a system of common fares and coordinated lines. To effect this, the Compagnie du Métropolitain de Paris and the Société des Transports en Commun de la Région Parisienne, which controlled the bus lines, were merged. Following the war, the Régie Autonome des Transports Parisiens (RATP) was created in 1948 to direct the unified system of public transportation. This organization was replaced in 1959 by the Syndicat des Transports Parisiens.

Although for a time government policies seemed to neglect public transport in favor of automobile facilities, it soon became apparent that no system of private transport could adequately serve the Paris region. The Métro system became subject, therefore, to a continuous program of modernization. New rolling stock was introduced, and during the 1960s station platforms were prolonged from seventy-five to ninety meters to accommodate longer trains. One of the most notable improvements came with the introduction of rubber tires for the cars. The tires were introduced experimentally in 1952, then employed on Line Eleven between Mairie des Lilas and Châtelet in 1956. In 1963 they were introduced on Line One, running from Vincennes to Neuilly. This line, the first portion of the Métro to be built, continues to be the most heavily traveled. Rubber tires were added to other lines during the 1960s and 1970s.

During the long period of controversy before the construction of the Métro, opponents of the underground system had made pessimistic predictions about the quality of the air in the tunnels. Once the system was built, it was maintained by André Berthelot, the first administrator of the Compagnie du Métropolitain, that the underground was among the healthiest places in Paris. He pointed out that 'the absence of dust is, indeed, almost complete in the underground. . . . The air of the Métro is incomparably less charged with bacteria in the morning at the beginning of service than in the majority of public places and streets.'[19] According to a contrasting view, however, the Métro was 'a badly ventilated cellar, recalling, from time to time, a sewer collector. One is hit in the throat, in descending the stairs, by an unending series of odors, of unbreathable emanations, a mixture of sweat, of tar, of carbonic acid, of metallic dust, etc., . . . and all of a warm heaviness like that of a day when a storm is about to break.'[20]

It had also been predicted, while the Métro was being contemplated, that Parisians would be repelled by the idea of traveling underground. Not only would the prospect of being denied light and air drive people away, but there was also the presumed threat of human passions unleashed in subterranean darkness. This assumption seems to have been inspired by the railroad tunnel of Batignolles, which took one minute to traverse. It had been suggested that 'only those condemned to death, and women who are sitting face to face with a ruffian in darkness, know how long a minute can be.'[21]

Once the Métro was in operation, however, with the almost frightening adaptability that characterizes the inhabitants of large cities, the Parisians simply accepted underground travel as a normal part of life. In 1960 it was found that one Parisian in four traveled daily in the Métro, and it was estimated that by the age of retirement the average Parisian would have spent two years of his life underground. [. . .]

Although the Métro is an important part of the Parisian transport system, it is only one aspect of a complex pattern of regional transportation. In 1972 over 12 million daily displacements took place in the Paris region, the largest concentration of movement naturally taking place in the commute between home and work. It was reported in 1971 that 850,000 people commuted daily to Paris from the suburbs, while 200,000 went from Paris to jobs outside. At the same time, 900,000 suburbanites traveled to employment in the suburbs and 700,000 inhabitants of Paris traveled to employment within the city. Inside Paris there were 7 million daily displacements, using public transport, with the greatest concentration during the evening rush hour. During a single hour at this time, more than 700,000 people were found to be using public transport, with 300,000 in the Métro and 230,000 passing through the railroad stations.[22]

The constant increase of commuting from the suburbs to Paris prompted consideration of an additional Métro system as early as 1929. Travelers arriving on the suburban railroad lines were compelled to transfer to the Métro or the bus system to reach the center of the city. An additional system was considered necessary, therefore, to link the suburbs directly to employment centers. The incorporation of the Ligne de Sceaux with the Métro in 1938 provided the first step in the creation of such a network.

It was not until the 1960s, however, that construction of the new rail system, called the Réseau Express Régional (RER), was begun (Figure 24.7). In outlining the proposed system in 1961, the Transport Commission explained why the problem could not be solved by further extensions of the Paris Métro. The Métro was considered to be already saturated and, with its numerous stops, unable to provide an efficient service to outlying areas. The commuter railroads were also considered to be overloaded and unable to penetrate the city center. The RER, it was suggested, would not only supplement existing transport systems, but open up new suburban areas for settlement.

The tunnels of the RER were designed to traverse Paris at a depth of ten to thirty meters, passing under existing Métro lines, but permitting transfer to the Métro at certain important stations (Figures 24.8 and 24.9). Trains were designed with the same height as railroad stock and intended to reach a speed of seventy to ninety kilometers per hour.

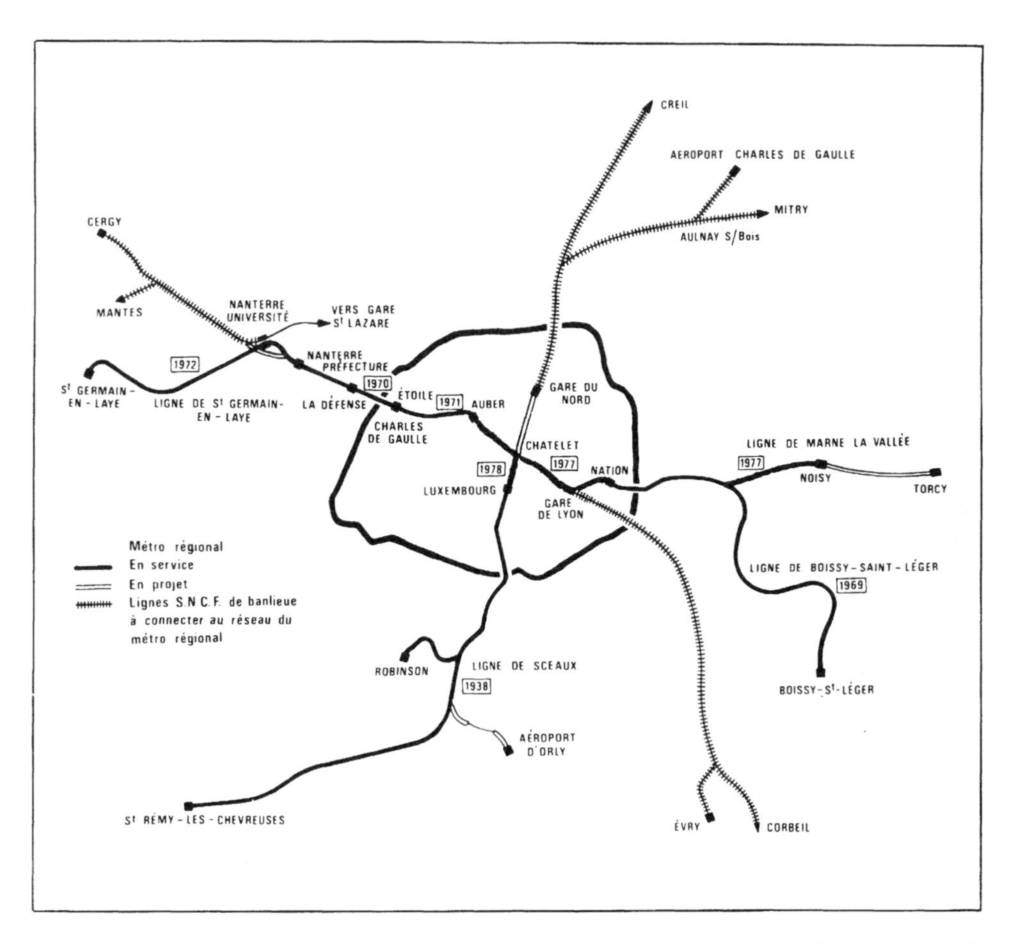

Figure 24.7 Routing of the Réseau Express Régional (RER), showing connections with the suburban railroad lines. *Note*: The RER lines are shown in black, and suburban railroads with hatched lines

The first section of the RER to be developed was an east–west line extending from Boissy-Saint Léger to Saint Germain en Laye. Construction began with the suburban sections, with the portion on the east linking Boissy-Saint Léger to Nation operating in 1969. On the west the line was first developed to provide a connection between the Étoile and a new business center at La Défense. This portion was opened in 1970, with an extension into the city to the Métro station of Auber added in 1971, and an extension westward to Saint Germain en Laye opened in 1972. By December 1977, the first RER system was complete. The east–west section, called Line A, now extended through the city, with connections to the Métro added at Châtelet-Les Halles and the Gare de Lyon. An eastern branch reached Noisy le Grand-Mont d'Est. A second line, Line B, incorporated the Ligne de Sceaux, leading southward from Châtelet-Les Halles to Saint Rémy-Les Chevreuse, with a branch to Robinson.

At this time, it was anticipated that by 1985 Line B would have been extended north to Charles de Gaulle airport at Roissy and to Mitry-Claye. Line A was intended to receive an eastern extension to Torcy (paralleling

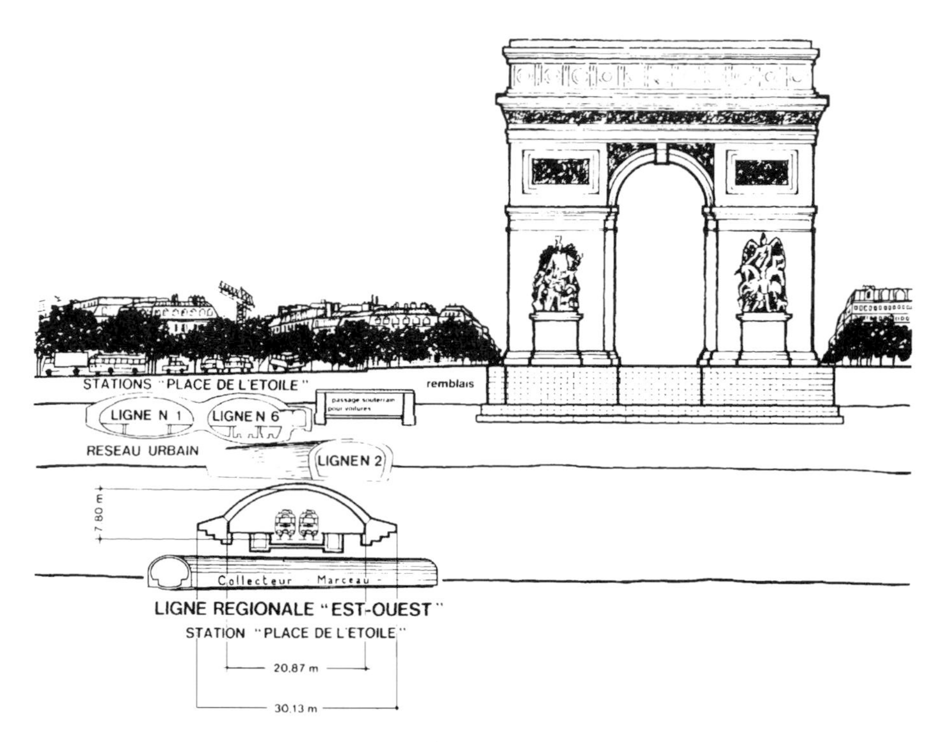

Figure 24.8 Subterranean section at the Place de l'Étoile, showing the relation of existing Métro lines to the new RER station

the axis of the new town of Marne la Vallée) and a western branch to the new town of Cergy-Pontoise. Additional RER lines were projected to serve such destinations as Montigny-Beauchamp and Argenteuil in the north, Versailles and the new town of Saint Quentin en Yvelines to the west, and Massy-Palaiseau, Dourdan, and Étampes to the south. An additional line was foreseen to extend to the new town of Melun.

Figure 24.9 RER station at the Étoile

As the region of Paris expanded during the period following the Second World War, the transportation system became subject to continual analysis. Patterns of movement, choice of transport, length and time of journey were dissected statistically, and the cost of commuting painstakingly evaluated in terms of money and time. As a technical problem, transportation in the Paris region has made constantly increasing demands on public resources, and as improvements in transport have often been accompanied by increases in the amount of travel, the capacities of the system seem always strained to the limit. [. . .]

In the view of many urbanists, problems of transport would be better solved by reducing travel than through increasingly costly and elaborate transportation systems. Yet, although the pedestrian-centered environment in which most citizens can live close to their work remains an ideal, the planning of the Paris region has incorporated the assumption that massive commuting will continue to characterize the Parisian way of life. In the view of some, much of the attraction of any major city lies in its wide range of employment, housing, recreation, and cultural facilities, and the ability to move about through the entire urban area is essential to maintaining freedom and variety of choice.

The present-day planning of the Paris region involves the creation of a series of so-called 'new towns' designed to contain both housing and employment. Although the development of such centers could reduce some commuting, they were not projected as self-contained or isolated communities but were instead designed to be linked to central Paris by rapid rail lines. The evident presupposition was that the inhabitants of such towns would in many cases travel to Paris to work. In the planning of central Paris, moreover, new poles of commercial activity have been developed adjacent to suburban railroad terminals, on the assumption that they will draw many of their employees from outside the city.

While the present regional plan reflects an attempt to develop a more coherent and unified pattern of urbanization for greater Paris, it offers no substantial alteration in existing commuting trends. Although public transport may become more rapid, convenient, and comfortable, the burden of a lengthy daily journey seems destined to remain part of the lives of many Parisians.

Notes

1 The position of the national government was based on the view that the railroads were of strategic importance to national defense and thus subject to the control of the central authorities. Exceptions to this rule existed, however, for transportation systems judged to be of purely local importance, and it was the contention of the Paris government that the Métropolitain system should be categorized as local.

2 Arsène-Olivier de Landreville, *Les grands travaux de Paris: le Métropolitain de Paris*, Paris, Baudry, 1887, p. 25.

3 Quoted in Georges Verpræt, *Paris: capitale souterraine*, Paris, Plon, 1964, pp. 216–17.

4 'Le sous-sol de Paris: l'héritage de l'histoire,' *Paris projet*, no. 3, 1970, p. 30.

5 Louis Heuzé, *Chemin de fer transversal à air libre dans une rue spéciale, passage couvert pour piétons*, Paris, A. Lévy, 1878, p. 5.

6 Charles Tellier, *Le véritable Métropolitain de Paris*, Paris, Schlaeber, 1885, p. 22.

7 Quoted in Jules Garnier, *Avant projet d'un chemin de fer aérien*, Paris, Chaix, 1884, p. 42.

8 Tellier, *Le véritable Métropolitain*, p. 29.

9 Verpræt, *Paris: capitale souterraine*, p. 217.

10 J. B. Berlier, *Les tramways tubulaires souterrains de Paris, 1887–1890*, Paris, Berlier, 1890, p. 7.

11 Roger Guerrand, *Le Métro*, Paris, Éditions du Temps, 1962, p. 16.

12 Quoted in Verpræt, *Paris: capital souterraine*, p. 212.

13 Quoted in Roger H. Guerrand, *Mémoires du Métro*, Paris, La Table Ronde, 1960, p. 60.

14 The story was dated July 20, 1900. Quoted in Verpræt, *Paris: capitale souterraine*, p. 215.

15 Bædeker, *Paris and Environs*, 14th edn, Leipzig, Bædeker, 1900, p. 28.

16 Guerrand, *Mémoires du Métro*, pp. 82–83.

17 Ibid., p. 84.

18 Pierre Lavedan, *La nouvelle histoire de Paris: histoire de l'urbanisme à Paris*, Paris, Hachette, 1975, p. 236. The statement was made by Martial Massioni, president of the Conseil Général de la Seine, in 1950.

19 Guerrand, *Mémoires du Métro*, p. 115.

20 Ibid., p. 116.

21 Verpræt, *Paris: capitale souterraine*, p. 217.

22 Statistics from Maurice Doublet, *Les transports dans la région parisienne*, Paris, Conférence des Ambassadeurs, November 1972, pp. 15–19, and Pierre Merlin, *Vivre à Paris 1980*, Paris: Hachette, 1971, p. 164.

25

THE METROPOLIS AS A CONSTRUCTION: ENGINEERING STRUCTURES IN BERLIN 1871–1914

Hans Kollhoff

Source: Josef Paul Kleihues and Christina Rathgeber (eds), *Berlin/New York: like and unlike: essays on architecture and art from 1870 to the present*, N.Y., Rizzoli, 1993, pp. 47–57

To celebrate the 750th year of its existence, in 1988, the city of Berlin gave itself a present: the conversion of the former Hamburg Station into an imposing new exhibition site. This gesture is symptomatic of an era that has begun to make itself comfortable in nineteenth-century structures and, under the pretext of preserving historical monuments, to sun itself in the borrowed light of the Wilhelmine era – in which a newly united Germany grew in economic and military power with Berlin as its capital. Berlin has again and again shown its inability to carry on the tradition of turn-of-the-century urban buildings and public spaces, resigning itself instead to architectural and urban design patterns that were little more than symbols of status and power even when they were new. Something very important has been lost from sight in the process: the impulses that made Berlin, with breathtaking rapidity, into a world center.

Ever since the 1920s, people have been talking about metropolitan architecture, seemingly unaware that such architecture had actually been built before the First World War. This was the period when factories, railway stations, shopping arcades, greenhouses, market halls, hotels, exhibition buildings, and fairgrounds emerged that established a typology of structures for the Industrial Age. Built by private investors and speculators, they provided the backdrop for the legendary 1920s, when the flight from the corrupt city set in and thousands dreamed of a better future out-of-doors, in the green belts and garden cities of cooperative projects.

The vision of a metropolitan architecture that emerged in Germany after World War I like an awakening from a troubled dream attracted enormous journalistic interest. Still, it remained largely retrospective in character and really can only be understood as an unconscious attempt to come to terms with the past. After the AEG buildings (Allgemeine Elektricitäts-Gesellschaft) in 1907, metropolitan architecture in Berlin came to a standstill. The brute force of the declining nineteenth century dwindled to nothingness as architectural volumes began to do without articulation and the lucid outlines of pared-down geometrical solids replaced decorative forms. It remained for the 1920s to speculate and philosophize about the big city and develop utopian ideas that in turn culminated in orderly, basically anti-big-city systems. Unlike art, literature, and film, modern architecture proved unable to accept the chaos of the metropolis of our day.

In the interest of a truly urban architecture,

then, it is time to consider the Wilhelmine era as more than the mere soil from which the flowering of the 1920s sprang. Rather, it was the period in which urban architecture in Berlin was born and grew to maturity. [. . .]

The period between 1879 and 1914 in Berlin was the period of a newly united Germany and soon thereafter a world center. By the end of this period Berlin had caught up with the enormous, well-nigh frightening technological lead enjoyed by London, Paris, and the great American cities. Berlin became the upstart among world centers, 'a colonial city,' as Karl Scheffler remarked in 1910, 'whose suddenness of development had more in common with American cities than with the old metropolises of Europe, recklessly expanding, violent, and established in a kind of no-man's land, far from the terrain of European culture.'[1]

Not surprisingly, the result was a cultural inferiority complex that led to a compulsive effort to create style and thus compensate for the city's respectable but internationally modest achievements in technology and the emergent field of engineering. Conversely, no effort was spared to bolster the self-confidence of the fledgling Reich with the aid of the most advanced technology available.

While swinging between technology and ostentation, Berlin produced nothing in the way of architecture or urban design that could be considered significant enough to include in an architectural history. In Leonardo Benevolo's *History of Nineteenth- and Twentieth-Century Architecture,* for instance. Schinkel's design for the Marschall Bridge (1818), appears in company with English and French bridges (all of which, apart from the Marschall, were designed by engineers). But this is the only mention of a Berlin structure in the entire first half of this voluminous book; in the second half another Berlin example appears – Peter Behrens's AEG turbine factory of 1907. [. . .]

There is a certain tragedy in the fact that this period, which more strongly and lastingly shaped the face of Berlin than any other, should have been artistically and technologically so insignificant, although it had forces at its disposal of which present-day Berlin can only dream. A closer look at the period between 1871 and 1914 becomes interesting, although – or perhaps because – we know our search for cornerstones of architectural history will be in vain. With extreme rapidity, indeed brutality, an urban topography was stamped out during these years that is both frightening and fascinating – residential blocks truncated by railroad tracks, an urban structure dissected by railroad lines, harsh juxtapositions of buildings, bridges, underpasses, embankments, colossal artifacts – gas meters, water towers – floating disembodied in the cityscapes. [. . .]

Gradually there came to be discovered in the works of engineers, especially in iron and steel construction, an aesthetic of a unique kind. Gottfried Semper's verdict that whoever accepted iron construction would find a meager soil for art made way for the insight that engineering structures were not intrinsically ugly, even when their utilitarian character was fully evident; in other words, there was no longer any reason to hide them behind a mask. [. . .]

The self-confidence of German engineers, however, was still relatively undeveloped. In the field of architecture they had no achievements to show that could match those of English, French, or American engineers. [. . .] Nor, of course, did the challenges exist – the bridging of great rivers, the world's fairs – that would have brought them forth. German engineers were able to unfold their skills only in areas where they were not blocked by architects or reduced to supernumeraries – in the field of machine building itself or in that of the technical facilities of the urban infrastructure.

Nevertheless, it was the accomplishments of engineers that shaped Berlin during the years from 1871 to 1914. By rapidly and directly reacting to the needs of the period they lent a lucid expression to the forces in society that commands our respect and admiration today. Not only the great urban spaces that still survive but also the anonymous components of the city possess a lasting fascination: the many miles of graceful archways supporting the interurban rail lines, the subway viaducts, the interurban embankments, the canals, the countless bridges that conduct these various transportation systems over and under one another, the giant railyards, and not least, the commercial and industrial yards that managed to survive war and renewal.

The low opinion in which such engineering accomplishments were still held in the late nineteenth century becomes evident from an entry in *Berlin und seine Bauten,* the standard work on the city's architecture. As late as 1877, the authors could still write, 'The earlier history of Berlin only rarely mentions construction projects to which we would apply the modern term, "engineering structures."'[2] Yet in its emergent phase the term engineering structure actually covered a broad range of buildings whose functions lacked the status or grandeur that would have made them worthy of an architect's attention. Until the end of the century these buildings included factories.

Friedrich Gilly already sensed the danger of separating architects from engineers; and Schinkel and the engineer Peter Beuth, with whom he had traveled to England, managed to reunite the diverging disciplines under the roof of the Academy of Architecture. It was not until 1879, however, when the faculties were split at the new Technical College in Charlottenburg, that engineers as specialists in the mathematical sciences could face architects as aesthetic specialists in a partnership of equals. In the meantime, methods of calculating highly complex types of steel construction had been found. In 1863, J. W. Schwedler had succeeded in computing spheroidal roofs for locomotive sheds and natural-gas tanks, and E. Winkler followed with double- and triple-articulated arches. Steel construction theory, which until then had limped behind practice, especially in the experimental field of greenhouse building, could now find creative application.

> The construction of gas container buildings of large diameter gave Schwedler the opportunity to free the rafter construction previously used for the roofs of these structures from their internal tie-rods, and to transfer by an arrangement of rings and groins all of the constructional elements into the spheroidal roof surface, thereby transforming the earlier beams system into a dome system. Since 1863, Schwedler's domed roof has been employed in every gas container building erected in Berlin. . . . As regards installation, the central section of the domes, measuring 23.66 meters in diameter, was fully assembled and riveted on the floor of the water basin, resting on several trestles. The raising of this approximately 12.000 kilogram section was carried out with the aid of twelve lever jacks which were installed on a projecting, gallery-like scaffolding.[3]

The raising took eight to ten hours, which meant that it could easily be accomplished in a day. The assembly process is comparable to that used to erect the roof of Berlin's New National Gallery in 1968 (Figure 25.1), one of the several reasons why we must attribute to

Figure 25.1 New National Gallery, Berlin, 1967. Hydraulic presses raise the steel-roof construction
Source: Landesbildstelle, Berlin

Mies's building a power that has more in common with the nineteenth century than with the Neues Bauen of the early twentieth – particularly in the girder supports, coffered ceilings, and not least the urbanity of its interior, which is the sole postwar example that can rival interiors produced before World War I.

In the National Gallery the conflict between industrial product and architectural detail was first settled, a conflict that had made life difficult for turn-of-the-century architects. As technical possibilities increased, building materials no longer presented sufficient resistance to the designer, and the result was an indecision or arbitrariness of expression. The first building to face this issue, and which came to embody it, was Peter Behrens's AEG turbine factory (1907). 'Behrens produced mass,' says Julius Posener, who also said he knows of only one large building in Germany that employed the new methods of steel construction as an architectural means, the Frankfurt Exhibition Hall by Friedrich Thiersch (1908). Berlin, Posener admits, has nothing to compare.

With the onset of industrialization, Berlin developed into the largest industrial metropolis on the European continent in the space of a few decades, assisted by modern technology, industrial capital, and an efficient transportation system. Stimulated by state subsides, private enterprise was at last able to take up England's lead, even though the Berlin region entirely lacked sources of raw materials.

With the establishment of the Royal Iron Foundry outside the New Gate in 1804, the stage was set for Berlin firms pioneering in machine building to move to Chausseestrasse, which came to be popularly known as Fireland. Instead of the enclosed industrial yards that predominated in the south and east, these factories in the northwestern part of the city, ranged as freestanding, solitary structures along the streets, providing the model for the emergent typology of large industrial spaces that came to maturity in the AEG assembly buildings of Peter Behrens (Figure 25.2). Yet Chausseestrasse was soon overcrowded, and in 1847 the Borsig Company was obliged to shift its operations to the district of Moabit on the Spree and eventually, in 1898, from there to Tegel. This migration to the periphery, which affected all of the expanding large enterprises, necessitated an expansion of mass transportation and municipal services, which in turn opened up new locations on the city's outskirts. When Berlin became the capital of the German Empire in 1871, these developments accelerated dramatically. By 1877, its population numbered one million, and by 1905, it had surpassed the two million mark.

The initial step in the expansion was the construction and improvement of highways and waterways: the first steamship appeared on the Spree River in 1815. By 1825, regular steamer traffic commenced, and the old Schafgraben was excavated, based on Peter Joseph Lenné's plans, to become the Landwehr Canal. In 1853, Schöneberg Harbor went into operation, and followed in 1859 by the Berlin-Spandau shipping canal.

Figure 25.2 Stator of a three-phase current generator, 1912
Source: Tilmann Buddensieg and Henning Rogge, *Industrie-Kultur, Peter Behrens und die AEG*, Berlin, 1981

The extension of the rail network began in 1838 with the inauguration of the Berlin-Potsdam Railway, allowing not only an increased flow of goods and raw materials into the city but a growing migration of labor. When Berlin became the capital, the need for transportation grew ever more urgent. In 1868, horse-drawn buses were introduced; building commenced on the belt railway in 1871, on the interurban railway in 1882, and on the subway system in 1902. The first electric trolley in the world went into operation in 1881 in the Lichterfelde district of Berlin.

Tracks proliferated and extended into every nook and cranny of the city until they could go no further, and came up short at the bumper of a terminal. Station after station was constructed, altering the face of Berlin entirely and driving every well meaning urban planner to distraction. It may well be that the impossibility of city planning was demonstrated here before it even had the chance to establish itself as a discipline. [. . .]

The interurban Stadtbahn, a four-track system that served as both a belt and a long-distance line, was the first viaduct railway in Europe. In the built-up areas of the city, its route was elevated on arched brick viaducts, with wooden truss bridges at river crossings and simple riveted steel girders at road crossings. Boasting 731 viaduct arches, 597 of which were open to business occupancy, the Stadtbahn represented the largest contiguous urban structure in Berlin, a chain of shops, storerooms, warehouses, workshops, and pubs that with the market halls guaranteed the provisioning of the city. Interurban and belt lines were designed to compensate for the deficits that had resulted from the configuration of competing terminals. Berlin never had a main or central station, only a number of terminal stations, each of which served a single main

Figure 25.3 Friedrichstrasse Train Station with taxi ramp, 1910
Source: Landesbildstelle, Berlin

line coming in from a different direction. By 1875, there existed eight such stations, arranged like the points of a star around the historical town core. Passengers who wanted to transfer from one line to another had to take cabs or walk through the city center to reach their connecting trains.

Alexanderplatz and Friedrichstrasse stations, glazed, triple-jointed arched halls that were huge by Berlin standards dominated the cityscape. Friedrichstrasse in particular (Figure 25.3), with its glazed walls and looming mass covering an area of 132 by 5,456 feet, became a favorite picture-postcard motif. [. . .]

With the invention of the dynamo in 1866 by Werner von Siemens, the generation and use of power now functioned independently of one another. Instead of a locomotive, a small electric motor could be installed in the first carriage of a passenger train, fed by power plants placed anywhere along the line. This development gave rise to the idea of a flexible inner-city transport system that, inspired by New York's example, could be built above the streets – an elevated train. After several plans had been rejected by the authorities, including a line through Friedrichstrasse, in 1891, Siemens received permission to construct a section of the belt line he envisioned between Gitschiner and Skalitzer Strasse, which was to connect the main-line terminals. In the meantime his company, Siemens & Halske, had built the subway system in Budapest and a network of electric streetcars in southern Berlin.

With the directness of engineering plans the tracks pushed themselves forward until they encountered some obstacle, which was duly tunneled through and the opening equipped with a frame – a ruthless procedure that gave rise to one of the most delightful and spectacular, but not untypical, situations along the trunk line of Berlin's elevated railway, the tenement tunnel on Bülowstrasse.

The junction point known as Gleisdreieck (Figure 25.4) received its characteristic triangular shape on account of the need to connect the east–west trunk line with Potsdamer Platz without the use of crossings, for on September

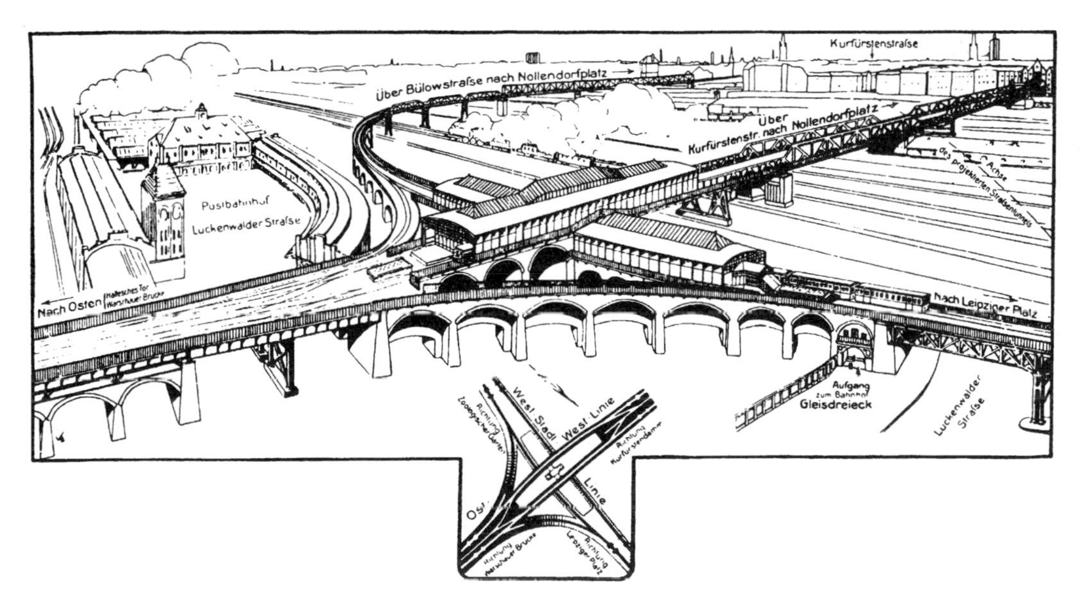

Figure 25.4 Gleisdreieck after redesign, c. 1915
Source: Landesbildstelle, Berlin

26, 1908, two trains had collided, casting one of their motor carriages off the viaduct and killing twenty-one people. To ensure adequate train frequency all three branches were provided with separated tracks so that each line could be operated independently of the others. The substructure, a stone viaduct, had to be replaced by five iron bridges. Differing from one another in length, direction, gradient, and height above the ground, they were combined in an extremely complex and apparently daredevil way into a filigree, three-dimensional grid.

The elevated-train viaducts of riveted rolled steel, designed by the engineer Heinrich Schwieger, soon drew harsh criticism because of their utilitarian plainness (Figure 25.5).

> Popular opinion initially turned against the elevated out of a latent aesthetic aversion to iron as an artistically expressive building material, an aversion that even affects experts and that probably has its source in the fact that people are used to considering iron an aesthetically inferior material that is employed for supports, train rails, and so forth. . . . This obliged the Association for Electric Elevated and Subway trains to earmark a considerable sum for the artistic design of the viaducts and stations, and from that point on, construction took place as if under continual public supervision.[4]

A competition was held; first prize went to Bruno Möhring. Since the constructive elements had to be made of wrought iron, freedom of design was limited to the cast parts at their base and top. The gradual evolution of the elevated viaduct eventually led to organically mature solutions in which the contributions of engineers could no longer be distinguished from those of architects. A merely utilitarian employment of materials made way for sophisticated architectural design.

Ornamentation was replaced by careful structural detailing, precisely placed rivets, and supports whose shape followed the play of forces within the viaduct. Necessity itself became a subject of artistic interpretation, with a view to finding an adequate expression of the building task based on its static function

Figure 25.5 Bridge for elevated train over the Landwehr Canal, c. 1900
Source: Landesbildstelle, Berlin

Figure 25.6 Central Hotel, Berlin, view of the conservatory, c. 1880
Source: Engraving from Georg Kohlmeier and Barna V. Sartory, *Das Glashaus* (Munich, 1981)

and the nature of the materials; it was a form of expression that avoided short cuts.

At Nollendorfplatz, the rails had to go underground in order not to disturb the peace of the posh West End. The cumbersome viaducts and noise would have disrupted Tauentzienstrasse. Technically, the U-Bahn posed little problem to Siemens and Halske after their experience in Budapest. The architectural demands were reduced as well, for the stations were invisible and therefore much easier to design than the obtrusive above-ground stations, which tended to dominate the cityscape.

Long ignored by architectural history, greenhouses evolved into vast iron translucent structures in which to cultivate and exhibit subtropical and tropical plants. Under a filigree shell of iron and glass an exotic climate could be produced in the midst of cold northern cities. The conservatories attached to hotels, restaurants, and dance halls became favorite gathering places for society (Figure 25.6). Unhampered by building codes and relatively free of the aesthetic obsessions of the Wilhelmine period, this building type could develop according to function alone, making it the predecessor of types of construction whose prime aim was to reduce mass to a minimum.

Even less hampered than greenhouses by questions of style were the structural experiments carried out in connection with man's first attempts to fly. The Airship Hall (Figure 25.7), built in 1911 on Johannistal Airfield, was a miracle of spidery iron construction comparable in every way to the flexible skeleton of Otto Lilienthal's flying machines.

If there was one single type of building, however, that characterized the waning nineteenth century in Berlin, it was the Panorama (Figure 25.8). No less a man than the young Schinkel introduced the Berliners to paintings arranged in panoramic form. The Sedan Panorama at Alexanderplatz Station, a seventeen-sided structure measuring 128 feet in diameter and 49 feet in height, had a Schwedler domed roof with a glazed rim 23 feet wide. The outside ring of the

Figure 25.7 Airship hangar under construction, c. 1918
Source: Landesbildstelle. Berlin

stepped platform, 36 feet in diameter, was rotatable, each rotation lasting twenty minutes.

> Located five-and-one-half meters above the floor of the platform is a walkway covered with vellums, which is suspended from the roof beams and which fulfills the regulations of ventilation and lighting. The thorny problem of illumination by night has been satisfactorily solved here for the first time, by means of seventeen differential lamps, or arc lamps, based on the Siemens and Halske system and installed over the walkway railing, for which electricity is provided by generators located in the basement.[5]

In 1892, when interest in panoramas was already on the wane, Berlin's most impressive example of this genre was built, the Hohenzollern Panorama. It was a massive, sixteen-sided stucco building, 131 feet in diameter, located on the axis of Moltke Bridge between the Packhof (Custom House) and Lehrte Station. Representing a quarter millenium of Brandenburg-Prussian history, the Hohenzollern Panorama apparently bored the Berliners despite - or perhaps because of - all its pomp and circumstance, and so it was replaced before the inaugural year was out by a depiction of 'A Lloyd Steamship Entering New York Harbor.'

> We actually feel as if we were on the deck of a great steamer just pulling into the huge basin of New York Harbor. . . . On our left we see the

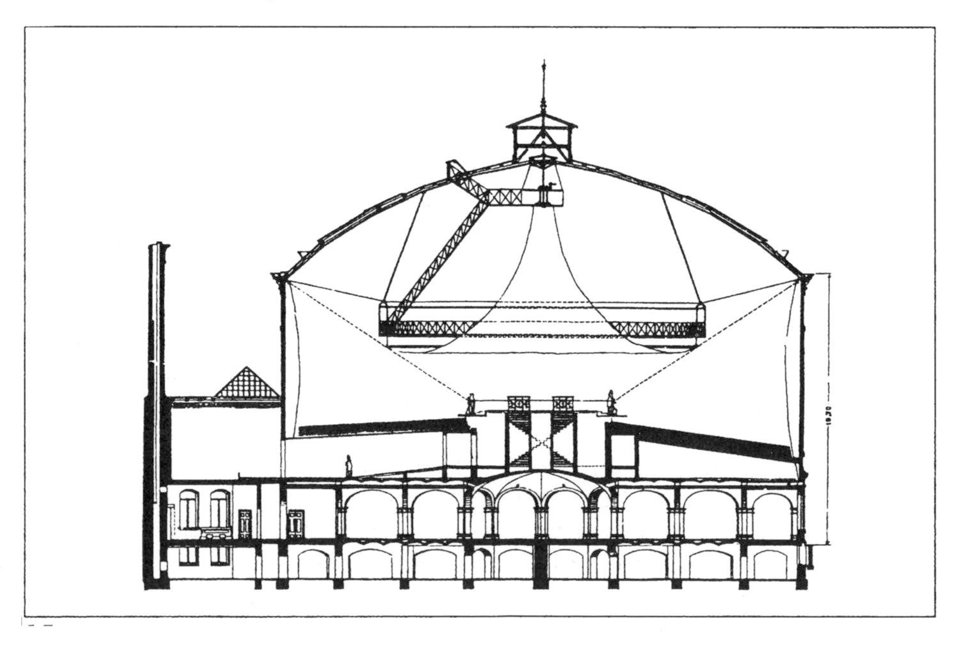

Figure 25.8 Sedan Panorama, designed c. 1810, cross-section
Source: *Berlin und seine Bauten*, vol II (Berlin, 1896)

> gigantic Statue of Liberty. . . . And there, right in front of us, to the right of the bow, New York's colossal suspension bridge arches at a dizzying height over the East River to the neighboring town. Like a fantastic, brightly shimmering mirage it soars through the sunny sky high above the whitish smoke of the steamers covering the water. At its left end, enormous red and white buildings bristle above the sea of New York's houses, which rise from the harbor shore to Broadway, the most magnificent of its streets.[6]

Their own city having nothing comparable, Berliners were reduced to goggling at the technological wonders of skyscrapers or the Brooklyn Bridge in a panorama or, as the author of the newspaper article suggested, to rushing down to Karl Stangen's travel agency at 10 Mohrenstrasse and buying a passage to New York. What they could boast, however, was an urban drainage system that was the most modern in the world in its day and one of the finest achievements of Berlin engineering. Until 1878, the disposal of all the city's sewage, including human wastes, was effected by means of cesspools that were manually emptied into stinking open carts. 'As a big city,' wrote August Bebel, 'Berlin did not really pass from a barbaric to a civilized state until after the year 1870.' The so-called reform of latrine facilities occupied engineers and local authorities for decades on end. The process began in 1859 when James Hobrecht, a thirty-two-year old master of waterway and railway construction, was made head of the Commission for the Preparation of Building Plans for the Environs of Berlin. Part of his job was to work out a 'drainage system for the projected streets, squares, and surrounding terrain.'

After an inspection trip in 1860 to Paris, London, and Hamburg, Hobrecht designed a system of drainage based on 'the principle of dividing the city into various separate districts or radial systems and disposing of sewage water on surrounding fields by means of irrigation. Machine power would be employed to pump the water up from the separate radial systems though pipes arranged in a radial manner to outlying properties which were to be purchased by the city.'[7] After the plan was approved by the magistrate and city council, the first phase of construction began in August 1872. The principal, technologically revolutionary feature of Hobrecht's conception of a closed system of water supply, sewers, and biological clarification of waste water was that instead of being emptied into the river, sewage was channeled away from it by means of pump stations and radial piping to fields on the extreme periphery of the city. Human feces were no longer treated as waste but as a valuable raw material for fertilizer, filtered through the soil and, thus purified, reintroduced into the ground water. The project involved dividing the city into twelve drainage districts each with its own pump station, which permitted the system to be expanded as the city grew. [. . .]

Perhaps the clearest idea of the apparently chaotic development of Wilhelmine Berlin is conveyed by the district of Moabit, as represented in 1888 in the Liebenow plan. The plan reveals Berlin as a city of engineers and at the same time a field of battle upon which countless architectural designs met their doom. Gathered in Moabit in a small space is the entire typology of large buildings, infrastructure, public housing, and places of amusement; the most progressive of industrial plants rubs shoulders with the most bizarre of exhibition buildings. Yet detectable in this chaos beyond the control of architects and city planners is an urban-design utopia of a kind that Schinkel had predicted. Schinkel was prepared to bring the confusion under control without resorting to baroque notions of the well-ordered city. As early as 1822, when he not only planned the museum but suggested moving the Packhof to the northern tip of Museum Island, he conceived of an urban space determined by relationships of tension among great, freestanding structures that would replace the former

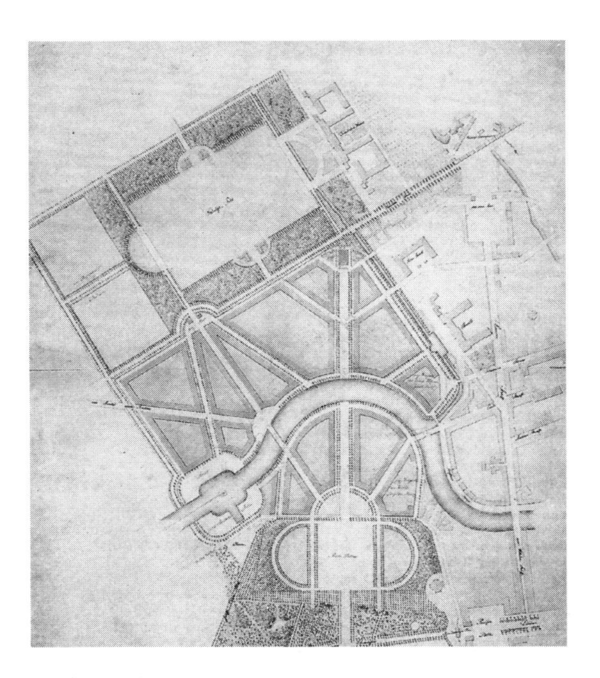

Figure 25.9 Peter Joseph Lenné, plan for the Pülvermühlengelände, Berlin, 1839
Source: Harri Günther, *Peter Joseph Lenné* (East Berlin. 1985)

Figure 25.10 Hans Kollhoff, Berlin-Moabit Design Seminar, ETH Zurich, 1988

axial relationships and the limited spatial typology of street, square, and courtyard (Figure 25.9). This was a vision of the city as a landscape, aiming at long, deep perspectives and consciously producing voids that would heighten the effect of mass in a controlled way. As the hesitant design process that was documented in his sketches indicates, Schinkel tried to establish fixed points with extreme precision (Figure 25.10). The Academy of Architecture in its uncompromising position within the frame of reference established by the palace, museum, armory, and Packhof buildings, embodies the quintessence of this planning approach. The formal Baroque dinner party has ended; the guests are standing and have begun to communicate with each other.

Notes

1 Walter Rathenau, *Die schönste Stadt der Welt*, Leipzig, 1902.

2 *Berlin und seine Bauten*, Berlin, 1877.

3 Julius Posener, *Berlin auf dem Wege zur einer neuen Architektur*, Munich, 1979.

4 Stephan Oettermann, *Das Panorama*, Frankfurt am Main, 1980.

5 *Ibid.*

6 *Ibid.*

7 *Berlin und seine Bauten*, Berlin, 1896.

26

BERLIN: TECHNOLOGICAL METROPOLIS

Thomas P. Hughes

Source: Thomas P. Hughes, 'The City as Creator and Creation', in Josef Paul Kleihues and Christina Rathgeber (eds), *Berlin/New York: like and unlike. Essays on architecture and art from 1870 to the present*, N.Y., Rizzoli, 1993, pp. 13–15, 17–18, 22–5, 27, 30–1

> *Technology cannot permanently be understood as an end in itself, but gains in value and significance just at that point at which it is recognized as the most defined means of achieving a culture. A mature culture, however, speaks only through the language of art.*
>
> Peter Behrens, 1910[1]

During the half century after 1870, Berlin and New York became technological metropolises. Earlier, London, Paris, Vienna, and Rome had become mother cities, although primarily government and commercial activities, not urban technology, stimulated and sustained their growth and prestige. [. . .] Berlin and New York became creatures and creators of a new technology, the technology of a second industrial revolution and urban development. A metropolis, in contrast to a large city, is an urban creation that expresses the history, the multifold social and cultural activities, and the aspirations of the larger national or regional society of which it is a part. Berlin and New York were alike in that both, among other characteristics, expressed the technological, that is a creative, exuberance of their countries during the second industrial revolution that began around 1870 and reached its apogee between the two world wars. [. . .]

[. . .] the second industrial revolution that shaped the rapid development of New York and Berlin involved the new power sources of electricity and internal combustion; new materials such as steel, glass, and reinforced concrete; the development of intraurban and highway transportation; and the organization of massive systems of production and communication. The revolution also depended on the inventions (especially in the United States) of independent inventors, the amassing of knowledge in technical colleges, universities, and private and government research institutions especially in Germany, as well as modern management techniques. Smoke, fire, and masses of workers comprise our image of the first industrial revolution; humming dynamos, city streets illuminated by electric arcs and incandescents, intraurban electric transit, city planning and urban housing developments of an international style, engineers bent over drawing boards, Edisonian inventiveness, scientific managers, and white-coated scientists working in industrial laboratories and research institutes characterize the second. Berlin and New York, not smoky Pennsylvania or Ruhr Valley towns, were the sites of the managerial, institutional, technical, and scientific developments of the second industrial revolution. Both cities had skilled workers who were experienced and gifted in practicing the high-technology arts of the day, such as fine machine work and complex electrical engineering.

Berlin and New York were creators of modern technology; they were also the creatures of it. The role of technology in the making of modern New York and Berlin has not yet been adequately recognized. The rapid growth of the two cities during the late nineteenth and early twentieth centuries has been conventionally attributed mainly to events such as the unification of Germany and the establishment of Berlin as its capital in 1871 and to the rise of the United States to world-power status and industrial preeminence. Relatively little attention has been given to the technology that has made possible the artifactual wonder - the modern metropolis. Except in specialist studies, for instance, the history of the electrification of transportation and communication in the two cities is rarely explored. Only infrequently has it been stressed that independent inventors, the source of much late nineteenth-century innovation, chose to do their creative work in New York and that the great research institutions of Germany found their home in Berlin. Too often the interweaving of financial potential and technological awareness by the investment banks and houses of New York and Berlin has been overlooked. Only a few historians have seen that power drawn from integrated networks of electrical transmission and distribution lines and applied in myriad ways in office, factory, and home have made possible modern Berlin and New York. Walther Rathenau, social critic, statesman, and head of Allgemeine Elektrizitats-Gesellschaft (AEG), the German electrical manufacturing company founded by his father, Emil, wrote of the electrical and other urban networks:

> In their structure and mechanical order all great cities of the white world are identical. Imbedded in a web of rails, they spread their net of highways over the countryside. Visible and invisible networks move streaming traffic underground and through city canyons, pumping masses of people from the suburbs into the heart of the city twice daily. A second, third, and fourth network distribute water, heat, and power; an electrical nerve system pulsates with life of the city. Food and goods glide on rails; water flows through the city; and waste empties through canals.[2] [. . .]

While the two cities were similar in many ways, they also had their differences. During the nineteenth and twentieth centuries, great cosmopolitan cities drew on an international pool in technology. Engineers, scientists, and managers in Berlin knew what their counterparts in New York were doing. Yet despite the common pool available, they did not create identical buildings, machines, processes, or technological systems for transportation, communication, and production. Each city's framework for creativity and invention was unique. Using technology to solve problems has not homogenized the world. Because, for instance, the geography, demography, politics, economics, and history of Berlin differed from those of New York, a Berlin style - and a New York style - of technology emerged. Style permits us to speak of the unlike; the common pool of technology engenders the like. [. . .]

Berlin and New York as electropolises

Berlin, a center of electrical manufacturing, has aptly been named an 'elektropolis.'[3] Modern New York, too, was an electropolis, but mainly because electrical technology made the modern New York physically possible, not because New York was a center of electrical manufacturing. In contrast, two of the world's largest electrical manufacturers, Siemens & Halske and AEG, had located in Berlin by 1890. In each city the presence of one of the world's leading suppliers of electricity - Berliner Elekricitats-Werke in Berlin and Consolidated Edison in New York - also contributed to the rise of these electropolises. The electrical supply networks of these utilities rapidly developed between 1880 and 1920 and made possible the communication, lighting, and horizontal and vertical transportation that were the identifying characteristics of these modern technological cities.

The history of electrical manufacturing and supply in Berlin had its beginning near and in New York. The independent inventor Thomas A. Edison and his associates, working at Menlo Park, New Jersey - only a short train ride from New York - invented and developed an incandescent lighting system that was installed on Pearl Street in the Wall Street district of New York in 1882. In 1881 Emil Rathenau, who had established himself in Berlin in mechanical engineering, visited the Paris International Electrical Exhibition and was so impressed by the Edison electrical lighting display that he acquired its patent rights in Germany. Financed by German banks, including Jacob Landau of Berlin, Rathenau founded in 1883 the Deutsche Edison Gesellschaft für angewandte Electricität (German Edison Company for Applied Electricity). There followed between the Rathenau interests and the long-established telegraph and electrical manufacturing firm of Siemens & Halske an agreement that, because of its cooperative character, was far more characteristic of Berlin and Germany than of New York and the United States. They concurred that Deutsche Edison would manufacture incandescent lamps and install the Edison system but that the Siemens company would manufacture the large equipment for the electrical systems. Deutsche Edison proceeded to install an Edison system at Friedrichstrasse 85 to supply the renowned Café Bauer in Berlin and then to construct, in 1885, a central lighting station on Markgraffenstrasse and a distribution system much like the one that Edison had installed in New York. Rathenau also established the Berliner Elekricitäts-Werke and transformed Deutsche Edison into Allgemeine Elektrizitats-Gesellschaft, which soon became, with Siemens, the best-known electrical enterprise in Europe. AEG and Siemens - creators of technology; employers of engineers, scientists, and thousands of workers; designers of factory and worker-housing settlements; and powerful economic and political movers - became the core of the Berlin electropolis. By contrast, the great electrical manufacturers in the United States, though deeply dependent on the inventions of Edison and other Americans, several of whom did much of their work in New York or its environs, established themselves in Schenectady, New York (General Electric Company), and Pittsburgh, Pennsylvania (Westinghouse Company).

Berlin and New York not only manufactured or invented electrical devices but by using them defined their own characteristics as cities. Before 1870, applied electricity was associated primarily with the telegraph. The electrical telegraph originated in the first half of the nineteenth century when it conveyed messages from one city center to another. Later, ingenious inventors showed how it could be used within cities to form a network of commercial communication, spread stock market reports, and carry fire alarms and police messages. After 1850, bold overland lines and transatlantic cables made cities nodal points of world communications. In 1876, Alexander Graham Bell, probably influenced by the earlier telephone experiments of the German Philipp Reis, introduced the speaking telegraph - or telephone - a communication system that within a decade was spreading a network of personal communication throughout large cities. The telegraph and the telephone changed the tempo of and brought precision to both communication and control activities. Communication at the speed of light transcended space and time. From New York or Berlin such investment bankers as J. P. Morgan in New York and George Siemens in Berlin, a cousin of the inventor Werner, oversaw, helped coordinate, and sometimes controlled far-flung commercial-industrial empires, including those involved in telephone and electrical manufacturing.

Electric lighting came to the cities first as brilliant arc lighting and then as soft incandescent lighting in the 1880s. Because distribution lines for electric current constituted a prime cost for the electrical supply utilities, such lines were established first in the densely populated cities. Berlin and New York quickly forged ahead of Paris and London in adopting the new lighting. Arc lighting in the streets and incandescent lighting in the interior spaces changed the rhythm of city life. Oil and gas lighting had shown the possibilities of nightlife, for pleasure and work, but electricity brought a dazzling illumination that suffused the evening hours with a glow and clarity that rivaled the stimulation of a sunlit day. Urbanites no longer ventured out into the dark; they strode along great white ways. [. . .] Offices, workshops, and factories that needed brilliantly lit workplaces for fine accounting and machine tending turned quickly to incandescent light. Newspaper publishers could readily set type and run presses for their morning editions. The bright new lights, however, were long in coming to dark tenements.

Electric motors came to the cities about a decade after incandescent lighting. Many industrial establishments in Berlin first drove their electric motors with current from their own power plants, but by 1900 the rapidly expanding Berliner Elekrizitäts-Werke could persuade industry that it was economically feasible to close down its isolated generating plants and take power from the urban network or grid of distribution lines that the utility was extending through Berlin. In New York, Consolidated Edison brought industry onto the network even more rapidly. Electric motors in factories made possible the reorganization of the workplace and of work. Previously, the central steam engines, with their geometrically extended labyrinth of transmission belts driving various machines, had severely constrained their placement and prevented subtle variations in factory layout and architecture. Once individual electric motors were attached to the various machines, experts in factory organization and managers of labor opened for themselves a plethora of redesign and reorganization opportunities. It is not coincidental that scientific management, assembly-line production, and industrial electrification rose in tandem during the early decades of the twentieth century.

Unlike steam power, electric light and power spread through cities as a great interconnected network, an urban system. An impressive sequence of inventions and developments made possible by 1900 the supplying of electric power and light to Berlin and New York from several large central stations generating alternating, or polyphase, power. Transformers, motor-generators, synchronous generators, high-voltage transmission lines, and low-voltage distribution lines interconnected arc lights, incandescent lights, home appliances, giant industrial motors, and the motors for streetcars, subways, elevated and later mainline trains. A few persons in several centrally located dispatching, or control, centers matched the generating capacity of the central stations to the consumption of the millions of diverse consumers. Berlin and New York might have seemed chaotic at times, but well-ordered systems of supply such as electricity, gas, and water bound people and things together in a great net of dependency.

Berlin and New York took early leads in developing electric power for transportation. Werner Siemens and his company took a historic step when they built an electric urban railroad in Berlin in 1879 on the occasion of the International Trades Exposition. The Siemens company operated the small electric railway with a locomotive, pulling several cars at four miles per hour around a circle of track. More than eighty-six thousand passengers were excited, even thrilled, by the experience, but the city did not build subways and elevated railroads with electric locomotives until 1896.[4] A year earlier, city-owned electric street-cars began operating in Berlin. In New York by 1902, electric locomotives had replaced the steam locomotives that drew trains along elevated railroads above Second, Third, Sixth, and Ninth avenues. In 1905, the elevated train transported 250 million passengers. Electric streetcars were common in New York in 1899 and by 1910 they were carrying 500 million passengers per year. [. . .]

Challenge and response: Martin Wagner and the Siedlungen

Following [World War I] the housing problem in Berlin became 'catastrophic.'[5] Shortages of materials had postponed building, and economic depression and deflation had further delayed the construction of housing acutely needed by returning soldiers and their growing families. Even before the war, housing for low-income Berliners was among the worst in Europe. In 1910, government statistics counted seven percent of the Berlin population living in twenty thousand apartments with at least five people per room. By 1925, the situation had worsened: more than seventy thousand Berliners lived in cellars.[6] Countless factory workers, artisans, and lower-middle-class office workers lived in *Mietskasernen,* large, multistory, square tenements with often only a single window on a dark, dank, inner court per apartment. Frequently only one room of a two-room apartment was heated. Many apartments had no running water or toilets, and residents were forced to share facilities. In these buildings there were sometimes four or more people in each room. With such housing conditions in Berlin and elsewhere in Europe after the war, it was not surprising that the French architect Le Corbusier had predicted architecture (housing) or revolution.[7]

In Berlin, architect and city planner Martin Wagner took action. He believed that new technology, especially American production technology, offered the most likely solution to the dire housing problem. Other avant-garde architects in Germany and Berlin shared his approach. Among them were Bruno Taut, also in Berlin, Walter Gropius at the Bauhaus in Dessau, and Ernst May in Frankfurt. In their enthusiasm for modern technology, they were like independent inventors who had flourished several decades earlier in the United States, especially in New York. The architects were among the principal founders of the modern, or international, style of architecture much as the independent inventors were originators of twentieth-century, or modern, technology. Wagner served as architect and *Stadtbaurat* (municipal city planner) in Berlin from 1919 to 1932. The 'leading engineer-architect of his era,'[8] he was politically and philosophically deeply committed to the Social Democratic Party and to socialism. He used technology and architecture to solve the pressing social problems of his day. Through housing, Wagner

Figure 26.1 For mass production of housing, Wagner advocated labor-saving machinery, rational materials layout, and orderly work procedures. Photograph shows concrete panel construction originating in the United States and introduced in Germany by a Dutch firm, Occident Gesellschaft, c. 1926
Source: Martin Wagner, 'Gross-Siedlungen: Der Weg zur Rationalisierung des Wohnungsbaues,' *Wohnungswirtschaft 3* (1926). Courtesy of Sammlung Baukunst, Akademie der Künste, Berlin

Figure 26.2 Labour-saving crane in use at Hufeisensiedlung Britz, c. 1924
Source: Sammlung Baukunst, Akademie der Kunste, Berlin

Figure 26.3 Earth-moving equipment and rails for efficient flow of work process at Hufeisensiedlung Britz, c. 1924
Source: Sammlung Baukunst, Akademie der Künste, Berlin

sought to bring 'light, air, and sun,' dignity, and order to persons of limited income.[9] He is best remembered in Berlin as the architect of several major *Siedlungen* (housing settlements), especially the first stage of the Hufeisensiedlung built at Britz, which he designed in association with Bruno Taut.

Enthusiastic about American production technology, Wagner despised American capitalism. In 1924, after an intensive three-week study tour of the American construction industry, he returned to Germany persuaded that the massive flow of traffic in American cities, the plethora of skyscrapers as tall as the Cologne cathedral, and the bright lights of New York's Broadway could not mask the greedy dehumanizing materialism of a civilization dedicated to the dollar. Yet his animadversion toward capitalism did not keep him from appreciating the American production techniques that he associated with the reforms of Frederick W. Taylor, the father of scientific management. He observed that in Germany there was much talk of Taylorism but little practice. In America, he heard relatively little discussion but witnessed widespread applications of scientific management doctrines. Like so many other Germans at the time, he also wanted to adopt the mass-production techniques associated with Henry Ford. Wagner believed that mass production of housing and consumer goods would ensure the loyalty of the working class to the newly established and fragile Weimar Republic. Leninist reformers in the Soviet Union held similar views at about the same time.

Drawing on American practices, Wagner articulated the ways in which costs could be lowered in the construction industry and a system of mass production introduced. He insisted that only through these economies could the financial means be found to solve the dire housing problem in Berlin. For the government to tax other sectors of a poor economy to subsidize the housing sector was, in his opinion, not in the national interest. He advocated mass production of housing through the specialization of labor and the substitution of machines for labor. He also wanted to standardize building components. He contrasted this method with the almost medieval construction techniques still used in Germany. The program for mass housing that he worked out with Gropius, Taut, and May specified the use of nontraditional and inexpensive building materials, the avoidance of labor-intensive practices such as building interior walls of small, hand-laid bricks, and the reduction of transportation costs by use of local materials and the newest materials-handling techniques. Perceptively observing that mass-production achievements depended on organizational and financial inventions and innovations as well as technical ones, he called for the vertical integration of organizations such as those involved in raw-materials production, building-components manufacture, and housing construction. Wagner realized that the cost of mechanization and administration had to be spread over many units of production (apartments or houses) in order to justify the investment in high-cost machinery and processes. From the managers of large-scale, industrial production enterprises, he also borrowed the cardinal principle of using capital and labor to capacity. He understood that idle labor and capital, especially that caused by poor scheduling and coordination of work, added greatly to unit costs. Work stoppages on the building site because of poor weather were a special problem of the housing industry. To counter it, he wanted more building components to be made in factories rather than on the open site.

Like many other inventive persons, Wagner relied heavily on the use of analogy. Before embarking on mass production, Wagner insisted that architect-engineer entrepreneurs, like the great manufacturers, had to analyze the market carefully, for housing needs were localized and houses were not transportable. Following the most recent trends in industrial practice, he wanted to establish a research laboratory to develop new materials and processes. He also recommended that architect-engineers, like automobile producers engaged in mass production, design and construct experimental models of housing before embarking on mass production. As in manufacturing, housing construction needed, Wagner maintained, large hierarchical organizations presided over by entrepreneurs to coordinate its numerous branches and phases. Henry Ford and his managers coordinating and controlling a massive system of production involving mines, rubber plantations, raw-materials processors, railways and ships, assembly lines, and dealer networks provided an example for those wishing to rationalize housing production.

Estimating that about eighty percent of the costs of an apartment in a housing settlement stemmed from construction and only about twenty percent from land, closing, interest, and administrative costs, Wagner concentrated in his reform of construction practices on replacing labor with machines. He offered as a prime example of wasteful labor the use of standard bricks. Wagner calculated that 24,500 bricks would be needed in each apartment in a housing development he was planning and, therefore, 24.5 million would be required for the thousand apartments to be constructed. He estimated that each brick had to be handled ten times, so workers would have to handle bricks 245 million times during construction. From the United States might come, he predicted, the solution to this labor-intensive process in the form of poured concrete or concrete-panel construction. He had reference, especially, to an American concrete-panel mode of construction that had been introduced in Berlin by the Dutch firm Occident. In 1926, however, he decided that further development of the process had to take place before it could be extensively employed.

Wagner's opportunity to respond on a large scale to the Berlin housing problem had come in 1925, with the commission for the Hufeisensiedlung. Capital shortage and conservative labor practices kept him from introducing much of the new building technology that he had seen in America and articulated for mass housing in Germany. Nevertheless, the Hufeisensiedlung is a milestone in the history of modern housing construction and Wagner's ideas served as a model for the rationalization of city planning in Germany. The layout of buildings and streets in the housing settlement facilitated the handling of materials, and a power excavator replaced hand labor. Traveling cranes moved materials to the workers. Apartment buildings were limited to four basic types. Coordination of the project by a centralized administration also provided an example of rationalization. Yet there were disappointing failures in this early rationalized *Siedlung* venture. Wagner complained that a thousand housing units were not enough to take full advantage of the economies of scale. In addition the newly designed mechanical excavator had numerous breakdowns. Changes in design of the four basic housing types during construction raised costs, which ultimately became higher than projected, so that few low-income workers could afford to live in the apartments.

Despite this, more apartments at the Hufeisensiedlung were added to those designed by Wagner and Taut. Over eight years, 2,317 housing units were constructed; Taut continued as architect for most of the period. These units served as a precedent and model for Onkel Toms Hutte (Waldsiedlung Zehlendorf), Siemensstadt, and Weisse Stadt, three other major housing developments begun in Berlin during the era of the Weimar Republic. Almost 2,000 apartments were constructed between 1926 and 1932 at Onkel Toms Hutte, for which Taut served as one of the architects; 1,370 for Siemensstadt in 1930–31, for which Walter Gropius, Hans Scharoun, and others were the architects; and in 1930–31, 1,286 for Weisse Stadt. Seventeen major *Siedlungen* were constructed in Berlin during the Weimar era. Construction of the Berlin *Siedlungen* was a substantial technological achievement. Not as dramatic as the New York response to the transportation problem, the Berlin solution to the housing problem demonstrated impressively how technology could be organized on a large scale and directed to social, even political, ends.

During the interwar years, the *Siedlungen* architecture of Wagner, Taut, Gropius, Scharoun, Hugo Haring, and other avant-garde architects introduced not only modern production technology, but also a modern, or international, style of design. As Gropius characterized the style, it embraced precisely impressed or cast forms, carefully controlled processes, pronounced contrasts of form and materials, orderly arrangement of building elements, and overall unity of design and color. The concepts were obviously borrowed from mechanical technology. The architects believed that the International Style, with its plain surfaces and rectilinear forms, was particularly well suited, like Henry Ford's Model T automobile, to construction using modern techniques.

Peter Behrens, architect and designer for AEG, also designed Berlin buildings expressing the order, control, and rationality of modern production technology. After 1908, Behrens designed a series of factory buildings for AEG in Berlin-Wedding. Included among them was the turbine construction building on Huttenstrasse now regarded as a major twentieth-century architectural statement. The characteristics of these buildings influenced the designs of other architects, including Gropius, Ludwig Mies van der Rohe, and Le Corbusier, all of whom worked in his atelier around 1910. Behrens insisted that the functionalism of engineering calculation would never fully express a modern style; the artistic sensibility of the architect was needed as well.

His factory buildings embodied new materials such as glass and steel. Offended by the nineteenth-century practice of cladding buildings with ornament to make historical references, he designed factories manifesting the technological processes taking place within them. He created appropriate spaces for the orderly process of manufacture and large expanses of glass and vertical supports of steel that provided proper lighting for work. Avoiding flat, banal functionalism by presenting walls

Figure 26.4 Hufeisensiedlung Britz, c. 1926
Source: Sammlung Baukunst, Akademie der Künste, Berlin

of steel, masonry cladding, and glass, he achieved volumetric effects with the play of light and shadow. His use of a rhythmic sequencing of prominent features, such as column supports, suggested the order of serial mass production. Behrens wrote of the 'articulation of large areas, a lucid contrast between prominent features and widely stretched flat planes, and a unified repeating sequence of essential features.'[10] [. . .]

Conclusion

During the early decades of the second industrial revolution, both Berlin and New York were creators as well as creations of modern technology. In Berlin, however, creative or inventive activity was mediated by influential institutions such as universities and manufacturing firms. In New York, such creative achievement was highly individualistic, as exemplified in the activities of the independent inventors. Berlin was also unlike New York in that *Siedlungen* builders not only utilized the modern technology of mass production, as did builders in New York, but made efforts to find new symbols and forms expressing their commitment to modern values such as rationality and economic democracy. The modern International Style first took root in Berlin, not in New York. By contrast, builders of great bridges, railroad stations, and high-rise buildings in New York cloaked their stark technical achievements in historical forms and symbols until the third or fourth decade of the twentieth century. On the other hand, New York displayed its remarkable technological prowess by creating an urban transportation system of unprecedented size in an unprecedentedly short time. There was little effort to find forms and symbols expressive of a new age; the raw and imposing nature of the technology said enough. New York's dramatic technology directly served the commercial life of the city; Berlin made efforts to integrate technology and culture, to find ways of relating technology to social and aesthetic, not simply economic, goals.

Notes

1 Peter Behrens, 'Art and Technology', in T. Buddensieg (ed.), *Industrielkultur: Peter Behrens and the AEG, 1907-1914*, Cambridge, Mass., M.I.T. Press, 1984, p. 215.

2 Walther Rathenau, *Zur Kritik der Zeit*, Berlin, Fisher Verlag, 1921, p. 15.

3 Sigfrid von Weiher, *Berlins Weg zur Elektropolis*, Berlin, Stapp Verlag, 1974.

4 *Ibid.*, p. 170.

5 Norbert Huse, 'Grossiedlungen der 20er Jahre-heute,' in *Vier Berliner Siedlungen der Weimarer Republic: Britz; Onkel Toms Hutte; Siemensstadt; Weisse Stadt*, Berlin, Bauhaus-Archiv, 1984, p. 9.

6 Hans Jorg Duvigneau, 'Die Bedeutung der Berliner Grossiedlungen für die Wohnungsversorgung-damals und heute,' in *Vier Berliner*, p. 13.

7 Le Corbusier, *Vers une architecture*, Paris, Editions Gres, 1923.

8 *Münchner Merkur*, May 28, 1957.

9 Sabine Schurer-Wagner, 'Architekten-Portrat,' *Der Architekt*, December 1985.

10 Behrens, 'Art and Technology,' p. 218.

27

PREFABRICATED HOUSES IN POST-WAR BRITAIN

R. B. White

Source: R. B. White, *Prefabrication: a history of its development in Great Britain*, London, HMSO, 1965, pp. 140–9

It would be inappropriate in a short history of trends in prefabrication to describe in detail the construction of each of the types of houses in this programme. Of the eleven, only five were built in numbers greater than 2,500. Of these five, the *Arcon,* the *Aluminium Bungalows* and the *Uni-Seco Structures* were probably the most historically interesting. Their numbers were finally about 40,000, 55,000 and 29,000, respectively. A very brief description of these three will serve to illustrate the technical background and the general inferences to be drawn.

The Arcon house

ARCON was the name of a group of architect-consultants who set out to design and sponsor a single-storey dwelling which could be assembled from stocks of prefabricated parts and thus contribute to the requirements of the programme. The collaboration and productive capacity of a number of industrial groups were enlisted, each of which could contribute steel, in light hot-rolled sections for the frame, in tubular sections for roof trusses and purlins, in light sections and sheet for windows, doors and internal equipment; asbestos-cement for the external cladding, including the roof covering; plasterboard for internal linings, ceilings and partitions; timber, plywood and wood-wool for prefabricated floor and ceiling panels and for thermal insulation. Thus the house became far less dependent on one or two critical materials than was, for instance, the Portal steel house. The Ministry of Works and a large firm of general contractors acted as agents and general co-ordinators for the group and managed the distribution of parts from various centres.

At the time of maximum production the output of bungalows may have reached as many as 500 per week.

The *Arcon* house (Figure 27.1), while not an example of complete prefabrication in the same sense as the *Aluminium Bungalow*, was simpler and less costly to produce and did not require special transport vehicles. It was a product of industrial and government organization by which ordinary, fabricated building materials were ordered, prepared, stocked, and erected by dry-assembly and without recourse to special plant or lifting-tackle. [. . .] The houses were fitted with standard Ministry of Works cupboards and kitchen-bathroom units.

The Aluminium Bungalow

The *Aluminium Bungalow* was much the most highly prefabricated house in the programme. Production of complete bungalows

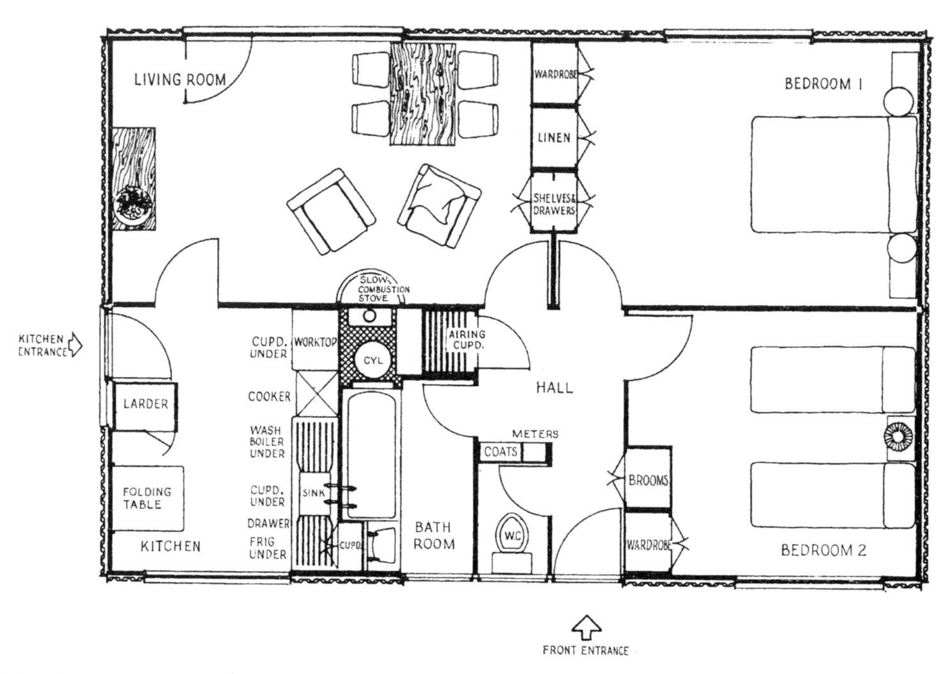

Figure 27.1 Plan of the *Arcon* house
Source: Courtesy Arcon, Chartered Architects

was concentrated in five large factories, well distributed over the country, and assembly-line methods were used to produce each house in four fully finished units which required mechanical handling and transport on special low-built trailers. There was therefore no need for centres of distribution for separate components.

Early in 1944, the possibilities of and problems involved in making use of the production capacity released by the inevitable end-of-war decline in fighter-aircraft production, to stimulate and make an important contribution to the post-war housing programme, was exercising the minds of ministers and industrialists. If a temporary house of which the principal parts were of aluminium alloy could be successfully prefabricated in these aircraft factories, then much of the labour employed there during the war could be retained until gradually re-absorbed, and stocks of alloy, including scrap from aircraft, could be used to great advantage to help solve another national problem. The proposition was attractive. As already mentioned, at the time of the passing of the first Housing (Temporary Accommodation) Act in 1944, the intention was to proceed with 50,000 steel houses, as well as other types, for the equipment of which some 200,000 prefabricated steel kitchen-bathroom units and sets of steel cupboards had been ordered. When it became clear that steel houses could not be produced in sufficient quantity before the middle of 1946, it was decided to concentrate instead on a temporary aluminium house, the production of which was considered feasible at an earlier date.

The development of the design for the *Aluminium Bungalow* was entrusted to a joint organization of the aircraft industries and the Ministry of Aircraft Production. This was known as the Aircraft Industries Research Organization on Housing (A.I.R.O.H.). The government departments concerned insisted that the house be designed specifically for the temporary housing programme, that is for a life of ten years. The designation 'temporary' was not very palatable for the designers, but the condition was nevertheless accepted.

The original intention was that practically all the temporary houses, of whatever type, should be fitted with the prefabricated steel kitchen-bathroom and cupboard sets designed by the Ministry of Works. Owing to difficulties of supply and production, however, only 28,500 of the former and 27,000 of the latter had been manufactured by January, 1948. Nevertheless a large number of both *Arcon* and *Aluminium* houses were eventually fitted with them (Figures 27.2 and 27.3). The cost of these sets, and especially of the refrigerators which were added to the equipment, far exceeded the original estimates (Cmd. 7304, pp. 10–11).

In the *Aluminium Bungalow* the entire plumbing system was contained in one of the four units, so that no plumbing joints had to be made on the site. The transport of the units imposed limitations of weight and size, but the lightness of the metal and of the air-entrained cement grout used to fill the panels of the external walls enabled the heaviest unit

Figure 27.2 Prefabricated kitchen-bathroom unit, kitchen side

Figure 27.3 Prefabricated kitchen-bathroom unit, bathroom side

of 2 tons $15\frac{1}{4}$ cwt (that containing the kitchen and bathroom assembly) to be kept within the maximum transportable load (Figure 27.4). The 7 ft 6 in. limit of width on many roads had perforce to be accepted as a fundamental design consideration. Figure 27.5 shows the plan of the bungalow with the plumbing assembly separating the bathroom and kitchen, at the living-room end of which is housed an openable, slow-burning stove with a back-boiler.[1]

The Ministry of Supply's specification for the bulk of alloy material to be used was DTD 479, which contained copper, manganese and magnesium, with a permissible addition of up to 10 per cent of pure aluminium for sheet and strip and up to 40 per cent for extrusions. The sheet and strip used for the external cladding of the walls and roof was coated on both sides with

Figure 27.4 A section of the *Aluminium Bungalow* being loaded onto a lorry
Source: Courtesy Aluminium Federation

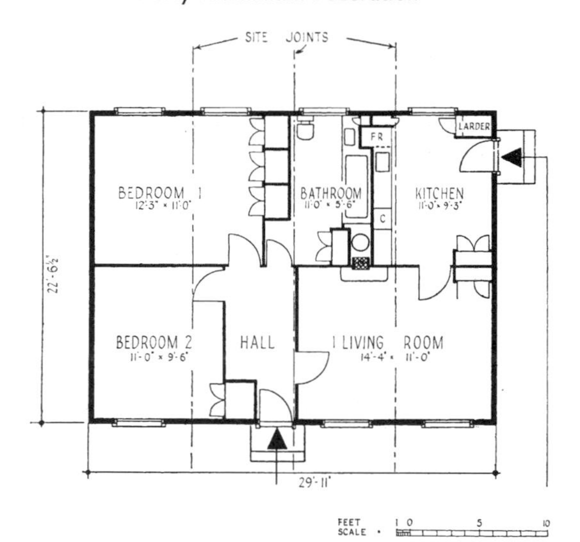

Figure 27.5 The plan of the *Aluminium Bungalow*, showing the four sections

99 per cent purity aluminium, and was known as *Alclad.*

With the exception of the nailing down of the floor-boards, which was done by hand, the entire production was a mechanized process. The wall frames, like shallow trays, were first sprayed on the inside with hot bitumen, then immediately passed, under a battery of cement pourers to be filled with air-entrained grout

Figure 27.6 Pouring of the air-entrained grout into the wall panels of the *Aluminium Bungalow*
Source: Courtesy Aluminium Federation

(Figure 27.6) which provided an insulating layer. The partly made wall panels were then passed through low-pressure steam drying ovens which enabled the grout to reach full strength in forty-eight hours, after which a further layer of hot bitumen was sprayed over the exposed cement face, and to which, while still hot, a $\frac{1}{4}$-in. plasterboard inner lining was pressed and secured by nails to wood fillets fixed to the frame. The layers of bitumen were intended to provide a water and vapour barrier and in addition to take up any relative movement that might be caused by different rates of thermal expansion. They did not however provide the continuous vapour barrier expected, because the first layer of bitumen became unevenly displaced over the aluminium sheeting when the grout was poured; moreover, the formation of small bubbles in the second layer also prevented continuity.

The finished wall was $2\frac{1}{4}$ in. thick and gave a thermal insulation equal to a brick cavity-wall of 11 in. with a plaster lining. The underside of the corrugated roof sheeting was likewise sprayed with bitumen, to which $\frac{1}{16}$-in. millboard was fixed.

The ceilings were of fibreboard on timber frames fixed to the aluminium trusses, and included additional insulation of glass-fibre, under which was a layer of building-paper, intended to act as a vapour barrier.

The final assembly of the components was done on the moving-belt system (Figure 27.7), during which the final paint spraying was carried out. Previously, the aluminium parts requiring special paint treatment, such as windows, had been passed through the appropriate processes of pickling, dipping and drying. Each unit of the house emerged fully wired for electricity, glazed and painted.

One of the five factories concerned had installed four assembly-belts, one for each unit of the bungalow; in another factory the four units travelled along a single belt. It was possible with this system ultimately to bring houses to completion at a rate of one for every twelve minutes of the working day. Careful regulation of the production of component parts was of course necessary to maintain an even flow; a shortage or a surplus at the wrong moment could cause conveyors to be halted, with crippling consequences.

The houses were contracted for and produced under the direction of the Ministry of Aircraft Production (later the Ministry of Supply). The preparation of the sites, concrete slabs and dwarf walls on which they were erected was the concern of the Ministry of Works, whose responsibility ended when the houses were finally accepted by the local authority in whose area they were built.

When the four low trailers, each carrying a unit of the house, arrived at the nearest spot convenient to the erection site, they were backed up between collapsible gantries which lifted the units clear and lowered them on to trolleys fitted with screw-jacks and rollers, enabling the units to be pushed along parallel sets of rails to their appropriate place on the foundation slab. For erection on sloping sites, however, a five-ton travelling crane had to be used. As soon as one unit was ready to be joined to the next, a wooden batten was placed in a vertical channel in the edge of the wall panel, ready to engage in a similar channel in the abutting wall panel. This batten thus acted rather like the tongue in matchboarding. The cavity in front of the battens was then packed with glass-fibre which was finally enclosed with spring cover-strips of aluminium (Figure 27.8); similar strips protected the roof joints.

At the eaves and base of the wall the units were locked by V-shaped interlocking connector blocks of aluminium through which a pin was driven horizontally (Figure 27.9). The main

Figure 27.7 Placing the roof sections on the *Aluminium Bungalow*
Source: Courtesy Aluminium Federation

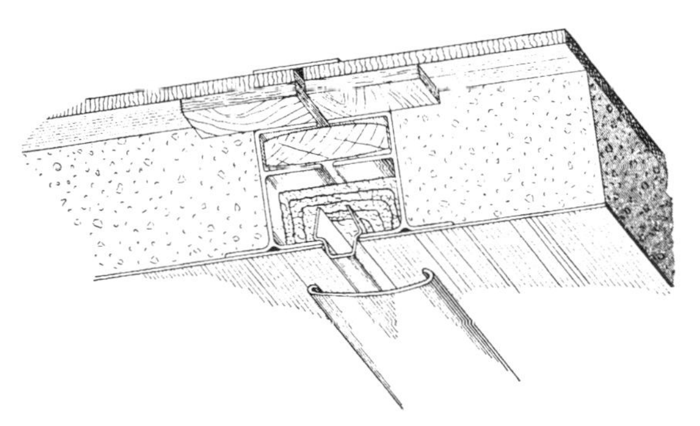

Figure 27.8 The *Aluminium Bungalow* – site joint between wall panels
Source: Courtesy Aluminium Federation

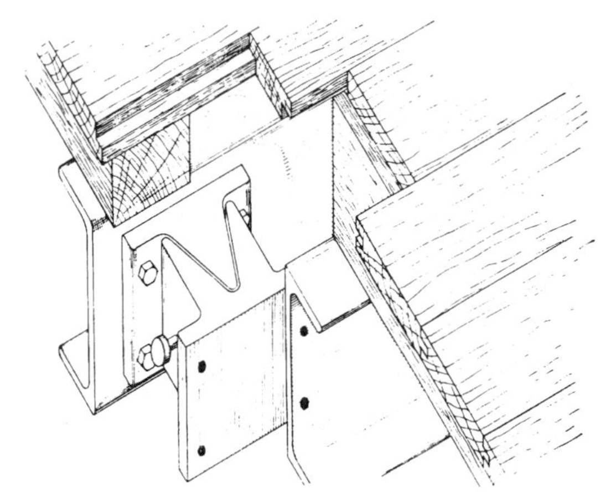

Figure 27.9 The *Aluminium Bungalow* – site joint at floor
Source: Courtesy Aluminium Federation

services were all carried in the kitchen-bathroom unit and work on joining them up could begin before the house was completely erected. Extension of electrical wiring to the remaining units was made by plugs and sockets. Erection time would vary a good deal according to the site; twenty man-hours has been commonly quoted.

The floor area of the *Aluminium Bungalow* was 629 sq. ft. In the White Paper of October, 1945, the cost was estimated at £1,365; by 1947 the cost was given as £1,610, and this did not include a proportion of the cost of tooling the factories for this programme, a cost which could not possibly be amortized over ten years of use. It was thus by far the most expensive of all the temporary houses, costing about £400 more than the *Arcon* house of similar size and equipment, despite the fact that some 20,000 more *Aluminium* than *Arcon* were produced. There is no doubt that this revelation of cost did much to shake the belief in total prefabrication, particularly as the houses were officially classified as temporary dwellings.

The *Aluminium Bungalows* were indeed designed as temporary dwellings; seen as such, structural deterioration did not render any house uninhabitable, even after twelve years, but from 1953 onwards incidences of corrosion of aluminium structural members were reported in increasing numbers, notably corrosion of those members at the base of wall panels where moisture was held for long periods. The prevention of further corrosion and traditional methods of repair were rendered extremely difficult owing to the factory method of construction, a factor to be considered in any design of prefabricated buildings.

In common with other types of temporary houses that included excellent fittings, many of these, such as cookers, refrigerators, sinks, drainers, cisterns and sheet-steel casings began to show steady deterioration after about fifteen years, requests for replacements becoming more and more frequent. Nevertheless, tenants generally expressed complete satisfaction with the living conditions and were loath to leave these houses for more permanent quarters. A more serious drawback from their point of view than incidences of corrosion of the aluminium was the occurrence of condensation. In a large number of instances this fault was found to be the result of cooking and washing with inadequate ventilation of the roof space. It had been hoped that the millboard under the roof sheeting, apart from providing some additional thermal insulation, would direct condensation occurring under the metal roof sheeting to the eaves, and allow it to drain away outside. However, owing to the large amount of vapour produced in some homes, and the lack of an effective vapour barrier in the ceiling, the millboard in some houses became saturated and detached, and condensate collected in the ceiling panels, which eventually leaked. The Building Research Station suggested providing a double layer of aluminium roof sheeting with a cork sandwich. The double layer was indeed provided in some later bungalows of similar type embodying a few alterations of detail, following which they were categorized as 'permanent' houses,[2] but the cork sandwich was omitted, and fibreboard was used under the sheeting instead. Even so, troubles from condensation appear to have been less frequent in these later, 'permanent', *B2* houses than in the original version (*B1*) intended for a ten-year life.

As also with many other houses of novel design which did not have brick chimneys, some over-heating of the casing round the flue-pipe occasionally occurred, as well as some corrosion of the chimney terminals.

In spite of these occurrences, and its probable complete disappearance from the scene in another ten years or so, the *Aluminium Bungalow* must be recorded as a great historical achievement in prefabrication. It made a timely, if costly, contribution to the mitigation of the serious housing shortage at the end of the war, and at the same time helped the large aircraft industry built up during the war to re-adjust itself to peace-time conditions. [. . .]

The Uni-Seco structures

The third temporary house in order of total numbers completed was the *Uni-Seco*, of which 29,000 were built, many of which are still occupied. Since it embodied a fairly high degree of prefabrication and since the system was continued for a time beyond the temporary housing programme to provide schools and houses for export, a brief description of the construction may be of interest.

Basically it was a system of standardized posts, beams and panels assembled on a grid pattern of 4 ft × 3 ft 6 in. The principal materials used were timber, asbestos-cement and wood-wool. Posts of box section were erected at 12-ft centres and connected by built-up timber and asbestos-cement roof-beams, sloped on top to give a series of low-pitch 'trusses' which were connected by secondary spars or purlins carrying roof slabs covered by roofing felt. The roof and wall panels consisted essentially of light frames of softwood covered on both sides with flat sheets of asbestos-cement which enclosed a sandwich of wood-wool. The upper part of the external walls consisted of a deep eaves-beam which served to connect the columns and roof-beams and to lock the heads of the wall panels. The feet of these panels were tongued to a continuous timber sole-plate bolted to the site concrete. Considerable use was made of a proprietary mastic to caulk external joints.

The Government's Temporary Housing Programme, admirable as it may have been in many ways, nevertheless caused a good deal of misgiving in many quarters; enough to show that the wholesale industrialization of house building to the extent envisaged, for instance, by the afore-mentioned Committee for the Industrial and Scientific Provision of Housing, would have occasioned vehement opposition.

Notes

1 Surplus heat from the flue was used to warm the bedrooms by means of air ducts carried above the ceiling, with an air intake in the hall.

2 *B2 Permanent Aluminium Bungalows* to distinguish them from the original, unaltered bungalows henceforth known as *B1*.

28

URBAN RECONSTRUCTION AND TRAFFIC PLANNING IN POST-WAR GERMANY

Jeffry M. Diefendorf

Source: *Journal of Urban History*, 15 (1989), pp. 131–58

Anyone familiar with West German cities cannot fail to be impressed by the role played by the automobile in German urban life. There are many clichés about Germans and their cars, ranging from the high speeds on the Autobahns to the spotless condition in which Germans supposedly keep their cars. Germany is an automotive society, and 1986 was the 100th anniversary of Karl Benz's patent of his three-wheeler. Nevertheless, when we think of German cities today, what we often recall is not the presence of autos but their absence: Inner-city pedestrian zones are ubiquitous in West Germany. To anyone aware of traffic conditions in Germany's cities prior to the 1960s, however, the success of the pedestrian zones implies the failure of most inner-city traffic planning of the postwar era. Most of the pedestrian zones, after all, used to be streets that carried cars, buses, and streetcars, often in great volume. What happened? What had been attempted with earlier traffic planning, and what went wrong?

In the following pages, I will discuss traffic planning for the inner cities and particularly the main arteries in the city centers. Much that German planners would include within a comprehensive traffic system must, for reasons of space, be excluded. I will refer only briefly to public transportation facilities such as subways and streetcars. The railroad system was and is independent of the city planning process, usually to the frustration of the city officials. Regional traffic planning – the connections between suburbs and city centers, and the highway and superhighway (*Autobahn*) network – must also lie beyond the scope of this essay.

A recurrent theme of the analysis will be a basic difference between modern and conservative planning ideas. The modernist, functionalist approach to planning was largely technical in nature. That is, rational, mechanical calculations about the functional needs of cities and their inhabitants lay at its center, and plans that contained green belts or called for skyscrapers were the result of these calculations. The conservative approach was more architectonic; it was more concerned with the design of structures (height, volume, surface materials, proximity to other buildings or spaces). It will be useful to distinguish between technical and architectonic conceptions of traffic planning.

In concentrating on inner-city traffic planning, I will explore some of the continuities that existed between the prewar, wartime, and postwar periods. Finally, I will try to evaluate the planning of the postwar years. Obviously I cannot present a comprehensive survey of events in all German cities. Instead I will draw on the experiences of just a few cities to illustrate my main points.

Traffic planning concepts

Prewar planning

It would be possible to present an almost unending list of citations in which major planners state that traffic planning was the heart, the *Kernstück,* of their efforts. I will limit myself to a statement by Herbert Jensen, the planner of Kiel, from his address to the 1953 meeting of the Deutsche Akademie für Städtebau und Landesplanung:

> To be a junction and assembly point of traffic, that is, to serve the exchange of goods, money, labor, ideas, and culture – that was the original and is today still the task of cities. . . . Traffic is for the city's body the life-giving flow of blood, the streets are the arteries. . . . We live in an age of traffic. Traffic is trump; we live in fact in an intoxication, an ecstasy of traffic, and of motorized traffic at that.[1]

This statement is entirely typical of the thinking of almost all planners of postwar reconstruction during the first 15 years after the war. Since traffic was the life blood of the city, and since traffic's dominance over other aspects of city life was incontestable, traffic planning had to have the highest priority.

The emphasis on traffic planning was of course not new to the postwar world, nor was it exclusively German. The sources of postwar concepts can be found in the planning theories of the 1920s, 1930s, and the Nazi era, when certain conservative ideas that were essentially antiurban, even antimodern, merged with progressive ideas on the subject of traffic. Conservatives, impressed by the garden-city movement, emphasized low-density housing in green suburbs with traffic arteries as links to the inner cities. The use of the term *artery* is of course also old; it derives from the use of an organic analogy to describe the growth of cities. Progressive planning ideas were stated most clearly in the Charter of Athens of 1932, the manifesto of Europe's (and Germany's) modernist architects. Thesis 77 of the Charter, for example, declared that planning must focus on a city's four essential functions: housing, work, recreation, and traffic – the system that tied the first three together.

The Nazis' ambitious plans for Autobahn construction and for the redesign of certain German cities had a major impact on German postwar planning. From the point of view of planning and engineering, the Autobahns embodied quite modern traffic ideas, even if the thrust was connections between cities and not traffic arteries in and through cities. The architectural projects of Hitler and Speer are well known.[2] The key point here is that Hitler's and Speer's interests in urban planning were architectonic: Both started with a concept of architecture and imposed it on the map of Berlin. The term used, *Neugestaltung,* was apt. Form was to dominate function. However, once the war began and planning for monumental buildings ceased, urban planning became more technical than architectonic in nature. Planners gave more attention to the functional requisites of urban life, especially traffic. This merits elaboration.

Wartime planning

Large-scale planning, though suspended by the outbreak of the war, was resumed and rethought in the light of the vast destruction in the bombed cities. Planning theory, moreover, continued to develop during the war, even if it was only published at length after the war ended. For example, Hans Bernhard Reichow's *Organische Baukunst,* published in 1949, was written during the war, as was *Die gegliederte und aufgelockerte Stadt* by Johannes Göderitz, Roland Rainer, and Herbert Hoffmann. The title of this latter book, which appeared in print in 1957, was a kind of *leitmotif* of postwar planning. In these works one finds old, crowded cities opened up, reorganized on functionalist lines, with residents living in garden suburbs, and with major traffic arteries serving to carry the life blood of the now healthy cities.

The tendency for Nazi planning to become more technical and less architectonic was undoubtedly also encouraged by the infusion of the experience of the Autobahn builders. Members of Speer's staff and members of the planning staffs of other cities designated for redesign as Nazi cities had worked on the Autobahn network in the 1930s. (Indeed, for some young architects, such projects were among the few actually available by which they had been able to earn a living.) Partly as a result, though Hitler's grand avenues did not lose their representative, ceremonial character, equal attention was given to new traffic arteries that would make the newly redesigned cities work. Examples are Konstanty Gutschow's plans for an east–west artery in Hamburg, Paul Bonatz's plans for new arteries in Stuttgart, and the east–west axis planned for Cologne.

The swing to technical planning took official form with an edict issued on 11 October 1943, by which Speer was ordered to begin preparations for planning the reconstruction of the destroyed cities. [. . .]

The instrument for the new planning was to be a 'Working Staff for Reconstruction Planning' (*Arbeitsstab zum Wieder-aufbau bombenzerstörter Städte*) headed by Rudolf Wolters and located administratively in Speer's ministry. Much of the work was done by this *Arbeitsstab's* inner circle, which numbered 10 or 12, with some projects farmed out to another two dozen or so architects and planners. Konstanty Gutschow, the architect for the redesign of Hamburg, was the real leader of the organization.

Major tasks of this group included preparation of standards and guidelines for future traffic planning and actual plans for 42 cities designated for initial reconstruction. In all cases the thrust was technical, not architectonic.[3] The guiding principles in the section on roadways were as follows:

> Clear separation according to function of through streets and access streets. As few as possible, but high-capacity through streets. Crucial for high capacity a minimum number and large size of crossings and exit and entry junctions. Through streets on a tangent to contained settlement cells.

The *Durchgangsstrassen,* or arteries, in built-up urban areas were to have a width of from 30 to 70 meters, depending upon the presence or absence of streetcars and center islands or dividing strips. The guidelines gave as examples of streets like Berlin's Unter den Linden, Vienna's Kärntner Ring, and Paris's Champs Elysées.

Using these guidelines, members of the *Arbeitsstab* and other architect/planners worked on plans for individual cities. The actual plans need not detain us, since none were implemented as drawn up in 1944–5, though it is noteworthy that the plans were all technical and not architectonic in nature. It is important that quite a number of men associated with the *Arbeitsstab* went on to play major roles in postwar reconstruction planning. [. . .]

At the same time, we should note that many postwar planners who had had no association with the *Arbeitsstab* based their planning concepts on the modernist ideas of the 1920s and early 1930s. This was especially true of Hans Scharoun in Berlin, Rudolf Schwarz in Cologne, and Walter Hoss and Richard Döcker in Stuttgart. Finally, whether we are talking about those who looked back to the Nazi era or before it for their planning ideas, the emphasis was on using the opportunities created by the bombing to address the functional needs of modern cities and provide technical solutions to perceived problems. Planners worried first about the layout of new streets, not the architectural style of the buildings that would be built along them. A commitment to an architectural style, whether modern or conservative, required a firm commitment to the values that underlie that style. The Germans, like many other Europeans, were suffering from ideological exhaustion after the war. Under the circumstances, technical solutions to pressing problems seemed the path to take.

Postwar planning

Now to the nature of postwar urban traffic planning. All over Germany planners argued that the bombing presented a unique opportunity to solve the problems caused by unplanned urban growth in the nineteenth century. High capacity *(leistungsfähige)* traffic networks, a functional layout, open areas of greenery, sun-lit housing and work places: now was the time to create healthy cities. Since city planners had no authority over the independent railroad system, traffic planning meant street planning, and the desire was to create what Hans Bernard Reichow later called 'die autogerechte Stadt.'[4] Everywhere planners made proposals for improving roads connecting suburbs with the inner cities. Suburban housing was in some cases the result of modernist, prewar developments like those of Ernst May in Frankfurt; in other cases suburbs of small garden cottages had been built during the Nazi era as part of the attempt to recapture mythic ties to nature. Where the destruction in the city centers had been especially severe, residents from the centers had sought housing in the outskirts, which also encouraged suburban growth and required efficient ties to the urban core. In any event, whether one was concerned with regional traffic flow or with rebuilding the city center, major changes had to be made in the old, inner cities if planners were to take advantage of the unprecedented opportunity for modernization.[5]

Widening streets was the easiest option, though it often required removing historic facades and demolishing damaged buildings that could have been repaired. Another option was undercutting existing facades to create arcades for pedestrians, while using former sidewalk space for roads or parking. Beyond street widening, it was necessary to create new streets by cutting through formerly built-up areas. Since these new arteries sounded all too much like the great ceremonial streets planned by the Nazis, postwar planners often went to great lengths to explain that their new roads were to move *traffic* efficiently, not marching hoards, and that they were not intended as 'representative' streets. Unlike the Nazi *Prachtstrassen,* magisterial boulevards that were to be lined with massive neoclassical buildings, the architecture on these postwar arteries was not considered very important.

Finally, the inner cities would be surrounded by high-volume rings. Feeder streets might come in from the suburbs like spokes of a wheel, or they might join the rings on tangents, reducing the number of crossings. Vehicles simply crossing the city on their way between points outside the city would normally be kept out of the inner city. This traffic would use the Autobahns or the ring systems. Some through traffic would, however, still cross the inner city on the new *Durchbruchstrassen*; the economic health of the *Altstadt* required that it not be an isolated island. It was expected that this redirection and channeling of traffic would relieve congestion in the inner cities, making life easier for pedestrians. Improved traffic conditions, plus a more functional layout of the city (i.e., relocation of industry) and better housing, would produce revitalized inner cities.

Some cases

Traffic planning in Hannover, [. . .] Hamburg, and Stuttgart can serve as concrete illustrations of these main trends. In all of these cities, planners chiefly used a technical rather than architectonic approach to planning, though there were differences in the ease with which planners were able to implement their ideas. Cologne is an example of a city that did not give technical traffic planning top priority, though, there too, street construction was nevertheless important.

Hannover

Of all of Germany's bombed cities, postwar Hannover long enjoyed the reputation of being a model of progressive reconstruction planning. Likewise, her chief planner after 1948, Rudolf Hillebrecht, established himself as probably West Germany's most influential city planner. The reputation of both city and planner rested on several things, including the remarkable cooperation between citizens and the planning department and a model inner-city housing project, but chief among them was Hannover's traffic planning.

Stimulated by Speer's Working Staff for Reconstruction Planning, preparation for Hannover's rebuilding began in the last year of the war. At a colloquium held on March 3, 1944, a spectrum of possibilities was presented. Professor Gerhard Graubner proposed putting all inner-city streets and rail lines underground, with only pedestrians and bicycles enjoying the privilege of surface transportation. Professor Friedrich Fischer proposed cutting through a major new axis straight across the city from the main train station to the Leineschloss, while Stadtbaurat Karl Elkart argued for a large-scale system of ring roads fed by arteries on a tangent to them. All wanted the old Hannover opened up, and all wanted greater physical separation of urban functions.

Some of these ideas were carried over into the postwar period, when traffic planning was always given the highest priority. In early 1948, the first postwar planner, Otto Meffert, submitted a plan that would relieve through-traffic by creating a double ring road, with part of the inner ring going along the west side of the Leine river. Several plans drawn up by a team led by Hans Högg at the Technical University kept variations of Meffert's rings but added a redesign of the Steintorplatz and a widening of the Georgstrasse to carry traffic through the city. Shortly after Hillebrecht's appointment to succeed Meffert in mid-1948, still another team of independent architects and officials set to work on a plan, one that kept some of the features of the earlier plans, but that enlarged the outer ring and added tangents rather than radial streets as feeders. (This feature resembled Elkart's 1944 proposal.)

Another source of planning ideas was the Aufbaugemeinschaft Hannover, a private organization of owners of inner-city property. Hillebrecht had encouraged the establishment of this organization, and at his suggestion it had engaged Konstanty Gutschow, under whom Hillebrecht had worked in Hamburg and while on the Working Staff for Reconstruction Planning during the war. Gutschow prepared a lengthy, technical analysis of the history of the inner city, of its likely future development, and of the various traffic plans prepared to date. Finally, Hillebrecht organized a public competition to solicit ideas for the redesign of the inner city. The competition addressed a number of issues, including land use and questions of form (*städtebauliche Gestaltung*), but traffic planning was given the most weight. From all of these sources, final street plans were drawn up by the planning office, in collaboration with the Tiefbauamt in charge of street construction and a representative of the planning department of the Technical University, with Hillebrecht mediating.

The result was a concept of Hannover, now the capital of the Land of Lower Saxony, as the hub of a large region. Indeed, Hannover became the primary example of good regional planning. Traffic would flow easily to and from the city, but not through the inner city. Rather it would use the new system of rings and tangents. To ensure further that high levels of

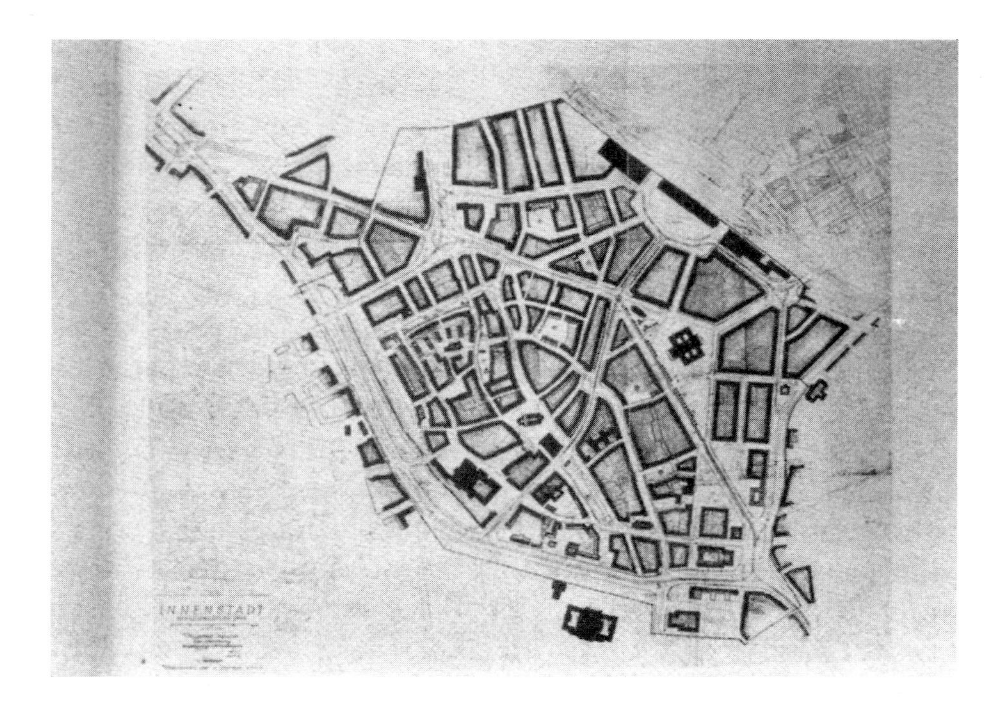

Figure 28.1 Hannover, plan for rebuilt inner city, 1949
Source: Der Aufbau der Stadt Hannover. Erste Denkschrift: Die Innenstadt, 'Bildteil.'

traffic would stay out of the inner city, skyscrapers would be allowed only outside the inner ring, while buildings inside that ring could not exceed the proportions typical of the historic core. (This was a decision made partly on technical and partly on aesthetic grounds.) Streets in the *Altstadt* were widened enough to facilitate the traffic that did flow through the inner city and to leave open the possibility of a subway system. Provision was made for a pedestrian shopping street (not a pedestrian zone, which was a concept of the 1960s) and for some mixture of functions. The inner city was not to become purely an administrative, cultural, or shopping center. It was important that people still live in the *Altstadt,* though under modern conditions. Hillebrecht was able not only to win approval for his ideas from the city government, but also to sell his ideas to the citizenry. Consequently planning and actual construction moved at a pace that won the admiration of planners from other cities.

Hamburg

[. . .] Hamburg's new traffic planning followed in many ways the wartime planning for the redesign of the city as Hitler's representative port. A competition had been held in 1940 for the design of an east-west axis on the southern edge of the inner city, along the Elbe. The competition was repeated in 1948, with some of the same architects submitting proposals. Building director Otto Meyer-Ottens argued that this Durchbruchstrasse was for traffic and not parades, like the Nazi axes, but in fact the arterial function of the new street was basically the same as in the wartime plans. Other key features were a new bridge alongside the old Lombardsbrücke over the Alster, a new tunnel

Figure 28.2 Hannover, part of a high-capacity inner ring road over the Aegidientor-platz, 1965
Source: Photo Heinz Koberg, Presseamt Hannover, 1/2

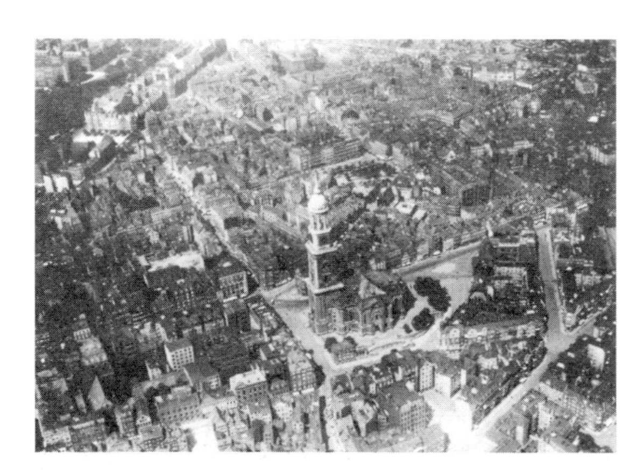

Figure 28.3 Hamburg, prewar conditions around the Michaeliskirche
Source: Baubehörde Hamburg, Lichtbildnerei, no. 29291

Figure 28.4 Hamburg, planners' projection of east–west artery past the Michaeliskirche, c. 1950
Source: Baubehörde Hamburg, Lichtbildnerei, no. 6368

Figure 28.5 Hamburg, east–west artery past the Michaeliskirche, 1960s
Source: Staatsarchiv Hamburg, Plankammer, 131-134/Ost 53

under the Elbe (Hitler had wanted a great bridge, but postwar Hamburg rejected this idea as too expensive), and improvements to the road along the old city wall. The purpose of all this, as Otto Sill, the later director of the Tiefbauamt put it, was 'high-capacity roadways and generously and lucidly designed squares for traffic.'[6]

Stuttgart

In Stuttgart, where once again traffic considerations lay at the heart of reconstruction planning, the building office had a more difficult time selling its ideas to the public. As had been the case elsewhere, planning had begun during the war. Paul Bonatz had had the leading role in planning a new layout for that city as one of Hitler's representative Nazi cities. In 1941 he prepared a major memo titled 'Stuttgart, Städtebau und Verkehrsfragen,' in which he called for a new, flow-through train station closer to Bad-Canstatt, a grand boulevard connecting the station with the old city, and a new 'city' with administrative and economic functions between the new station and the old center, which would remain the cultural center.

Some of these ideas remained as goals for Stuttgart's postwar planners. In April 1947, a group from the Technical University under Professor Carl Pirath - who had also participated in the wartime planning - prepared a general traffic plan for the city. This was largely adopted by the reconstruction office headed by Walter Hoss and submitted to the city council for approval that summer. This *Verkehrsgerippeplan* still called for moving the train station, though at some future date when more resources would be available. Otherwise, the most important features of the plan were first, the creation of a new, high capacity artery to the northwest of the existing train station (today the Theodor-Heuss Strasse, Friedrichstrasse, Heilbronner Strasse) to relieve traffic on the Konigstrasse and improve traffic flow in front of the train station. The existing bottleneck was the relatively narrow Rote

Figure 28.6 Stuttgart, still-standing façade of Kronprinzenpalais, 1949, demolished for enlargement of the Planie (view from the Planie)
Source: Landesbildstelle Württemberg, Stuttgart, no. F93123

Figure 28.8 Stuttgart, view from the Planie, 1984 (streetcars and most auto traffic now underground, and this part of the Planie is a pedestrian zone)
Source: Landesbildstelle Württemberg, Stuttgart, no. 62857

Figure 28.7 Stuttgart, view from the Planie, 1974, of the buildings that replaced the Kronprinzenpalais (auto traffic flows under the building with the Württembergische Bank sign)
Source: Landesbildstelle Württemberg, Stuttgart, no. K429125

Strasse. Second, the Neckarstrasse on the southeast side of the train station and Schlossgarten was to be improved (today, part of it is named the Konrad-Adenauer-Strasse). Finally, an artery was to be built connecting these two streets by widening the Planie, which ran between the Neues Schloss and Altes Schloss. This would require demolition of the Kronprinzenpalais, a bombed building whose facade still stood and could have been restored. This traffic plan, it should be noted, was approved long before any general town plan was ready for discussion.

There was dissatisfaction in several quarters. The city had imposed a ban on new construction in the inner city while planning was underway, and property owners who wanted to start building became increasingly angry as planning dragged on. Property owners were unhappy also about the amount of land that might be expropriated for street construction. Many citizens protested what they felt was both an excessive modernization of their city and what they saw as arrogance and secrecy in the building administration. Finally, conservationists objected to the demolition of the Kronprinzenpalais. For example, the art historian Hans Fegers, a member of the reconstruction advisory council stated: 'I would consider it irresponsible to neglect today's tasks out of historical considerations and interests.' However, what troubled Fegers was 'the fact, that in many places pure traffic questions are obviously accorded a superior place to architectonic form. I want to say very strongly that the architectonic form of our cities should not only and not even in the first instance be made dependent on considerations of traffic needs.' Even Bonatz, when asked for his opinion, abandoned some of his earlier views and came out in favor of rebuilding the Kronprinzenpalais and using one-way streets to solve the traffic problem.'

Hoss's position was that Stuttgart had to rebuild as a modern city, not one tied to a romanticized past.

> We must think like modern men and seek a vital connection between content and form. . . . I do not believe that we need musty and narrow passages like closed-in Italian cities. We do not want to look into a fairyland around every corner. Our sense of life demands free and open spaces, filled with air and light - which are in any case self-evident demands of modern city planning.

Hoss's colleague Richard Döcker, who had helped start the reconstruction office before accepting a teaching position at the Technical University and who was also a member of the reconstruction advisory council, argued that 'We must use our chance and think ahead and plan for the next 50 to 100 years. Our descendants expect that of us. Should we take a crooked path and waste a clean, clear solution merely to save a few stones of the Kronprinzenpalais?' Hoss and his supporters eventually won out, and the new traffic network was constructed, but the whole debate left behind bitter feelings about reconstruction planning in Stuttgart.

Cologne

As mentioned earlier, the approach to inner-city traffic planning in Cologne differed from that in most other cities, where traffic planning enjoyed the highest priority. Here the greatest emphasis was put on the preservation of the unique qualities of the historic *Altstadt*. This focus on quality meant a focus on *Gestalt*, on architecture and form, more than on functionalism or technical issues. In general, Cologne's reconstruction planning aimed at preserving the historic street pattern - since most historic buildings had been destroyed in the bombing - rather than introducing radical changes.

Rudolf Schwarz, Cologne's chief planner until 1952, clearly rejected understanding the role of city planning as primarily technical. 'City planning,' he said, 'is not a matter of science but of history.'[7] He argued that planners of the Nazi era had done a favor by revealing the fallacy underlying most modern planning. If, for example, all one wanted to do was to construct a major east–west artery through the inner city, then the 72-meter wide axis planned under the Nazis was in fact the way to do it. If one built such a street, then all kinds of traffic would be sucked into it, and 72 meters would not be too wide. Yet most people after the war found the idea of such huge axes abhorrent. 'We must actually thank this mistaken planning for having once proven, in the form of an enormous proposal, that one cannot develop a city like ours out of technical standards and the ideology of technology. That has once and for all led to an absurdity, it won't work.'[8] Technical planning, whether Nazi or modern, naturally led to results that Schwarz found inhuman. Human values, particularly those derived from culture and religion, ought to come first in planning. The individual human had to be the standard, thus setting limits on building size and placing the individual pedestrian before masses of people in mechanized means of transportation.

In the end, Schwarz also had to make concessions to the demands of traffic. His conception of traffic planning was unusual. Even though Cologne in fact already had ring streets dating from the 1920s, he rejected the idea that either a ring system, a radial system, or a tangent system of arteries was appropriate for Cologne. Instead he argued that because traffic followed the flow of the Rhine, and because the old ring streets terminated at the Rhine, what really suited Cologne was an S-shaped *Band* or ribbon system, connecting the old city to newer suburbs to the north and south.

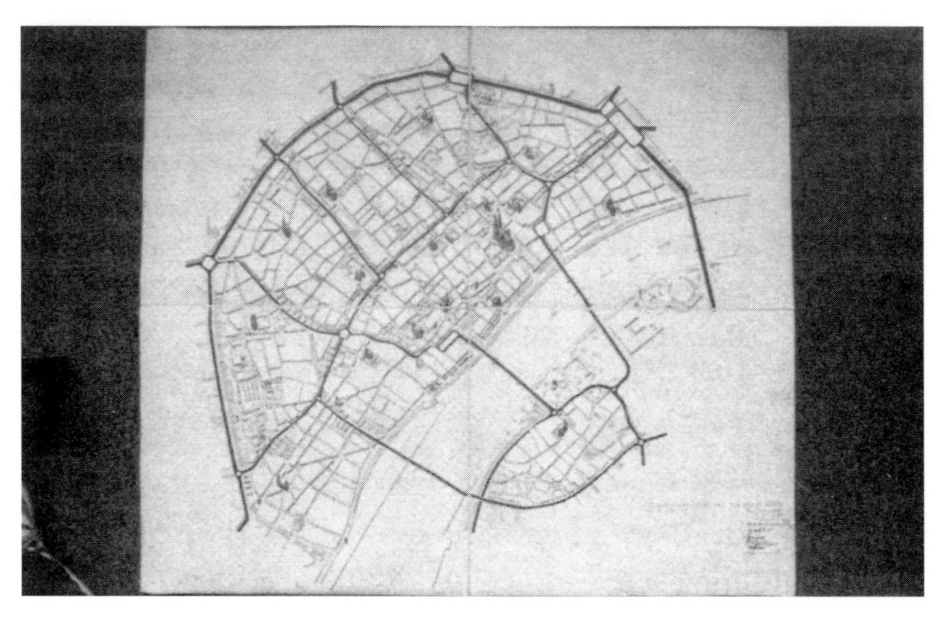

Figure 28.9 Cologne, reconstruction plan of 1948, showing the main arteries to be built through the old city (the plan shows the proposal to move the railroad station.)
Source: Historisches Archiv der Stadt Köln, Plankammer 2/1233

Nevertheless, the east-west artery was retained, though not on quite as large a scale as had been planned by the Nazis. He also planned a north-south artery, a road that came close to a similar artery planned under the Nazis. Eventually completed as today's Nord-Süd-Fahrt, this artery came under heavy fire, and I think it fair to say that Schwarz's lack of enthusiasm for traffic planning delayed both planning and construction of the road.

Conclusion

What were some of the consequences of postwar German traffic planning? There is no doubt that the widening of streets and the creation of new arteries meant a considerable loss of historical substance; damaged buildings that could have been restored were torn down, and the character of others radically changed by the opening up of space around them. Since many of these street alterations proved inadequate and were subsequently abandoned in favor of the pedestrian zones, that loss must be seen as most unfortunate. The debate over new traffic arteries was always also a debate over the retention of the historic street plan, the *Stadtgrundriss,* which, along with existing historic buildings and the historic *Gestalt* (height, roof line, building materials), constituted the essence of the *Altstadt.* As it happens, the small, twisted streets of the historic street plan proved more suitable for pedestrian zones than did the broad traffic arteries that had to be abandoned when they proved incapable of handling the volume of traffic that developed. And even where the *Durchbrüche* still function as main arteries serving the automobile, they have sometimes had a damaging impact on the character of the inner city. Thus, for example, Cologne's Nord-Süd-Fahrt, Hamburg's Ost-West-Strasse, and Stuttgart's Konrad Adenauer Strasse-Neckar Strasse all constitute unpleasant breaks in the continuity of the urban landscape. Finally, the early concentration on technical traffic planning sometimes meant that

Figure 28.10 Cologne, plan of 1964, showing north–south artery through the inner city
Source: Historisches Archiv der Stadt Köln, Plankammer

Figure 28.11 Cologne, north–south artery, 1963, a clear break in urban continuity
Source: Photo Walter Moog, Historisches Archiv der Stadt Koln, ZSB 4/66, no. 05/1679, Freigegeben Reg. Präs. Düsseldorf

architectural form was neglected. Hannover, for instance, has been criticized for undistinguished buildings, however progressive the layout of the streets.

Why were these mistakes made? Some have argued that the concept of modern traffic planning was basically sound, but the refusal of state and national legislators to make possible extensive expropriations of private property made it impossible for planners to change the inner cities in the way they wanted. Moreover, because so much of the underground capital investment in sewers and in water, power, and gas lines was relatively undamaged after the war, responsible planners in Germany's impoverished cities had no choice but to retain most of the old street structure, even though they knew that those streets would prove inadequate to handle future automobile traffic.

These arguments are not very convincing. In fact, city planners usually were able to implement new plans that they thought would work. The difficulties of expropriation and the need to use existing utilities were not decisive, though in some cases they may have kept planners from seeing their plans fully carried out.

More important is the fact that those in charge of overseeing reconstruction grossly underestimated both the speed of German recovery and the growth in the use of motor vehicles. It was common right after the war to speak as if two or three generations would be necessary for reconstruction. Thus planners thought they would have plenty of time to plan and build. Also the market for inner-city land, especially parcels on which stood destroyed buildings, was very poor right after the war, and planners expected that totally destroyed areas would remain empty, which would make them relatively easy to use. These expectations were proved false. By the mid-1950s, however, everyone was talking about new building (Aufbau or Neubau) and not anymore about reconstruction (*Wiederaufbau*).

And of course the use of private motor vehicles grew much, much faster than anticipated. Hannover's Rudolf Hillebrecht, for example, has said that he believed the poverty of the German cities and the German people would mean that decades would be needed before many Germans could afford their own cars. Only at some distant date would Germany perhaps reach American proportions of one auto for every 10 inhabitants.[9] When Friedrich Tamms completed Düsseldorf's reconstruction plans in 1949, he noted that in 1939 there had been one motor vehicle for every 28 residents. He expected that 'in the future,' Germany could expect one vehicle for every 10 residents – an approximate mean between America's satiation and Europe's prewar condition.[10]

The 'future' arrived quickly. Consider these numbers from Stuttgart, where in 1937 there had been one motor vehicle for every 16.2 residents. In 1949, the year in which the city council was debating a major artery through the inner city, motor vehicle registration increased 46 per cent compared to 1948, and there was the equivalent of one vehicle for every 20.7 residents. The following year there were 15.6 residents per vehicle, in 1951, 13 per vehicle, and in 1952, 11.3 residents per vehicle.[11] Obviously these increases outstripped the expectations of the planners. Munich in 1950 had only 16 automatic traffic signals. It is no wonder that in another decade motor vehicles were overwhelming the new arteries that were supposed to handle traffic growth for years to come.

Finally, one can argue that postwar traffic planning, even though seeming quite modern in a technical sense, *overvalued* the role to be played by automobiles in the inner cities. As traffic came to exceed the capacity of the new arteries, the functional, technical nature of postwar planning was called into question, and this happened at the same time that the planners' ideas about the quality of urban life was being challenged by critics like Jane Jacobs and Alexander Mitscherlich. Soon old narrow streets, the cores of building blocks, even some of the larger avenues, were being converted into pedestrian zones from which all vehicular traffic was excluded except that necessary for deliveries to merchants. This process was facilitated by the construction or extension of subway systems to move people to and from the pedestrian zones. The proliferation of pedestrian zones reflected both the failure of postwar traffic planning and a turn toward a more architectonic concept of the inner cities.

Notes

1 Herbert Jensen, 'Der Wettlauf zwischen Verkehr und Städtebau,' in *Deutsche Akademie für Städtebau und Landesplanung: Jahresversammlung*, Munich, 1953, 16–18. Jensen was partly responsible for Munich's famous pedestrian zone of the 1960s.

2 Barbara Miller Lane, *Architecture and Politics in Germany, 1918–1945*, Cambridge, 1968; Robert R. Taylor, *The Word in Stone. The Role of Architecture in the National Socialist Ideology*, Berkeley, 1974.

3 Most architects were technocrats, not ideologues, and technical planning was an opportunity to work while avoiding moral, value-laden questions.

4 [. . .] Reichow's book was titled *Die autogerechte Stadt. Ein Weg aus dem Verkehrs-Chaos*, Ravensburg, 1959.

5 It should also be noted that urban planning and reconstruction in postwar Germany was the responsibility of

the towns. There was no national government until 1949 and no national building law until 1960. The role of the federal government was thus very small. Likewise, the influence of the occupation governments on traffic planning was also minimal, though of course the Germans were well aware of the rapid growth of automobile use in the United States.

6 Otto Sill, 'Der Verkehr,' *Hamburg und seine Bauten 1929-1953*, Hamburg, 1953, 246.

7 Schwarz, 'Gedanken über die kunftige Grossstadt,' *Der Tagesspiegel*, 4 April 1947.

8 *Ibid.*, p. 13.

9 Interview with author in Hannover, 25 June 1983.

10 Tamms, 'Planungsaufgaben in Düsseldorf,' *Stadtplanung Düsseldorf. Ausstellung Ehrenhof*, Düsseldorf, 1949, 25.

11 Kurt Leipner, *Chronik der Stadt Stuttgart, 1949-1953*, Stuttgart, 1977, 167-171.

Part 3
URBAN TECHNOLOGY TRANSFER

This final section consists of texts that illustrate the transfer of technology from the developed West to the East. Reading 29 reveals why new forms of urban transport were so slow to come to St Petersburg. There the city council presented decades of resistance to late-nineteenth-century proposals for improving urban transport, first by a network of horse-trams, and later by steam-railways. Why the council eventually changed its attitude is here made clear, as are the reasons for the delays in electrification. The consequences of the new transport for the rate of suburbanization and distribution of social classes in St Petersburg are also discussed.

Local opposition was also effective in preventing the building of Moscow's Metro for over 30 years. When it finally opened in 1935, there was official reluctance to recognize any influence of the capitalist West's technological know-how, a view that still prevailed in Moscow in the 1990s. What was the truth behind the propaganda? Reading 30 uncovers new evidence and comes to a clear conclusion.

Reinforced concrete and techniques of prefabrication, previously used in the West, were adopted in Soviet Russia from the 1950s for a rapid solution to the country's dreadful urban housing conditions. Reading 31 shows how Western technology brought very different results under the influence of the Soviet regime. The text assesses the achievements and failures of the greatest housing programme ever implemented.

Moving further east, to British India, reading 32 again finds local resistance to Western technology, this time in the shape of sanitary reform. Focusing on the holy city of Allahabad, this text compares the Victorian public health measures introduced in metropolitan Britain and colonial India. Outstanding here is the difference in motivation: sanitary improvements in India were driven by the priority placed on the health of that foundation of imperial power, the Indian Army. The changes in Allahabad, an army headquarters, are analysed, considering the influence of contemporary legislation in Britain relating to water supply and overcrowding in cities.

29

THE DEVELOPMENT OF PUBLIC TRANSPORTATION IN ST PETERSBURG, 1860–1914

James H. Bater

Source: *Journal of Transport History*, New Series, 2 (1973), pp. 85–102

In St Petersburg, 'everything is new but the pavement and the droskies [sic]'.[1] In describing the *droshky,* the one-horse, two-passenger cab which was the principal mode of public conveyance at mid-nineteenth century, the Englishman Laurence Oliphant was comparatively charitable: 'if locomotion by the latter be not enjoyable it has the merit of being, in the first place, singular, from the manner in which the passenger seats himself across a cushion behind the driver; and, secondly, exciting, from the extreme difficulty he finds retaining his seat there, which is considerably increased when a wheel comes off – an incident of not unfrequent occurrence.'[2] On this topic, foreign travellers frequently expressed more caustic opinions.[3] The ensuing decades witnessed a number of important changes in public transportation facilities; however, even as late as the First World War many features remained 'singular'. To be sure, the street tram had been added to the sundry types of vehicles which crowded mid-nineteenth-century St Petersburg thoroughfares, but it remained horse-drawn long after electric traction had been introduced in cities of comparable size. Railways serviced the peripheral regions of the city and the surrounding settlements, but daily commuting to the city was extremely limited. Even the motor car was conspicuous by its absence,[4] and one observer in 1910 ranked St Petersburg last among major European cities to be 'invaded' by the automobile.[5] Vehicular transportation in the early 1900s remained very much the province of conveyances 'of the most non-descript and antedeluvian kind.'[6]

Yet this was not the state of affairs in a small provincial town. In 1852, St Petersburg, capital of the Russian Empire, was still primarily a court-administrative centre, albeit one with more than 550,000 inhabitants. Subsequent events, many of which were precipitated by the emancipation of the serfs in 1861, metamorphosed the city. At the turn of the century, in addition to a vastly proliferated government bureaucracy, St Petersburg was host to industrial and commercial activities of no mean proportion by international, not just national, standards. A voluminous immigration, especially following the depression of the 1880s, substantially increased the population so that by 1914 it had reached 2.2 million.

The inauguration of horse tram service in North American and European cities after the middle of the last century, the subsequent technological improvements and their rapid, widespread adoption, particularly electric traction in the 1890s, certainly improved the mobility of sizeable numbers of urban dwellers. And increased mobility, as reflected for instance in the lengthening of journeys to work, was an important factor in the transformation of the fabric of nineteenth-century cities. In this regard, the extension of electric or steam railway service beyond the confines of the densely settled city centre into the suburbs did much to effect changes in urban morphology and socio-economic structure. While there is a growing body of literature concerned with the role of public transportation in the metamorphosis of nineteenth-century cities, for St Petersburg, indeed for Russian cities in general, the relationship between public transportation and mobility is an important topic which has not before been investigated. [. . .]

Public transportation in St Petersburg

The economic and social changes unleashed by the emancipation of 1861 had a perceptible impact on St Petersburg even during the 1860s. Data do not permit a precise calculation, but approximately 175,000 of the 720,000 total population in 1869 had been added since 1861. Many new factories had appeared giving employment to thousands, and expanded commercial activities, mirrored by an increase in port traffic and the numerous new financial, wholesale and retail establishments, required the services of even greater numbers (to say nothing of the burgeoning bureaucracy). To accommodate the influx of people, housing construction was augmented, but less rapidly than minimum sanitary and space requirements would have dictated, and thus, the shortage of working-class housing, so critical a problem after 1890, was already in evidence. During the 1860s the combination of long working hours and limited disposable income obliged the working classes to seek accommodation proximate to the place of work. Since the opportunities for employment were still greatest in the central part of the city, it was there, rather than to the suburbs, that the flow of migrants was directed. In this regard, A. Kopanev's description of a spatially intermingled nobility, bourgeoisie, merchants, traders, handicraftsmen and working classes during the second quarter of the nineteenth century still had much validity when applied to St Petersburg in the 1860s, although W. Blackwell suggests that Kopanev overestimated 'the proximity of the various classes'.[7] The increasing population density, expanded economic activity and resultant considerable congestion had by the early 1860s prodded the City Council into exploring ways and means of facilitating movement within the city. Inasmuch as the sundry conveyances available for personal hire, as well as the omnibus, were beyond the means of the bulk of the city's population for use on a regular basis, the intra-urban mobility of the general public was limited. One possibility for improving this situation was the horse tram, already operating in many European and North American cities.

The emergence of the street tram system

The development of a horse tram system in St Petersburg was comparatively retarded. The first line, on Vasil'yevskiy Island, did not begin operation until the summer of 1860 and was primarily intended to expedite the movement of goods 'from pier to customs storehouse'.[8] In 1862 the Nevskaya, Sadovaya, & Vasileostrovskaya Horse Railroad Company was granted permission by the City Council to construct and operate three horse tram-lines in the city.[9] A year later a tram-line was operating along part of the principal thoroughfare, the Nevskiy Prospect, and shortly thereafter, the system had been extended to include a line on Sadovaya Street and one on Konno-Gvardeyskiy Boulevard, which was linked with the original line on Vasil'yevskiy Island (Figures 29.1 and 29.2). While less than 10 kilometres in total length, the existing network did connect the port and customs complex located at the south-eastern extremity of Vasil'yevskiy Island, and the Nikolayevskiy railway station on the Nevskiy

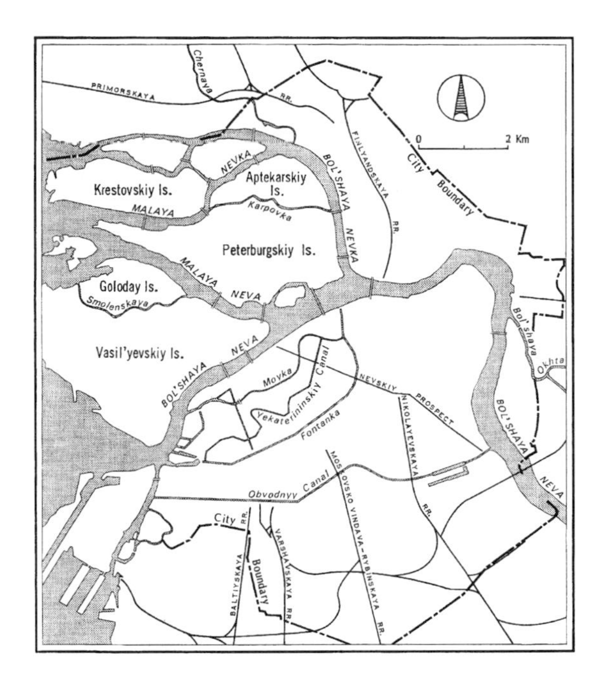

Figure 29.1 St Petersburg, 1914

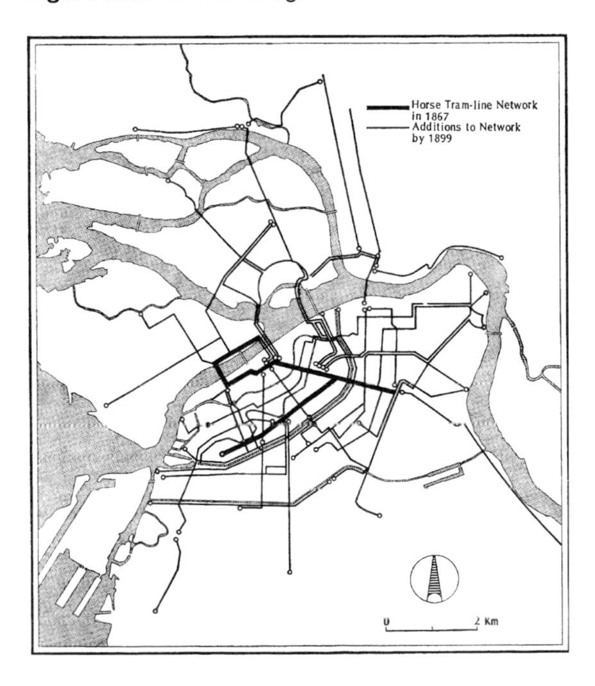

Figure 29.2 Horse-tramway system, 1867 and 1899

Prospect. As the Nikolayevskaya railway provided the only rail link with Moscow and the interior provinces, its importance as an artery of commerce was enormous. During the 1860s for example, it accounted for more than 80 per cent of all traffic into and out of the city.

By the late 1860s, the policies of the company operating the tram system had changed even if the system itself had not. One of the original intentions behind the company's application to the City Council to build and operate a tramway network, perhaps even the principal one, had been that of providing a goods haulage service through the city from railway station to port. While granted permission to operate the trams on city streets, the company had to comply with several restrictions imposed by the City Council. The most important restriction was that goods haulage had to take place between 11 p.m. and 9 a.m., ostensibly 'to minimize the daytime congestion'.[10] Night-time use of the service was far less than originally contemplated. In the first full year of operation, 1864, only 1.5 million tons of goods were hauled, whereas two years earlier the total tonnage of goods traffic through the Nikolayevskiy railway station exceeded 18 million. Since application to the City Council to have the restriction lifted was unsuccessful, two alternative courses of action were adopted. In the first place, more frequent service was provided on the existing network in order to generate increased passenger traffic. Secondly, submissions were made to the City Council during the period 1867 to 1868 on behalf of the company for authorization to extend the tram system. The discussion which this started is revealing of the conservative attitude of many members of the City Council to the horse tram.

The first application, in 1867, sought permission to extend the network into the peripheral regions of the city. The Council's initial reaction was negative, and the need to expand the system in the central city first was noted. Nonetheless, a committee was set up to examine the proposal further and it reported back to the Council in late 1868. It also recommended that the proposal be rejected, and in so doing, put forward a number of arguments against horse tram networks which, directly or indirectly, were related to the financial well-being of the city. Foremost in the argument against extension of the service was the potentially adverse effect on existing modes of intra-urban horse-drawn public transportation. Since thousands of horses were involved, each of which was subject to a tax of 10 roubles per annum levied by the city (in addition to the licence fees paid by the *izvochiki,* or drivers), a rapid diminution in their number owing to competition from the horse tram would reduce city revenue. Moreover, the horse tram with its smaller operations, service and maintenance work-force requirements could seriously jeopardize the many thousands of jobs created by the transportation industry as it then existed. The committee also suggested that the rails not only used valuable land, but were a hazard to other vehicles, that the horse tram was less flexible, and in winter less effective than the customary horse-drawn sleigh, and that further extension of the system would simply serve to inconvenience the population. For good measure, the committee also recommended that the existing horse tram network should be closed for both passenger and goods movement from 1 December to 1 April.

Despite a detailed rebuttal of the foregoing arguments by the Nevskaya, Sadovaya, & Vasileostrovskaya Horse Railroad Company, the negative view of the Council prevailed, although the committee's recommendation for complete closure during the winter months was not endorsed. The perceived threat to the economic well-being of this sector of the city's economy cannot be assumed to have had general currency outside the City Council. The reason for this was that the Council was scarcely a body representative of the city's population. During most of the 1860 to 1914 period its electorate comprised less than 3 per cent of the population of St Petersburg. But whether representative or not, it remained the considered judgment of the body with decision-making responsibility in this matter. Aside from improvements in the service on the existing network, major development in public transportation was stopped for nearly six years.

In 1874 a newly formed horse railway company obtained from the City Council authorization to expand the horse tram-line network in the central city. Within four years, 90 kilometres of track had been added and in the process established the horse tram as a major element in public transportation. The success of this company and the gradual acceptance of the horse tram by successive City Councils resulted in a modest articulation of the system during the next two decades. The only other company authorized to operate a horse tramline in the city was the Nevskaya Prigorodnaya Railroad Company, which, in 1878, began servicing the industrial district along the southeastern reaches of the Bol'shaya Neva River (Figures 29.1 and 29.2).

Concurrent with the articulation of the network, from 10 kilometres in 1865 to 133 in 1898, was a substantial rise in passenger traffic, from less than 2 million to over 85 million passengers for these dates respectively. As might be expected, most of this traffic was concentrated on the main routes in the central city. From Figure 29.3, it is evident that the Liteynaya line had the greatest volume of traffic in 1899, 10.5 million passengers, with the Nevskaya, 7.8 million, and the Sadovaya, 6 million, following close behind. Traffic on the lines to the south of the Obvodnyy Canal was particularly light, despite rapid urban-industrial growth in this region.

The sizeable earnings of the horse tram companies and the comparatively little financial benefit to the city from the privately owned system were significant factors in transforming City Council attitudes and policies. In 1893,

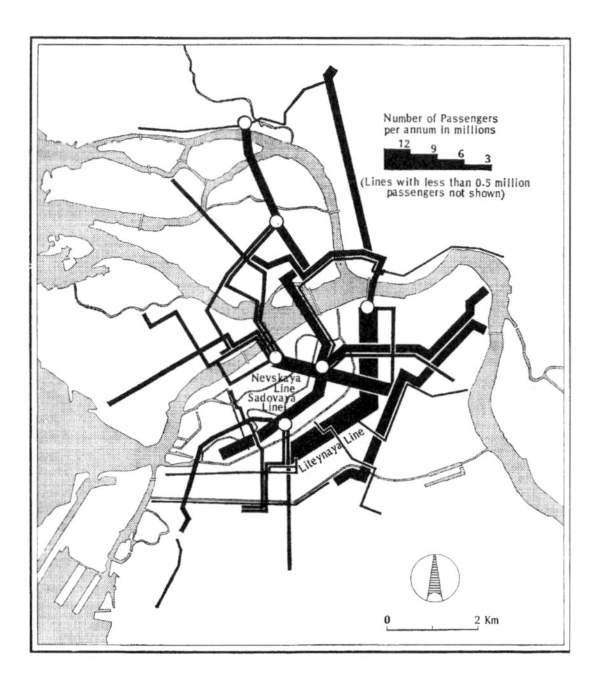

Figure 29.3 Annual horse-drawn tram traffic, 1899

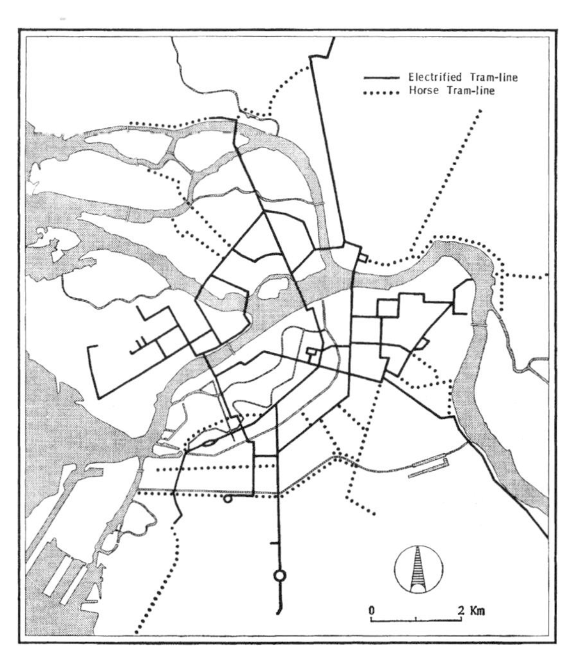

Figure 29.4 Tram system, 1914

gross receipts for the three private companies were 2.8 million roubles, 86,000 of which found their way into city coffers. Owing primarily to the rising volume of traffic, receipts over the next few years climbed steadily. Traffic expanded from 57 million in 1893 to 85 million in 1898; gross receipts during the same period rose from 2.8 to 4.2 million roubles. Thus, when the concession granted the Nevskaya, Sadovaya, & Vasileostrovskaya Horse Railroad Company expired in 1898 the City Council did not renew it and the facilities were taken over by the city. Given the substantial increase in gross receipts and the customary high profit levels, it is not surprising that attempts by the city to gain control of the principal company operating in the city were stalled by litigation for several years. However, by 1902 the two principal horse tram-lines had been acquired and plans for vastly extending the system were drawn up. Total track was to be increased 150 per cent, new trams were to be added and service improved, and electric traction was to be introduced. As in the case of virtually all municipal services, necessary expansion was constrained by a shortage of capital. The realized expansion in track was scarcely a quarter of that planned, moderate additions were made to the rolling stock, and electrification was introduced, but not until 1907, a rather late date even by comparison with other Russian cities. Recourse to the bond market enabled the city to raise 9.5 million roubles to further electrification and by 1914 the principal lines were electrified. But as Figure 29.4 reveals, many of the suburban lines at this date were still serviced by horse-drawn trams.

Once started, the conversion to electric-locomotion was expeditious and its impact on trip generation perceptible. Traffic in 1909 had reached 193 million compared with 85 million passengers in 1898. However, in 1914 the system carried over 300 million persons, 90 per cent of them in electric trams. The impressive growth between 1909 and 1914 in tram traffic is the result of a complex of socio-economic, as well as technological, changes, but the improved service owing to the increase in rolling stock and somewhat greater average speed of the electric tram was undoubtedly important. The corollary of increasing passenger traffic was ever greater gross receipts. These exceeded 17 million roubles in 1914 and available information suggests that net profit to the city was over 5 million roubles.

The part played by the City Council in the development of the street tram system in St Petersburg of course is very easy to misinterpret when the principal source for assessment is published minutes. Inevitably, vested interests and 'back-room' negotiations influenced developments and in this respect, it is unlikely that the extremely restricted electoral base of Council was a wholly positive element in municipal affairs. On the evidence available, it would appear that the actions of the City Council with respect to street tram matters fall into three general periods. Up to the early 1870s, policies adopted reflected considerable reservation on the part of the Council as to the benefits of a horse tram network. The authorization granted a private company in 1874 to expand the system in the city signalled a modification in attitude and policies. From this time on, the need to expedite intra-urban transit and the potential of the horse tram in this regard were gradually acknowledged by municipal authorities and limited expansion of the system was accommodated. After 1898 the city became directly involved in the business of mass public transit, and by 1914 controlled all but a small part of the total tram system.

Competing modes of public transit

Although it carried by far the greatest volume of traffic by 1914, the street tram of course was not the only facility available to the public. Within the city, hordes of *droshkies* and similar conveyances crowded the thoroughfares, passenger craft of various types plied the canals and rivers during the ice-free season, and for the few who commuted regularly from the outlying regions the steam railways offered a limited service.

A large-scale daily migration via the railway to St Petersburg from the outlying towns and villages was earnestly desired by those who saw in such a movement a possible cure for the 'eternal scab of our city', as the housing shortage was metaphorically described in 1910.[11] Such a pattern of commuting, however, simply did not evolve in the pre-war period. Moreover, annual traffic (Figure 29.5) showed a marked seasonal variation, which the data presented in Table 1 for 1896 typify.

This seasonal variation was a consequence of the traditional exodus to the *dacha* (summer house) during the summer months by the wealthier classes. For the mass of the population suffering from the perennial crisis in housing,

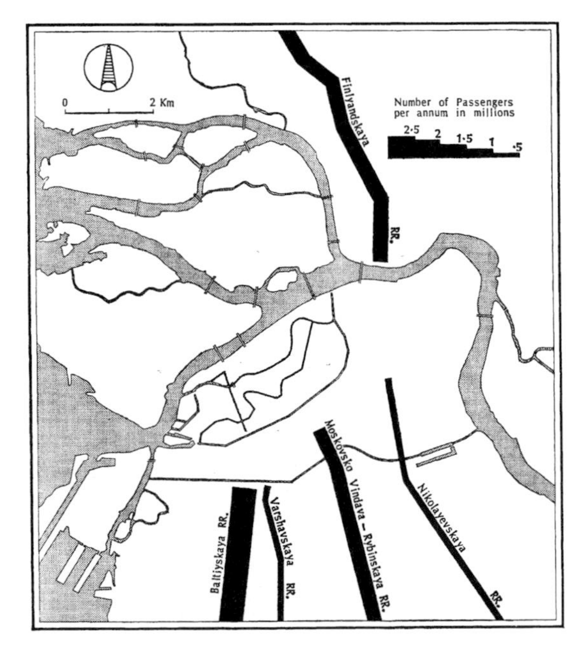

Figure 29.5 Annual suburban passenger-railway traffic, 1896

daily commuting was out of the question if only because of the length of the workday, the low remuneration and the continuous erosion of purchasing power, reasons little changed from the 1860s.

The potential of the railway in assisting intra-urban communication was also recognized, and various proposals for constructing an integrated steam railway network were put to the City Council during the period from 1880 to 1914. One of the most probable schemes, and one that on several occasions was discussed by the Council, involved covering the Yekatermininskiy Canal (Figure 29.1), laying a rail line along its route, and linking this line with a ring rail line extending from the southern reaches of the city to the north-east. In the early 1890s promoters could legitimately point out that passenger traffic along the canal was rapidly declining and that goods traffic was unimportant, comprising primarily firewood and construction materials. They argued that the advantages of using the canal in the manner proposed were overwhelming, since as a route for a railroad it would 'penetrate into the centre of the city', as well as obviate the need for clearing a right of way with its attendant economic and social costs.[12] The rail line was viewed as greatly assisting goods movement within the city as well as permitting direct, inexpensive and frequent commuter service. Of existing railways only the Nikolayevskaya and Moskovsko Vindava-Rybinskaya penetrated very far into the central city and neither accounted for a sizeable share of the existing commuter traffic (Figure 29.5). This proposal, along with many others having similar intent, was rejected by the Council on each occasion it was put forward. Even the critical problem of providing a ring rail line to link the north-eastern regions of the city with the southern was not resolved until 1913. After at least two decades of complaint by industrialists to all levels of government, a group of entrepreneurs in the metal-working industry finally despaired of any financial assistance and gaining City Council approval for construction of a ring railway that would skirt the built-up area, raised the necessary capital, 62 million roubles, themselves. In 1914, the long-discussed ring railway, intended to facilitate goods movement, was completed. For rail commuters, the facilities remained basically unchanged.

The rapid upsurge in tram traffic after the mid-1870s had the predicted impact on the traditional modes of public transit. Water-borne traffic was severely affected, especially on the canals within the central city. In 1882 for example, a concession was granted by the city for steamboat passenger service along the Yekatermininskiy Canal. But the steamboat could not compete with the horse-drawn tram and was dropped with the expiration of the ten-year concession. Although service was later re-established along sections of the Yekaterininskiy Canal, and continued on the Fontanka and the Moyka canals, most of the total water-borne traffic in 1898, 8.3 million passengers,

Table 29.1 Railway passenger traffic – suburban region,* 1896

Month	*Incoming*	*Outgoing*
January	96,693	98,199
February	89,217	92,773
March	115,495	120,456
April	132,969	139,520
May	253,181	274,776
June	301,965	313,299
July	315,338	308,686
August	287,366	248,061
September	162,831	145,018
October	121,978	118,786
November	113,671	129,873
December	125,980	129,393
Total	2,116,684	2,118,840

Note: * These data are for the Nikolayevskaya, Varshavskaya, and Baltiyskaya railroads only (see Figure 29.5)
Source: *Entsiklopedicheskiy Slovar'* (St Petersburg: Brokgauz-Yefron, 1900), vol. XXVIII, 329

was accounted for by journeys *across*, rather than along the waterways. The reason for this situation was the relative shortage of bridges, which had been a long-standing obstacle to both vehicular and pedestrian intra-urban movement (Figure 29.1). The scores of small vessels which provided a means of crossing the water-ways, particularly the Bol'shaya Neva and its tributaries, continued to be an essential component in the total transportation system despite a steadily declining share of traffic. As a result of this service, more direct journeys were possible.

With the advent of the electric streetcar, a diminution in number of horse-drawn conveyances for personal hire was inevitable, and between 1909 and 1911 alone approximately 2,600 departed permanently from the streets of St Petersburg. This sizeable attrition notwithstanding, some 14,000 still remained in operation, and these vehicles remained an integral part of the urban milieu until 1914.

Public transportation and mobility

Although the rapid increase in tram passenger traffic during the period from 1860 to 1914 has been established, before concluding it is essential that the question of intra-urban mobility be considered. Unlike many other cities of comparable or smaller size, the 'city fathers' of St Petersburg did not welcome the horse tram during the 1860s with unbridled enthusiasm. As noted above, the one company which did acquire a concession from the City Council entitling it to operate three horse tram-lines had no success in its bid to gain approval for extension of the system. The extent to which the horse car in the mid-1860s served the general public is clearly revealed when it is realized that total traffic in 1865 averaged scarcely three trips *per annum* per urban inhabitant. On the same basis, horse tram traffic in New York in 1864 was the equivalent of 87 trips per person. In 1893, the average number of horse tram journeys *per capita* had reached 55, still much below the figure of 111 for Toronto in 1890. Owing to the phenomenal rate of population growth after 1890 the average for 1905 had scarcely altered and at 58 trips *per annum* per person St Petersburg compared unfavourably with Berlin where there were 119 journeys. Even in 1913, several years after the introduction of electric traction, the capital lagged behind Moscow, 150 as compared to 165. Moscow, although also boasting of an electric tram system, could not be considered a forerunner in the provision of urban transportation facilities.

On the basis of the data presented, tram usage in St Petersburg would seem to have been comparatively low during the 1860–1914 period. Moreover, when it is recalled that regular commuting from the suburban regions via the railroads was restricted to the economically privileged classes, and that there was no intra-urban railway service, the question of the level of mobility within St Petersburg assumes even greater significance. In the preceding discussion the length of workday, low remuneration and steady erosion of purchasing power owing to inflation were suggested as possible reasons for the seemingly low level of usage of public transportation facilities. At this point it might be useful to put into rough perspective the cost of public transportation.

On the eve of the First World War, the official tariff for *droshkies,* the traditional one-horse, two-passenger cabs, was 20 kopeks for the first quarter of an hour, and 5 kopeks for each additional 5 minutes up to three-quarters of an hour. Night-time fares, that is from 12 a.m. until 7 a.m., started at 30 kopeks for a quarter of an hour and increased similarly. For a trip of an hour's duration day-time and night-time fares were 60 and 90 kopeks respectively. If it is assumed that the official minimum speed of 10 kilometres per hour was maintained, then the minimum fare would entitle the passenger to a journey of $2\frac{1}{2}$ kilometres in distance. Given the congestion which was common to the thoroughfares of the central city, this assumption is certainly open to question. In fact, 60 kopeks is suggested as the appropriate fare for a trip of just this distance along the Nevskiy Prospect, the busiest of all the city's avenues. To traverse the city would involve a charge of considerably more than a rouble (100 kopeks). Hiring any of the other types of horse-drawn vehicles, such as the *likhach,* the *kareta,* or in the winter the *troyka,* was substantially more costly. At 20 kopeks per kilometre, a motorized taxi cab was a no less expensive proposition. By comparison with vehicles hired for personal use, the fares on the tram were quite cheap. Depending on the seat, whether 'imperial' (roof-top) and exposed, or inside the car, first stage horse tram fares at 3 or 4 kopeks had scarcely changed from the time horse cars were introduced. At 5 kopeks, first stage fares on the electric streetcar were competitive. A journey by electric tram across the city, for example from the Finlandskiy to the Baltiyskiy railroad station, a distance of approximately 10 kilometres, was a multi-stage trip and would cost 15 kopeks. On the few remaining canal steamboats and omnibuses, first stage fares were usually 4 kopeks. Of all modes of intra-urban public transportation, the lowest fares were for the shuttle ferry service across the major arteries of the Bol'shaya Neva system. During the day, 2 kopeks, and at night, 3 kopeks, were charged for these extremely short, convenience journeys.

In spite of the fact that the wages in St Petersburg were above the national average, the inordinately high cost of living meant that for most citizens there was little surplus after expenditures on basic necessities. Budget studies of the highest paid of factory hands, the metal workers, indicate that salary increases during the early 1900s scarcely kept abreast of rising general price indexes in the city. In view of this situation, it is perhaps not surprising that one budget study carried out in St Petersburg revealed that as late as 1908 expenditures on streetcar fares were regarded as a luxury (along with bath-houses and postage) and amounted to 14 kopeks in one month. At prevailing fares this represented at most four single stage trips on the horse tram.

In many cities innovations in transportation technology combined with social legislation facilitated the intra-urban mobility of the working classes. The early introduction of the electric streetcar, the initiation of suburban services, either electric or steam, at times and at fares geared to the needs of the working class were important in this regard even if they were not always entirely successful in meeting the objectives initially set. In St Petersburg little in this respect was done to accommodate the working classes. As late as 1904, the poor service provided by the horse tram was such that a much-needed low-rent housing development on Goloday Island (Figure 29.1) was deemed a failure because it was effectively too remote. The numerous schemes which sought to alleviate the very serious housing shortage by building low-cost, working-class accommodation on the periphery of the city, and in the outlaying suburbs and towns, all assumed daily commuting via workmen's trains comparable to those operating in other cities of Europe and North America. The provision of such service did not materialize and in 1914 press reports still continued to stress the long-standing urgency of encouraging and facilitating the process of suburbanization. Even though by 1914 more than 90 per cent of all tram traffic was accounted for by the electric street-car, and 113 out of 190 kilometres of the total track were electrified, service on these lines did not begin until 8 a.m. and therefore was not especially suited to the needs of the working class. Moreover, with the advent of electric traction, a number of horse car routes were discontinued, particularly in the central city (see Figures 29.2 and 29.4).

With regard to the journey to work, and prevailing perceptions of distance within the city, an article in the *Torgova-Promyshlennaya Gazeta* of 1912 is instructive. At this late date it could be stated with an air of surprise that in the workers' district of Gavan, on the western extremity of Vasil'yevskiy Island, were to be found employees 'from distant [*otdalennykh*] factories, . . . the Laferm [tobacco] factory on Tenth Line, the Bekker [musical instrument] factory on Eighth Line . . . and even one [factory] close to the Tuchkov bridge' (leading to Peterburgskiy Island near the Malyy Prospect).[13] In the latter instance, the plant was still less than 3 kilometres straight-line distance away and would involve less than a 30-minute walk!

Summary

The period from 1860 to 1914 witnessed the adoption of many of the major technological improvements in public transportation facilities found in the larger centres of North America and Europe. In the adoption of innovations however, St Petersburg cannot be counted as one of the fore-runners. Indeed, change in the public transportation system was slow to be realized as the late introduction of electric traction indicates. It was remarked earlier in the paper, though not substantiated, that the need to be close to the place of work during the 1860s owing to limited mobility was in part responsible for the spatially heterogeneous distribution of social classes. The concluding general comments on mobility suggest that the role of public transportation in the journey to work of a large segment of St Petersburg's population was minor throughout the 50-odd years under review. This is an important issue and requires detailed investigation. Only by establishing, even if roughly, differences in levels of mobility amongst the various groups comprising the city's population, will the patterns of economic and social activity at the turn of the century be comprehensible, and hence the process of urban growth be better understood.

Notes

1 Laurence Oliphant, *The Russian Shores of the Black Sea* (2nd edn), Edinburgh, 1853, 2.

2 *Ibid.*, 2–3.

3 See, for example, Richard S. Bourke, *St. Petersburg and Moscow*, New York, 1970; reprinted from original edition of 1846, 178–9.

4 In 1913 there were only 2,585 motor vehicles. N. N. Petrov, 'Gorodskoye Upravleniye i Gorodskoye Khozyaystvo Peterburga', in *Ocherki Istorii Leningrada*, Moscow-Leningrad, 1957, vol. III, 910.

5 G. Dobson, *St Petersburg* (1910), 119.

6 *Ibid.*, 131.

7 William L. Blackwell, *The Beginnings of Russian Industrialization 1800–1860*, Princeton, 1968.

8 'Obozreniye Promyshlennosti i Torgovli v Rossii', *Vestnik Promyshlennosti*, No. 10 (1860), 2.

9 'Konno-Zheleznaya Doroga', *Izvestiya S. Peterburgskoy Gorodskoy Obshchey Dumy*, No. 5 (1867), 234 (hereafter cited as *I.S.P.G.O.D.*).

10 'Po Khodataystvu Tovarishchestva Konno-Zheleznykh Dorog o Razreshenii Onomu Prolozhit' Neskol'ko Novykh Liniy', *I.S.P.G.O.D.*, No. 15 (1868), 795.

11 'Oraniyenbaumskaya Elektricheskaya Zhel. Dor.,' *Ekonomist Rossii*, No. 5 (1910), 8–9.

12 See 'Po Proyektu Tsentral'noy Stantsii Dlya Zagorodnykh Zheleznykh-Dorog i Gorodskoy Opoyasyvayushchey Zheleznoy Dorogi', *I.S.P.G.O.D.*, No. 21 (1893), 8.

13 B. Kalinskiy, 'Bor'ba s Zhilishchnoy Nuzhdoy v S. Peterburge', *Torgova-Promyshlennaya Gazeta*, N. 283 (1912), 1.

30

THE MOSCOW METRO: PRODUCT OF INDIGENOUS TECHNOLOGY OR WESTERN INFLUENCE?

Michael Robbins

Source: Michael Robbins, 'London Underground and Moscow Metro', *Journal of Transport History,* 18 (1997), pp. 45–53

When the first section of the Moscow Metro was officially opened on 15 May 1935 it attracted much attention in the world's press. It seemed astonishing to some that there should be an urban underground railway in the Soviet Union at all; to others it came as a vindication of communist ideals. Pictures of the ornate architectural features of the stations were widely reproduced, and there was much congratulation. It was not then known how much technical bumbling lay behind the glittering facades. The project was originally supposed by the Soviet authorities to be a matter of straightforward mining technique, for which coal-mining experts from the Don basin would supply all the necessary skills, but it was briefly acknowledged that the experience of Berlin, Paris and London had contributed to decisions taken during the design process. By the opening date Berlin had disappeared from the list; New York, cited once, does not seem to have taken any active part.[1] Increasingly the impression was conveyed by the Soviets that socialist technology had solved the problems independently of any foreign capitalist ideas; and this is still the prevalent view in Moscow.

Information can now be put together from sources in documents held in Moscow and London which makes it possible, in spite of tantalising gaps in the evidence that remains, to come to a probable conclusion as to the contribution made by foreign sources. The conclusion seems inescapable that the London Underground made far and away the largest contribution to the designs and operating practices (though not the architectural treatment) that were adopted in Moscow.

I

This article attempts to provide neither a history nor a description of the first Moscow Metro, but a short historical introduction is called for, to provide a framework for the story of its 'foreign relations'.

A proposal for a Moscow city underground railway was put to the municipality by Balinsky, an engineer, in 1902. Opposition from the Archbishop of Moscow and the Russian Imperial Archaeological Society, based on the probable danger to historic buildings created by tunnelling work (presumably at shallow level) and vibration from the running of trains, appears to have blocked the scheme.

By the end of the 1920s conditions for passenger movement within the city and its fast-growing suburbs had become bad enough to cause the Moscow soviet and the supreme central government authorities to consider building an underground railway. From 1932 N. S. Khrushchev of the Moscow soviet was closely associated with the work, under L. M. Kaganovitch, People's Commissar of Transport, and the account in the following paragraph is based on his published memoirs. Whatever the authenticity of the political and personal parts of his book - and there was much comment, not to say controversy, outside the USSR after its publication - there seems no reason to doubt the essential correctness of the statements made there about the Metro scheme.[2]

'When we started building the Moscow Metro,' Khrushchev wrote, 'we had only the vaguest idea of what the job would entail.' P. P. Rotert (also spelt Rotter, Roter or Rottert), construction engineer, was put in charge, but things did not go well, and Kaganovitch told Khrushchev to take charge, on the ground that he had some experience of mining. Y. T. Abakumov, head of the coal works in the Don basin, was made deputy to Rotert, and Khrushchev, while remaining secretary of the Moscow soviet, says that he spent 80 per cent of his time on the Metro job, going down the shafts and through the completed tunnels every day. A young engineer, V. I. Makovsky, put forward a startling proposal - to abandon the 'cut and cover' (so-called 'German') method and to excavate deep-level tunnels on the 'English' method. It would be more expensive, but the tunnels would make excellent bomb shelters, and the alignment need not follow the lines of the main streets. It would mean using escalators, not lifts (which Rotert was proposing). Khrushchev had to be told the meaning of 'escalator', a word he had not heard before. Discussion, citing the example of Piccadilly Circus station in London, rebuilt and equipped with escalators in 1928 (which Makovsky had visited), went right up to the Politburo, where Stalin cut short the argument about cost and ordered deep-level tunnels with escalator access. Khrushchev implies that the civil defence factor justified the extra expense in Stalin's mind, and it is hinted that if Piccadilly Circus, 'the best station in London, right in the heart of the most aristocratic section of the city', had escalators, then Moscow would show that it could match it. No date is given for this meeting, but it presumably took place in early 1932. It is clearly implied that technical information from abroad was available, and here there are useful indications from British sources.

II

The course of political relations between Britain and the Soviet Union during the 1920s was, to say the least, uneven. The first Labour government had recognised the Soviets in 1924, but diplomatic relations were broken off in 1927, to be resumed in 1929. Trade relations, on the other hand, had been continuous since 1921. British firms like Metropolitan Vickers competed successfully with German firms for contracts to equip Soviet power stations, and the government gave export credit guarantees in approved cases. The British embassy in Moscow regularly reported on developments in this field, noting in 1930 that Ralph Budd, of the (US) Great Northern Railway had been invited to act as adviser to the Soviet government on 'the rebuilding of its railway system on American lines and standards', and it was minuted that Japanese engineers had also been invited to advise on railways. Budd was in Moscow in the summer of 1930 and advised on the Moscow Junction Railway, but there is no suggestion that he took part in any discussions on the Metro. Main-line railways were then considered to be a quite different kind of thing.

In June 1931 Sir Esmond Ovey, the British ambassador, in the course of a report on housing and town planning in the Moscow district, noted that among the recommendations of a planning report was the construction of an underground railway, to be begun in 1932, as well as peripheral (main-line) connections.[3]

On 18 August William Strang, of the Moscow embassy, reported a conversation with Major Ralph Glyn, MP, who was in the USSR looking at transport matters, including an underground railway system. Glyn found 'the authorities singularly ignorant about underground problems' and tried to disabuse them of the idea that tunnels could be made big enough to take main-line rolling stock. (Here he was going too far; the thing had been done with the Great Northern & City Railway in London, opened in 1904 from Finsbury Park to Moorgate for through running of main-line stock, though not so used until 1976.) He suggested a visit to London to inspect the Underground system, 'which is recognised here to be the best in the world', and he recommended an arrangement for providing British expert advice and supervision, the Soviet authorities being free to place orders for material as they wished.[4]

More than six months earlier, in December 1931, these matters were already being discussed and decided in London. On 5 December Dr Leslie Burgin, a solicitor, and MP for Luton, wrote to Ramsay MacDonald, the Prime Minister:

> An important group of English electrical engineers and consultant advisers has been asked to undertake the construction of the whole of the underground railways of the city of Moscow . . . Considerable orders for electrical material would be placed here, and British engineers would be given the entire charge and supervision of the work. [This appears to have been an optimistic overstatement.] I have been consulted on the legal side and as a preliminary am enquiring whether Great Britain is desirous of assisting in such an enterprise. The matter from its size and importance is urgent.

No more than three days after the date of Burgin's letter the Prime Minister minuted, 'Would warmly welcome chance to obtain contracts for British industry,' while noting that there were questions regarding credit terms. So there was encouragement from the highest level of government.[5]

Exactly when and how the London Underground became involved in the enterprise is not yet clear from the documents available. The decision must have been taken during the winter of 1931–2, for by June 1932 a fairly detailed report on the scheme, with recommendations, had been prepared by Underground officers, and was sent to Moscow. Construction had been under way there since November 1931, but, not surprisingly, had run into difficulties. The first line, when it was opened, had stretches of open-air, shallow 'cut and cover' and deep-level tube alignment.

III

A sixty-three-page paper entitled 'Moscow: Report on the proposed underground Railway System and General Traffic Conditions', by three London Underground officers, E. T. Brook (Rolling Stock Superintendent), Evan Evans (Assistant Superintendent of the Line) and J. C. Martin (Resident Engineer), who all visited Moscow, was completed in July 1932. It was accompanied by seven double foolscap pages containing answers to specific questions put by Rotert, the engineer in charge of construction.

The report went straight into a survey of the existing traffic conditions in Moscow and the proposals for improvements. There were than 1,944 trams in service (1,033 of them motor cars) and 200 buses (out of a fleet of 250). Metrostroi, the organisation set up to build an underground railway system, had prepared a comprehensive scheme, providing for five lines (i.e. ten radial routes), with two 'circle' or peripheral lines to follow. Town planning was already based on the proposed lines, which 'appear in general to be suitably placed', although 'the interchange stations in the centre should be almost entirely rearranged' and the siting of stations should be modified, as they were too close together. Consideration should be given to dividing lines at their outer ends, and circle lines should not be built – two recommendations derived straight from London experience: the Hampstead and Highgate branches of the subsequently named Northern Line and the western extension of the Piccadilly, and the recurrent attempts of London operators to free themselves from the constraint of a continuous Circle service, an object not yet attained.

Moscow was planning a service of eight-car trains (six cars at the outset, uncoupling to four cars in off-peak periods) at minute-and-a-quarter intervals. The Londoners urged them not to try a closer headway than a minute and a half. On station design Moscow's ideas were indefinite, but escalators had been agreed on. Island platforms 525 ft (160 m) long were proposed, and a flat fare was recommended.

Part II of the report contained detailed proposals for the construction and operation of the system. On the basis of a geological section provided by Moscow, the report found that 'the construction of underground railways under Moscow will be a work fraught with very great difficulties . . . about as difficult as it could possibly be' – water and running sand would be encountered. Moscow had plans for sub-surface tunnelling, which had gradients as steep as 1 in 25 and 1 in 30 and a very long construction time and which would affect streets, tramways and sewers. There was also a deep-level scheme, generally in carboniferous clay or 'some other hard ground', and this was to be preferred. Moscow insisted that a shortage of iron prevented its use for tunnel lining rings, so it favoured the sub-surface line built with concrete and stone; if they were to go to deep level, concrete lining blocks would have to be used. The report recommended deep-level tubes with cast-iron lining. (Concrete lining blocks were not used on the London Tubes until the 1960s.) All the tubes and escalator shafts (mostly with rises between 22 ft and 43 ft, except one of 60 ft and one of 75 ft) should be driven under compressed air.

For rolling stock, on the standard Russian 5 ft gauge, the report recommended eight-car trains, all steel: motor cars with forty-eight seats and standing capacity for 140, trailers with fifty-two seats and 146 passengers standing. A general arrangement drawing was supplied. There would be a positive current by third rail, with return through the running rails. Welding of rails was recommended (with some hesitation – it was not regularly adopted by the Underground until 1939.)

Frank Pick, the managing director, added some general and very pointed comments of his own, based on his London experience. This six-page paper shows Pick's mastery both of concept and of detail, at his management best. Without preamble he went straight to his first point: the spacing of stations was too close. Stations must of course be located right at the accepted traffic centres, but outside the central area they should be at least 1,400 m apart; surface transport would deal with journeys up to three miles. Next, extensions should be planned, in conjunction with housing development, into undeveloped territory. Escalators should be provided at stations where flights of thirty-six to forty fixed steps were proposed. A two-zone system of charging should be adopted, not the flat-fare system proposed or 'the complicated system of differential fares' as in London. All lines should emerge into 'free air' at each end. At the outer ends, branches were not satisfactory; buses and trams should be the feeders. Trams were not necessarily objectionable in themselves, but the development of bus services was to be preferred. Deep-level construction should be adopted in the centre, with island platforms at stations. Working shafts should be used for ventilation. Stations should be regarded as centres 'at which other public services might be rendered' – post offices, telephones, police call boxes, and public lavatories – at surface level.[6]

IV

Such were the recommendations made in July–August 1932. From fragments of correspondence that survive, it seems that there was pressure to deliver them quickly, for fear of competition from other countries. Markham & Co., of Chesterfield, supplied a tunnelling shield: only one, for further examples were manufactured in the USSR. The 'preliminary general arrangement' drawing of rolling stock had been prepared by the Underground Chief mechanical Engineer's office and stamped 'WSGB' (W. S. Graff-Baker), 16 June 1932. The Underground officers' report was completed while David Anderson, of the consulting engineers Mott Hay & Anderson, was in Moscow (26–30 July 1932), drawing up his construction report. It went into some detail, based on short visits, about the tunnel lining to be adopted to ensure a fully watertight structure, recommending cast iron lining with lead-caulked joints; reinforced concrete lining should be used only for running tunnels (not station or escalator tunnels) in layers of good clay. Gradients in running tunnels should be no steeper than 1 in 50.[7]

V

Documentary evidence in Moscow throws some light on the development of the Metro's foreign relations. Rotert, the construction chief, wrote on 3 March 1932 to the USSR trade delegation in London (Arcos), asking for technical information. On 5 April J. Janson, the Arcos head in London, wrote that he had made enquiries of Frank Pick, of the London Underground, who had replied on 1 April asking for information on the electricity supply and main-line railway characteristics. On 25 May the Soviet of People's Commissars, under V. M. Molotov, decided to build without delay the first section of Metro from Sokolniki Square to Sverdlov Square in deep tunnel, and to call upon foreign experts for advice. In September (the exact date is not clear) a letter signed by N. A. Bulganin, chairman of the Moscow soviet, was sent to Pick, stating that Messrs Brook, Evans and Martin and the consultant Anderson had been in Moscow and had spent much time giving explanations and advice. Their visit lasted from 18 until 30 July.

By June 1934 construction work was far enough advanced for the presidium of the Moscow soviet, with L. M. Kaganovitch in the chair, to invite further consultations with foreign specialists and to propose sending a number of their own operators abroad (5 June 1934). On 14 May 1935, the day before the line opened for traffic, the Moscow soviet resolved to award commemorative medals to Messrs Pick, Cooper, Brook, Evans, Anderson and Martin. On 15 May there was a long discussion with Cooper, Brook and Evans on technical questions concerning maintenance and operating. Some matters of design, such as the use of check rails on curves, were also aired, as well as the respective duties of train crews and rolling-stock maintenance staff. From the notes of this discussion it would appear that many operational issues had yet to be decided, but it may merely have been that the Moscow people were running through the whole of their procedures to invite comment or that some of the senior party men were showing off their concern for the workers by getting themselves primed with basic, down-to-earth questions – questions that must surely have been settled by the date of opening for traffic. It would seem surprising, for example, if procedures for track inspection and switching current off and on were still open to discussion.

On 16 May there was a further, larger meeting with a 'floor audience'. Messrs Cooper, Brook and Evans went over the ground again, this time dealing also with emergency procedures and equipment and with the need for a training school. The meeting concluded before all questions had been dealt with – thirty-seven were taken away for written reply. On 17 May there was another meeting, with L. M. Kaganovitch present, mainly about accidents and safety measures. [. . .]

VI

It is clear that, in 1935 at any rate, the builders and operators of the Moscow Metro felt they had learned much from the advice they had received from the London Underground. But any hopes the British government may have had of important orders for equipment coming to Britain were disappointed – after the prototype shield, everything was manufactured in the Soviet Union, which was not then inhibited by international agreements on copyright and patenting. In one way, which had been hinted at in the planning discussions, the deep-level tunnels proved their worth: when the German air raids on Moscow began in July 1941 Stalin and some of the general staff worked in bunkers at the Metro's Belorussia and Kirovskaya stations.

The Metro went on to develop in its own way and to set the standards for post-war construction in Leningrad (1955), Kiev (1960) and a string of other cities, including Budapest and Prague. By 1971, with Moscow's population over 7 million, its Metro was carrying over 2,000 million passengers a year, which made it the most intensively used urban underground railway system in the world. Visitors from London Transport – and indeed many non-professional London visitors too – could perceive, beneath the lush decoration of the first generation of stations, the familiar lineaments of their own Underground. The rolling stock was so close to London designs that when W. W. Maxwell, the assistant production engineer at Acton works, visiting Moscow in 1955, saw the underside of a car from an inspection pit he immediately exclaimed, 'District 1930s stock!'[8]

In return Moscow gave London an idea: a concourse at low level with high vaulted roof, with short side passages to the platforms. It was to be adopted at Gants Hill, Ilford, on the Central Line eastern extension. Construction with five intersecting cast-iron tunnels was begun in 1937, suspended during the war, and completed for use on 4 April 1948. Similar in design to the sub-surface concourse at King's Cross on the Metropolitan Line, opened in 1941, it remains the only one of its kind at deep level in London.

VII

The foregoing had already been written when a communication was received in June 1995 regarding a series of twelve photocopies of drawings prepared in London in 1932 which had been sent to Moscow in 1993. Comments were contained in a letter from Metrogiprotrans, the State Institute for the Design and Planning of Underground Transport, Moscow.[9] The letter stresses the high quality of the preliminary designs of Metrostroi (the planning agency) and the 'serious, substantial work organised by the Soviet government to study foreign experience'. It included visits by groups of experts to Paris, Berlin and Hamburg in 1926 and by V. I. Makovsky to London and other cities (published 1935). The 'English Expert Examination' confirmed decisions 'already expressed in one or other version of the preliminary design'. (This implies that alternatives were still open.) The recommendation of cast-iron tunnel-lining could not, however, be generally adopted at the time (it was widely used in the second and later stages) because of a shortage of the material; locally produced cast-iron tubing was used only in escalator tunnels at three stations. On the other hand, London recommendations which did not appear in the preliminary design were adopted in respect of step size on escalators, escalator speeds, 'continuous rating' and emergency lighting in cars. As for the line network, gradients and station layouts, foreign experience of constructing and operating underground railways was drawn upon.

The view in Moscow is that the Metro grew up 'without doctors and medicines of foreign technical firms, without foreign-made equipment and materials. Moscow Underground is a 100 per cent Soviet product.' Taken literally, about construction and equipment, this is perfectly true. But it implies no discredit to the authors of the Metro - indeed, rather the reverse - that they were careful to ensure that they had the fullest possible information about the 'state of the art' in the outside world of underground railways, and they drew fully on the experience of those already in the business. In the process their ideas matured rapidly from the somewhat simplistic notion that the thing was essentially a matter of mining technology into a fully developed passenger underground railway that could stand comparison with that of any of the world's great cities when it was opened in 1935.

Notes

1 O. S. Nock, *Underground Railways of the World* (1973: the most conveniently accessible account in English), pp. 175–83.

2 Nikita Khrushchev, *Khrushchev Remembers,* intro. C. Grankshaw (1971), pp. 52–6, 43, 485.

3 PRO, FO 371/15621, N4541/4541/38, 22 June 1931.

4 PRO, FO 371/15616, N5785/950/38, Strang (Moscow) to H. J. Seymour (London), 18 August 1931.

5 PRO, FO 371/15616, N7867/950/38, Burgin to Prime Minister, 5 December 1931; minute initialled 'J.R.M.', 8 December 1931.

6 London Transport Museum library, Ga.473.61.1; Pick's comments, 10 August 1932, LT document IG.947.

7 Mott Hay & Anderson, letter to M. A. Ozersky, USSR trade delegation, London, 6 August 1932 (copy in LT documents); letter from R. Beresford, Mott Macdonald Group, 8 January 1992.

8 Information from Antony Bull (who was there), 15 January 1992.

9 Letter to Michael Robbins, 14 June 1995, signed by the president, S. Seslavinsky, and the manager of the technical department, V. N. Kiselev.

31

HOUSING THE CITIZENS IN SOVIET RUSSIA, 1955–90: THE TYRANNY OF TECHNOLOGY

Blair A. Ruble

Source: Blair A. Ruble, 'From *Khrushcheby* to *Korobki*,' in William Craft Brumfield and Blair A. Ruble (eds), *Russian Housing in the Modern Age*, N.Y. and Cambridge, Woodrow Wilson Center Press and Cambridge University Press, 1993, pp. 232, 234–44, 248–51, 256–7, 259–60, 263–7, 269

[. . .] The massive housing construction program launched by Nikita Khrushchev during the late 1950s succeeded in turning over nearly 70 million apartments to just under 300 million Soviet citizens by the late 1980s. The average living-space available to each Soviet urban dweller doubled between 1956 and 1989, increasing from 7.7 square meters to 15.8 square meters. Indeed, as the 1990s begin, only about 12 per cent of the Russian Republic's population lives in dwellings smaller than the average national per capita allocation of thirty years ago. Qualitative indicators also demonstrate improvement, showing that around 90 per cent of all Soviet urban residences had running water, central heating, and indoor plumbing by the late 1980s.

Yet it is not clear precisely what is being measured by Soviet aggregate housing data. As the popular weekly newspaper *Argumenty i fakty* observed in September 1990, 40 per cent of the residents in Moscow's Oktiabrskii district lived in communal apartments and some 22,000 inhabitants languished on the district's waiting lists for new housing, while official statistics proudly proclaimed an average living-space allotment of between 18 and 20 square meters.[1] Concentrations of communal apartments linger on in run-down center city areas, leaving many students, young families, and pensioners trapped in slumlike conditions. After nearly three-quarters of a century of Soviet power, in fact, hundreds of thousands, if not millions of Soviet citizens resided in log cabins (*balki*), low-grade mobile homes (*vagonchiki*), and murky basements (*podvaly*). The last round of Communist Party promises to provide a separate apartment unit to each family by century's end was greeted with profound distrust and cynicism.[2] Now that the Party is gone, even empty promises of a brighter housing future have disappeared as well.

For all the remaining problems, it is important to recognize that skid-row warrens are no longer the urban norm. According to nearly every *statistical* measurement employed by Soviet housing authorities, the housing program launched during the 1950s – perhaps the most ambitious governmental housing program in human history – has been a tremendous success. The great failing in this formulation favored by the housing bureaucrats was that most of the beneficiaries of this enormous project did not much like the new apartments into which they were moving. Families liberated from communal dwellings quickly found their new abodes to be poorly planned and shabbily constructed. Years of labor by residents are frequently required to correct a familiar litany of irksome deficiencies in the superficially modern Soviet high-rise apartment building of the 1990s: persistent elevator breakdowns, plummeting water pressure, electrical surges, and upper-story windows that shatter in 'high winds' even though all is calm at ground level; the list goes on. [. . .]

Whatever the routine difficulties of Soviet reality, it is clear that the 1980s and 1990s witnessed a resurgence in what may be called domesticity: the space and opportunity for devotion to family and personal life, values, and tastes. This revival was reflected both in abstract architectural journals and in the lives of Soviet citizens. The family and the individual were fighting back. To understand against whom, one must turn back to 1953 to the gargantuan Soviet housing industry that was spawned by the struggle for supreme political power which followed the death of Joseph Stalin.

Housing the people

The battle for political dominance began even as Stalin struggled to draw his last breath. A lengthy and pressing agenda for change had long been accumulating despite the surface calm of the dictator's last years. Included on it were an end to mass political terror; a new openness to the outside world; greater flexibility in economic management; renewed investment in agriculture; expansion of consumer goods production; and an improved standard of living for the Soviet Union's poorest citizens. Together these constituted a ready-made political program for any potential leader with the courage and savvy to grasp for it. After some initial hesitation, Nikita Khrushchev moved to make this program his own, wielding it as a political club to drive other contenders off the political stage.

Housing lay at the heart of any attempt to gain broad-based popular support.[3] Soviet housing conditions were shocking. The collectivization and rapid industrialization campaigns of the 1930s had forced tens of thousands of peasant families off the land into already substandard urban housing left over from before World War I. The devastation of World War II, which had consumed vast expanses of the Soviet heartland, only compounded an already appalling situation. The total destruction of frontline cities such as Kiev and Minsk, and extensive damage to other battered communities such as Leningrad, insured widespread dislocation and homelessness. A dearth of postwar construction did nothing to alleviate suffering. Perhaps a single statistic best illustrates the common misery of that era: an average of 3.3 families lived in each Leningrad apartment in 1951 - and Leningraders reputedly were much better housed than other Soviet citizens.[4]

Khrushchev moved into action during mid-1955. He prompted the Communist Party and the Soviet architectural community to convene several conferences on housing, including a national gathering at the Moscow Architectural Institute and then another in the hallowed halls of the Kremlin. By the end of the next year, three distinct bodies - the Twentieth Communist Party Congress, the Party's Central Committee, and the all-union Council of Ministers - had issued proclamations announcing a housing program intended to provide each and every Soviet family with its own apartment over the course of the next three Five-Year Plans.

The origins of both the housing program's quantitative successes and its ruinous artistic and social deficiencies are to be found in the critical years immediately following Stalin's death. Elite housing projects of the late Stalin regime, such as the wedding-cake apartment towers on Insurrection Square (*ploshchad' Vosstaniia*) and along the Boiler-Maker

Figure 31.1 Apartment house on Kotel'nicheskaia Embankment, Moscow
Source: *Oltarzhevskii, Stroitel'stvo vysotnykh zdanii v Moskve*

Figure 31.2 Apartment house on Kotel'nicheskaia Embankment, Moscow: lobby
Source: *Oltarzhevskii, Stroitel'stvo vysotnykh zdanii v Moskve*

Embankment (*Kotel'nicheskaia naberezhnaia*) (Figures 31.1 and 31.2), reflected an elegance and attention to human needs that rivaled luxury buildings of the period in the West. A housing program that sought to reproduce the more obtainable features of the Stalinist high-rises on a mass scale would have held out the promise of sustaining a modicum of domesticity in Soviet life. Instead, the administrators controlling the country's centralized command economy turned housing into simply one more category of industrial output measurable in terms of gross production (in this case, thousands of square meters of living space).

In the end, Khrushchev's housing program worked about as well as every other sector of the Soviet economy. An initial decade of success turned sour as the Soviet housing industry produced thousands of desperately needed apartments that few residents would find comfortable by the time of the Brezhnev 'stagnation.' Indeed, much of the housing stock constructed at the beginning of the housing drive was deteriorating so quickly that it was simply uninhabitable twenty years later. Recent visitors to one Muscovite encampment dating from the early 1960s had to march around piles of rusting bathtubs and under emergency scaffolding (to protect them from collapsing balconies) before they could make their way into desolate 'Beirut-nouveau' entryways covered with shards of glass. The low-quality construction of the period was but one of many consequences of policies that treated habitation as just another production 'problem.' This supposition defeated any attempt to retain a human quality in housing.

The technological imperative

Boris Yeltsin recounts in his autobiography a story that is reminiscent of the attempt to perfect a 'centrifugal pipe-casting machine' by Vladimir Dudintsev's hero Lopatkin in the venerable 'thaw'-era novel *Not by Bread Alone.* Like Lopatkin, Yeltsin's father spent much of his adult life trying to invent a labor-saving device, in this case a machine that would lay bricks. And like Lopatkin, he was ultimately stymied by the system. Interestingly, Yeltsin also reports that entire research institutes are still trying to come up with a machine that can mix mortar, lay bricks, and clean off surface mortar, moving forward all the while.[5]

Soviet builders, like Yeltsin's father, have long been mesmerized by the idea of creating a construction system that would eliminate the need for human labor. Although no one has discovered a system that will lay bricks, a number of other 'industrialized' construction systems have been devised in response to the Communist Party's decision to rehouse the entire Soviet population in a matter of years. A laudable end, perhaps, but one that quickly became reduced to absurd means. Witness, for example, a pronouncement that the prime indicator of the effectiveness of construction was the weight of the building (lighter being better).[6]

The Soviet construction industry abandoned concern for design under pressure from Communist Party leaders to build housing as quickly as possible. Builders concentrated their efforts on developing prefabricated construction techniques. Such initiatives moved on two fronts. First was the effort to reduce the range of options available in housing to a bare minimum. As we shall see, this search for prototype plans and blueprints would last for years. Second, Soviet authorities attempted to develop industrial prefabrication methods that would enable housing to be built as rapidly as possible while reducing costs. The program's first decade saw relatively modest savings in total construction costs (between 10 and 15 per cent). The largest reduction in expenditures – between 35 and 40 per cent – resulted from the use of standardized design packages for apartment buildings. This statistic suggests that the high aesthetic costs of innovation in construction technology were not necessarily being offset in accounting ledgers.

Having failed to develop brick-laying machines, Soviet engineers turned to reinforced concrete. The search for new construction techniques began immediately after various Communist Party and state proclamations on the goal of providing new housing for all Soviet citizens within fifteen years. Engineers and architects understood from the outset that construction technology had to change in order for the new demands to be met. Industrialization of the building site became inevitable.

The furious pursuit of faster construction techniques led, in 1958, to widespread endorsement of large panel construction as the preferred industrialized construction technique (Figure 31.3). The Moscow city soviet established its Special Architectural-Construction Bureau (*Spetsial'noe arkhitekturno-konstruktorskoe biuro* [*SAKB*]) to study precisely how plumbing and window fixtures might be integrated into designs for prefabricated construction. This move by Moscow authorities was merely one local manifestation of a national trend begun with the creation of the State Committee on Construction Affairs (Gosstroi) in 1955 as the Soviet Union's principal national institution involved in the regional and city planning and construction process. Gosstroi began operations as the successor organization to the Committee on Architectural Affairs. It was involved in one way or another with every phase of the construction process from planning to general contracting, and directed a centralized national administrative and support network for construction and related planning efforts. This system included more than a dozen specialized research and design institutes, as well as scores of local design centers such as SAKB.

The State Committee on Civil Construction

Figure 31.3 Large panel construction
Source: Stroitel'stvo i arkhitektura Moskvy, 1970, no. 10

and Architecture (Gosgrazhdanstroi) was also subordinate to Gosstroi. As with their economic counterpart Gosplan, Gosstroi and Gosgrazhdanstroi were replicated at the level of the Soviet federal republics by fifteen state construction committees (one for each republic). These same functions were performed at the local level by the construction and architectural departments and the architectural planning administrations of various city and regional soviets as well as by planning institutes subordinate to Gosgrazhdanstroi. (This national system collapsed together with the Soviet Union itself in 1990 and 1991, although many remnant local organizations remain active.) An extensive network of contractors headed by city and regional construction administrations operated dozens of smaller trusts that became, in effect, construction firms. A series of specialized construction agencies were charged with such specific tasks as facade reconstruction, subway construction, bridge construction, and university development throughout the USSR.

Meanwhile, Leningrad construction teams transformed the Soviet housing industry forever with their successful completion of several experimental reinforced concrete large panel apartment houses in the city's Shchemilovka district during 1956 to 1958. By the early 1960s, the Leningrad construction techniques were being employed in such diverse areas as Kiev and Sverdlovsk. After Shchemilovka's debut, construction specialists around the Soviet Union worked to perfect and advance large panel construction systems. The fruits of this labor were to be seen in small and large towns throughout the Soviet Union, on newly set-down streets named for local shock workers and minor revolutionary or war heroes (major heroes having long before donated their names to the central arteries).

Discussions continued to rage over the effectiveness of new techniques as well as over the precise cost of differing construction practices. Such debates largely ignored the needs of the families that were scheduled to inhabit the new apartments. Proponents of new building methods paid even less attention to aesthetic considerations. Meeting production quotas and schedules mattered most of all: Soviet housing firms, naturally, were behaving just like any other Soviet industrial enterprise of the period.

The passage of time – and the perfection of large panel construction systems – prompted fierce debates over the optimal height of apartment buildings from a purely cost-efficiency point of view. The squat five-story structures of the late 1950s (Figures 31.4 and 31.5) – the *khrushcheby,* which is a play on *trushchoby,* the Russian word for slum – gave way by the mid-1960s to nine-story prototype apartment buildings. Taller structures were now thought to be more efficient as a result of improving technology, especially in the development of elevator, gas, and water systems. The reinforced concrete panels then in use were considered to be less expensive than brick, thereby permitting more economical construction of larger structures. Sixteen-story apartments would become the norm within a decade. Engineers continued to trumpet the glories of the tall building, paying little or no attention to the needs and desires of the

Figure 31.4 Housing project no. 8A, Moscow
Source: *Arkhitektura SSSR*, 1956, no. 5

residents. By the mid-1980s, a prototype apartment house emerged from Soviet drawing boards numbering no fewer than thirty-seven stories.

The continuously growing scale of apartment buildings had an overwhelming impact on the urban environment. The old khrushcheby of the 1950s were downright cozy compared with what followed, as the bogus economic criteria of the centrally planned economy came to drive architectural and urban planning and decision-making. Standardization powered this new Soviet machine age, much as it had propelled other machine ages earlier in the century. Enormous 'boxes,' or *korobki,* blighted the urban landscape.

The prefabricated elysium

Architects struggled valiantly to perfect design standards for the new apartment towers as the engineers moved to refine their large panel construction methods. National competitions were held during the late 1950s to develop

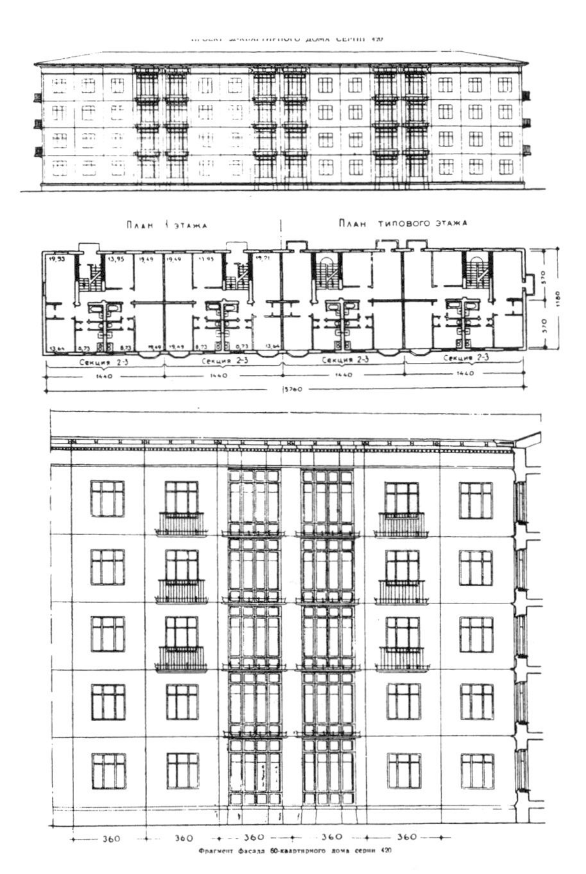

Figure 31.5 Standardized apartment house designs
Source: *Arkhitektura SSSR*, 1964, no. 1

prototype building designs. Construction authorities anticipated that the competitions would lead to the generation of a limited number of apartment types, which would in turn facilitate their own efforts to reduce the number of options available to residents. Fewer options, it followed, would facilitate standardization, thereby advancing industrialized construction techniques. Initially, some architects believed that as few as eight standard apartments could accommodate all possible housing needs.

The Soviet construction industry experimented with various designs for three- to five-story apartment buildings throughout the last years of the 1950s. Architects and engineers sought uniformity of design, with but a few alterations to accommodate varied climatic conditions. Some designers, for example, proposed that hallways and passageways be left open to the elements in more moderate climates so as to conserve on materials and facilitate speedy construction. The average apartment unit size of the period hovered around 33 square meters, or the size of many efficiency apartments in American east coast cities. This space allocation would come to be seen as somewhat luxurious in light of subsequent housing.

Construction trusts attempted to move beyond the architectural sketches of design competitions by integrating their growing experience with large panel construction methods into town planning. Large-scale projects, such as Moscow's Novye Cheremushki and Severnoe Chertanovo districts, Malaia Okhta in Leningrad, and Akademgorodok near Novosibirsk, offered more opportunities to discover new ways of reducing construction costs. Housing officials were never satisfied by cost-cutting innovations. The drive to reduce expenditures and improve efficiency dominated discussions of housing during this period, a rather curious development given the current ideological claims as to the superiority of central planning over profit-driven markets in addressing basic human needs. By the early 1960s, the desire for faster, more economical construction was exerting pressure over the entire Soviet housing industry to trim back living space. Average housing allocations in the newly built apartments declined during the early 1960s from between 16 and 17 square meters per inhabitant to just 12 to 13. Residents confronted rather severe constraints as a consequence of these cutbacks.

The 1960s also witnessed new national competitions for prototype apartment designs. Technological innovation and improved building materials accelerated housing construction beyond all previous levels. Apartment buildings grew in height, while the size of individual units dwindled even further. The average newly built apartment in Moscow continued to shrink until 1965, when the trend toward smaller units finally was reversed.

The history of Russian housing is replete with attempts to determine the optimal space allocation for each individual resident. Unlike the 1920s, when socialist theory emphasized the need to view apartments as spare 'machines for living' (or the 1930s, for that matter, when rapid industrialization prompted families to double, triple, and quadruple up in each apartment), the reduction in the available housing space assigned to the average citizen of the 1960s was motivated by the requirements of an increasingly inflexible construction industry.

The speed with which new apartment buildings dotted – or blighted – the urban landscape disguised the down-sizing of individual apartment units.

This shrinkage, so obvious from the inside of the new cottage-sized rooms, was invisible in aggregate data which continued to record improvement in housing norms. Each family member moving from a communal apartment to an individual apartment in some massive new housing project gained greater space per capita even though the rooms being inhabited were often actually smaller. Such gains became less dramatic with the passage of time as increasing numbers of Russians found themselves adjusting to their new domiciles.

Residents reacted no more sympathetically to the khrushcheby than they had to the communal experiments of the 1920s. The poor quality of building construction eventually presented severe difficulties to municipal housing authorities. The khrushcheby, which today require massive renovation, account for perhaps as much as a quarter of all housing in Moscow. Large panel construction continued to replace more traditional (and more flexible) building forms, such as brick construction.

Over one-quarter of all new Soviet housing in 1965 was built with large panel methods, a figure that increased to nearly one-half a decade later. Renovation of reinforced concrete buildings – together with disposal of obsolete prefabricated panels that are not as readily recyclable as brick or wood – presents yet another vexing problem for those charged with upgrading Russian living standards.

The spread of prefabricated towers across the Russian landscape prompted a backlash among the general populace, although their ire was not publicly articulated for some years. Early expressions of anger toward housing arrangements included Vladimir Voinovich's *The Ivankiad,* a scathing and hilarious account of the mortal struggle over one apartment, which appeared in the West and circulated widely within the Soviet Union.[7] This novella contributed to its author's banishment from the Soviet Union. The 1977 film *An Irony of Fate,* which began with cartoon housing-blocks springing out of the soil to march threateningly toward a meek and cowering human figure, won wide attention and acclaim. These look-alike towers grew like mushrooms after a rain shower as every feature became standardized throughout the USSR. [. . .]

Two professional communities were particularly outspoken about the deficiencies of the Soviet housing effort. Sociologists pointed to the folly of forcing all Soviet households into approximately a dozen standardized floor plans. The architectural community, for its part, challenged the aesthetic merits of panel construction. Controversy would erupt by the later 1960s over the aesthetic and socio-demographic price of large panel housing construction. [. . .]

A district, not a home

Architects and planners [. . .] [argued] that individual units need not provide every service required for human habitation. Instead, apartments were to be viewed as but one component in a new residential system that brings each resident into sustained contact with an entire urban neighborhood. The micro-district (*mikroraion*), or superblock, arrived in the Soviet Union together with large panel construction methods in the 1950s.

The superblock has a distinguished history in Modernist thought.[8] Part of the retreat from the industrial city advocated by radical architects and planners of the early twentieth century, superblocks sought to remove community life from the presumed tyranny of the street. Functional zoning separated housing from services and industry, with green areas serving as vital passageways between each area. In theory, the blocks would serve as totally integrated neighborhoods, with all social and commercial services available to residents within easy walking distance of individual apartments. The blocks were thought to be a sort of urban garden city in which residents would live, study, and work safely removed from the dangers of congested, dirty, and noisy urban districts of the old industrial city. [. . .]

The superblock concept had two additional advantages for planning and construction agencies of the 1960s. First, superblocks were easily integrated into industrialized housing production. Second, superblocks were readily transformed into impressive cardboard models and mock-ups. This second characteristic permitted the Soviet city planner to satisfy bureaucratic and political patrons who desired to show off models of future developments more than they sought to display actual housing projects (which, somehow, never quite managed to live up to the promise of the cardboard paste-ups). The architectural press of the period abounds with eloquent briefs in favor of the socialist micro-district, while talented draftspeople supplemented words with eye-catching sketches of new projects. Apologists for the micro-district argued that individual apartment units were merely the smallest part of an

Figure 31.6 Contemporary Moscow
Source: *Stroitel'stvo i arkhitektura Moskvy*, 1982, no. 7

all-encompassing system. Few members of the housing field during the early 1960s acknowledged that the actual micro-districts under construction fell far short of the images projected by design plans and models. [. . .]

Most new micro-districts became socialist bedroom communities with the result that more and more Soviet citizens must now commute farther and farther to their jobs on increasingly overloaded public transportation systems.[9] Hours are spent each week by Soviets packed onto undercapitalized yet overused subways, buses, and trams so crowded at times that one can 'stand' without touching the floor.

Consumer and public services similarly have failed to move with the population to the new micro-districts. Long-standing difficulties in coordinating the service sector with residential development within a centrally planned economy have meant that new districts have been served particularly poorly by consumer, health, educational, and recreational services.

Aesthetic concerns should not be ignored at this point, as large panel construction methods have combined with the micro-district approach to city planning to produce urban vistas that exude Kafkaesque anomie (Figure 31.6). Leningrad authors were among the first to complain about the unattractive cityscape arising about them. Articles throughout the 1970s by Leningrad architects asked why it was that new districts had to be so displeasing, and implored readers to think about the kind of city in which they wanted to live. Such complaints were hardly limited to Leningrad. The narrator in *An Irony of Fate* sardonically observes that one of the great achievements of recent Soviet construction has been to make folks feel at ease when traveling to an unknown city. After all, travelers will feel perfectly at home anywhere since every city looks exactly alike, with identical street names and streetscapes right down to the local 'Rocket' movie house.

The fusion of an inadequate service infrastructure with aesthetic deprivation has produced an environment that is acknowledged to be both psychologically debilitating and socially disruptive. Crime in the large panel micro-districts has become a significant problem, especially where cultural amenities (and even food) are deficient or lacking altogether. Interestingly, criminality in urban areas in Russia follows global patterns, with crime rates increasing in large high-rise housing projects just as in North America and Europe. [. . .]

Gain without equality

Russians at the end of the 1980s were housed much differently from the way they were at the end of the 1950s. In contrast to overcrowded communal apartments in central urban areas and the squat, cramped khrushcheby on the outskirts, Soviet cities have been surrounded by massive apartment blocks clustered in bleak mega-districts. Individual apartment units house multiple generations, usually of the same family. Nearly all urban housing built during the past quarter-century offers indoor plumbing, hot and cold running water, and electricity. The housing program launched by Khrushchev has been both a resounding success and a woeful failure. All of these elements are visible in Strogino, a rather typical development built at the end of the 1970s on the banks of the Moscow River nearly 20 kilometers northwest of the Kremlin.

Strogino sits on a picturesque bluff on the Moscow River's right bank, not very far from the Khimki Reservoir and Moscow's Ring Road. A variety of parks and recreational facilities are located within easy reach. Built on a scale that makes Co-op City in the Bronx, New York, appear meager and snug, the area features minimal landscaping which accentuates the vast spaces surrounding each building. Open space has been configured so as to create powerful wind tunnels during the winter. All public spaces appear to be studiously uncared for, especially the graffiti-covered entryways and corridors strewn with shards of broken glass. Strogino provides ample proof of a frequently repeated Gorbachev-era maxim that property which belongs to everyone belongs to no one. It is for this reason, among many, that the radical Moscow city soviet has been looking for ways to encourage private ownership of housing.[10] The proposed transfer of title for state housing to its tenants would dramatically expand a cooperative housing market that had already remained quite robust throughout the Brezhnev era.

Moscow's planners initially conceived of Strogino as a largely self-contained community, with a number of local employers including various industrial sites. Environmental activists were able to bring an end to industrial construction so that only a very few enterprises have been able to open for business. Transportation facilities have failed to keep pace with housing construction, forcing residents either to confront long, uncomfortable bus rides to the nearest metro (subway) stations or, in a strikingly large number of instances, to rely on private automobiles. Consumer services have remained inadequate ever since the complex's initial residents moved in over a decade ago. Strogino, in short, is ugly, depressing, and profoundly inconvenient for a majority of its residents. Despite all of these seemingly serious failings, Strogino has become a much sought-after micro-district. Once inside individual apartments, a visitor discovers rather spacious and thoughtfully planned residences. Domesticity has reappeared in Strogino, overcoming many of the impediments placed in its path by an overly rigid and unimaginative construction industry. [. . .]

Khrushchev's massive housing drive has given Russians the worst of all possible housing worlds. The tyranny of technology has proven every bit as restrictive as the tyranny of the market. Social gains have been substantial only when measured against the norms of three decades ago. Glasnost-driven media now expose the existence of an army of homeless people found throughout urban Russia. Central planning neither addressed the human concerns so prominent in socialist ideology nor provided the flexibility and resilience so apparent in capitalist housing markets. [. . .]

The nature of public space is a direct consequence of the tyranny of the technology. Engineers concluded early in the development of industrialized construction techniques that the design of public spaces was too burdensome and wasteful a task. Entryways and corridors were kept to a minimum so as not to complicate the prefabricated design and the block section construction methods the engineers so adored. Merely recalling the grand entrance of a Stalin-era elite apartment building suggests that numerous Russians might have reached a different conclusion.

The history of the European and North American apartment building demonstrates the value of decorous public space for sustaining congenial domesticity. Although bourgeois pretensions are easily mocked, it is clear that the ennobling experience born of certain middle-class affectations in Western apartments built over the past century is sorely absent from today's Soviet housing. The abandonment of a commitment to pleasant public space during the late 1950s has contributed to the dehumanized impression given by so much Soviet housing in the 1990s. [. . .]

Notes

1 'Kto v tereme zhivet?' *Argumenty i fakty*, 1990, no. 37 (15–27 September 1990): 5–6.

2 Gregory D. Andrusz, 'A Note on the Financing of Housing in the Soviet Union,' *Soviet Studies* 42 (July 1990): 555–70.

3 A point subsequently emphasized by Khrushchev himself, in Nikita Khrushchev, *Khrushchev Remembers: The Last Testament*, trans. and ed. Strobe Talbott, Boston, Little, Brown, 1974, pp. 100–5.

4 Denis J. B. Shaw, 'Planning Leningrad,' *Geographical Review* 68 (April 1978): 189.

5 Boris Yeltsin, *Against the Grain: An Autobiography*, trans. Michael Glenny, New York, Summit Books, 1990, pp. 24–5.

6 S. Khotchinskii, 'Puti realizatsii stroitel'noi programmy,' *Arkhitektura i Stroitel'stvo Moskvy*, 1957, no. 2 p. 12.

7 Vladimir Voinovich, *The Ivankiad*, New York, Farrar, Straus, and Giroux, 1976.

8 See, for example, the discussions in Robert Fishman, *Urban Utopias in the Twentieth Century*, New York, Basic Books, 1977; Peter Hall, *Cities of Tomorrow*, New York, Basil Blackwell, 1988.

9 Soviet planners have managed to produce horrendous commuting conditions even in as relatively modest a community as Iaroslavl', an industrial center of approximately 700,000 residents about 150 miles to the northeast of Moscow on the Volga. Nearly all of the housing constructed during the past two decades in Iaroslavl' has been built either across the Volga in a district connected to employment sites by a single two-lane bridge, or in a new mini-district complex to the north of the city some 45 minutes by crowded tram from the city center, and a good 30 minutes from most of the city's major factories.

10 Francis X. Clines, 'Moscow Acts to Give Tenants the Apartment They Live In,' *New York Times*, 8 July 1990, pp. 1, 4.

32

ALLAHABAD: A SANITARY HISTORY

J. B. Harrison

Source: K. Ballhatchet and J. Harrison (eds), *The City in South Asia,* London, Curzon Press, 1980, pp. 167–190

[In 1885 Dr C. Planck, sanitary commissioner of India's North-West Provinces, looked back on his experience over two decades:] 'The centres of population as first known to me presented an appearance of much neglect. The local officials of supreme authority, as a rule, gave little thought to sanitary conditions, often spoke of their occasional visits to the less public quarters of a town or city as dangerous to health, a duty most safely performed under the influence of a lighted cheroot, or with the aid of an occasional pinch of powdered camphor, as a defence against the prevailing unwholesome atmosphere. In the old days, indeed, I have walked with many authorities in unwholesome lanes of a city site - to their display of much disgust, holding of noses, and rapid departure to a place of purer air. But never could anything of evidence be discovered, favouring the idea that the authorities were in any measure to blame for this condition of things. Plainly, indeed, the view was expressed that, if the natives chose to live amidst such insanitary surroundings, it was their own concern. And how they managed to do it without greater penalty of death, than seemed apparent, was a frequent cause of expressed surprise.

As a fact, however, the local authorities of those days had as little knowledge, or thought, of the true death penalty of a city or town, as of their true position as the people's defenders against insanitary conditions. Conditions not born of any desire or fault of the people, but inseparable - until the authorities shall interfere with decision for their remedy or prevention - from the life circumstances of many families closely associated.

At the present day I have good reason to believe that much of this spirit of apathy, or unwillingness to acquiesce in long-established malpractices or insanitary evils, has given place to an ardent desire for improvement . . .'[1]

The purpose of this paper is to sketch an answer to the questions raised by Dr Planck's valedictory survey - under what impulses and under whose pressure did attitudes to sanitary matters change, to what evils did awakened official zeal address itself, with what understanding, and with what effect, and finally, how the general public responded to this concern and activity - using Allahabad as principal source in estimating 'the true death penalty of the city'.

In England the pressure for sanitary and public health reform can be traced back to the early nineteenth century and to the quite unprecedented growth of towns - growth ranging from 47 per cent for Manchester and Salford, 60 per cent for West Bromwich, to 70 per cent for Bradford in the single decade 1821 to 1831. The resultant breakdown of sanitary arrangements, and of efficient water supply, coupled with often gross over-crowding led to a reversal

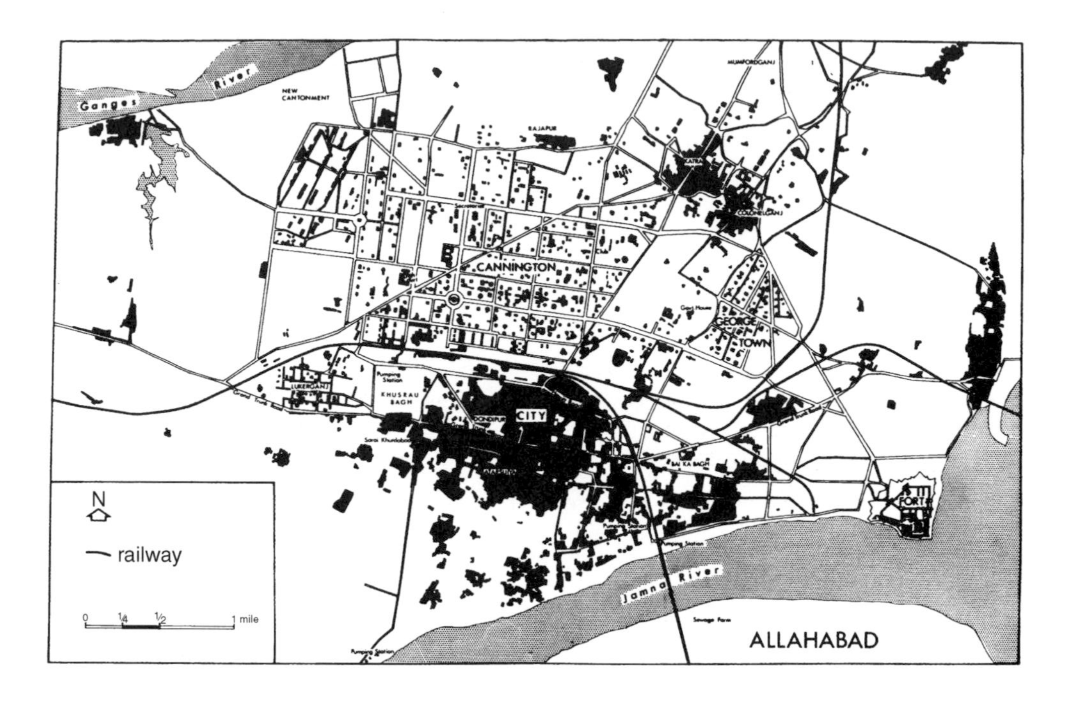

Figure 32.1 Map of the city of Allahabad

of the downward trend in mortality rates. Though the reversal was only slight, it was dramatically peaked in 1831–2, 1848–9, 1854 and 1867 by outbreaks of cholera, which 'struck down many hundreds of thousands of victims, killing tens of thousands',[2] and did so with an unpleasant disregard of social class. Among the poorer classes, moreover, typhus, always endemic, also assumed alarming epidemic proportions, in 1826–7, 1831–2, 1837 and 1846. One response was a flurry of temporary activity by local authorities in years of alarm. Another, of more permanent importance, was the prosecution of a series of major enquiries: in 1838 by Drs Southwood Smith, Arnott and Kay, employed by the new Poor Law Commission to study the 'constantly acting causes of destitution and death' in London;[3] in 1842, on a national scale by Chadwick, who to the 1838 material added reports from his Assistant Commissioners, from the medical officers to Boards of Guardians, and from local doctors, to construct his Report on the Sanitary Condition of the Labouring Population of Great Britain; in 1844 a Royal Commission on the Health of Towns; and 1869, in response to medical criticism, the Royal Sanitary Commission.

The conclusions set out by Chadwick were

> That the various forms of epidemic, endemic, and other disease caused . . . by atmospheric impurities produced by decomposing animal and vegetable substances, by damp and filth, and close and overcrowded crowded dwellings prevail among the population in every part of the kingdom . . . .
>
> That the formation of all habits of cleanliness is obstructed by defective supplies of water.

> That the annual loss of life from filth and bad ventilation are greater than the loss from death or wounds in any wars in which the country has been engaged in modern times.
>
> That the ravages of epidemics and other diseases do not diminish but tend to increase the pressure of population.
>
> That these adverse circumstances tend to produce an adult population short-lived, improvident, reckless, and intemperate, and with habitual avidity for sensual gratifications.[4]

Having skilfully played upon pocket, sympathy and fear Chadwick proceeded from diagnosis to remedy: 'The primary and most important measures, and at the same time the most practicable and within the recognised province of public administration, are drainage, the removal of all refuse of habitations, streets and roads, and the improvement of the supplies of water.'

And to this statement of objectives he added the riders that sanitation should be entrusted to 'responsible officers qualified by the possession of the science and skill of civil engineers', and to 'a district medical officer independent of private practice', both supported by 'uniformity in legislation and in the executive machinery'.[5]

Chadwick's findings were confirmed by the Royal Commission on the Health of Towns, and his pleas by the Royal Sanitary Commission's proposals that the unsystematised sanitary laws and jurisdictions should be made 'uniform, universal and imperative'.[6]

With varying degrees of delay and of acceptance of Chadwick's bland statement that public interest should override private property rights where health and sanitation were concerned, legislation followed. In 1848 the first comprehensive Public Health Act was passed and a General Board of Health was established. It was made imperative upon local bodies to appoint a surveyor and an inspector of nuisances, to make public sewers and require owners to provide house drains, to cleanse streets and fill up offensive ditches, and to provide adequate supplies of water, if necessary from local authority water-works. London had to wait until 1855 for its Act for the better Local Management of the Metropolis, the establishment of a central Board of Works, and provision for the appointment of Medical Officers of Health and Inspectors of Nuisance by vestries and local boards. But in the next decade further acts were passed against the adulteration of food and drink, to prevent pollution of the Thames, to make vaccination compulsory, to establish isolation hospitals, and to control Artisans' Dwellings while in 1871 and 1875 the Local Government Board Act and the Public Health Act provided for expert central review of local enforcement of a consolidated body of health regulations.

The Public Health Act of 1875 established bye-laws of general application, prescribing minimum housing standards. The Royal Commission on Housing in 1885 led in 1890 to a Housing Act which not only raised those standards, but gave local authorities power to recondition existing houses and to clear away slums. Together with legislation covering working conditions in factories, shops and homes, these acts sought to provide an adequate environment. The Boer War, however, revealed an alarmingly poor physique among army recruits, and the report of the Duke of Devonshire's inter-departmental enquiry into physical deterioration, published in 1904, led to a wider range of legislation - the Open Spaces Act in 1906, the Act establishing the School Medical Service in 1907 and the Maternity and Child Welfare Act of 1919 - which added a new dimension to the idea of public health.

What had been a late element in the evolution of sanitary policy in Britain - the spur of military needs - was in India the starting point of serious concern. The importance of sanitation to the army, established and publicised by the Crimean War, was confirmed by the Mutiny in which the losses by ill-health had vastly outnumbered those incurred in combat. As a result of that experience, and of the great increase in the size and cost of the British element in the post-Mutiny army in India a Royal Commission on the Health of the Army in India was appointed. It reported in 1863, not only upon directly military matters such as the duties and diet of European troops in India, their barracks, hospitals and cantonments, but also upon the sanitary state of Indian towns and cities. As the Report put it, 'It is indeed impossible to separate the question of health, as it relates to troops, from the sanitary condition of the native population, especially as it regards the occurrence of epidemics . . .'[7]

The 1863 report led directly to the creation of a Sanitary branch within the Home Department of the Government of India, and to the appointment of Sanitary Commissioners and Assistant Commissioners to the subordinate governments under a Sanitary Commissioner to the Government of India. The impulse behind these appointments is made manifest in the form of the Sanitary Commissioners' annual reports which opened always with sections on the European Army and the Native Army before dealing with the General Population. And when submitted to Parliament, the Report on Sanitary Measures in India was accompanied not by a review from the Local Government Board, but by a memorandum from the Army Sanitary Commission at the War Office. Nor was the military bias thus imparted redressed, as it would have been in Britain, by powerful professional bodies, such as the Institute of District Surveyors, the Institute of Civil Engineers, or the British Medical Association, for in India these professions were largely in the military hands of officers of the Royal Engineers and the Indian Medical Service. The administration of civil hospitals and dispensaries, and of jails, medical attendance on government servants and responsibility for public health was entrusted to I.M.S. Officers on loan to Government. Even the 'civil surgeon' to be found in every district was an officer liable at any moment to recall to military duty. The direction taken by medical research in India was long influenced by military needs, whence the attention to typhoid, which killed European troops, rather than to consumption which carried off the sepoy, or the treatment of V.D., which kept the equivalent of three regiments permanently in hospital, as a military but not a civilian problem.

Military interests were not, of course, the sole determinants of public health policies in India. In the North-Western Provinces and later in the Punjab, as the annual sanitary reports make very clear, the mortality and illhealth which overtook the cultivators in districts brought under canal irrigation, quickly became a matter of deep concern to revenue officials, and to the medical authorities. No less clearly, in Bombay, Karachi and Calcutta the interests of trade - and of the world community - were brought to bear when plague broke out in these ports. And in every town with a considerable body of European civilians, official or non-official, some sort of station committee came into existence to secure a minimum standard of health and comfort, for themselves at least. Calcutta with its Justices of the Peace appointed in 1794, its Lottery and its Fever Hospital Committees, and from 1847 its Board of Improvement Commissioners might lead the way - both in the size of its sanitary problems and the amount of taxation and effort deployed for their solution - but by 1853–54 the Station Committee at Allahabad had four dispensaries in action in the native city, Katra and Daraganj, the cost defrayed from the Magh Mela fund,[8] and two years later it reported an expenditure of Rs. 700 a month on conservancy from the proceeds of the Chaukidari Tax. Indeed, as early as 1827, Fanny Parkes had described Allahabad as 'pretty and well-ordered, the roads the best in India', with station ice-pits, but also with both a leper and a blind asylum supported by public subscription.[9]

Such local effort lacked funds, lacked direction, continuity and professionalism, and failed to involve even the 'respectable natives' of the town, except by way of an occasional consultation or appeal for subscriptions. In 1861, however, post-Mutiny financial embarrassment led Wilson, Finance Member of the Government of India, to propose a transfer of responsibility for roads and public works to local bodies and the grant of a corresponding power to raise taxes locally, a shift towards decentralization which Lawrence was to follow by transferring the costs of town police, and Mayo of local education and medical services to local authorities. In 1861 the Allahabad magistrate accordingly sounded out leading townsmen on the subject of local taxation – and found 'an almost universal feeling in favour of octroi. The only objectors were a few of the large traders'.[10] In 1863 a municipal committee was appointed, under Act XXV of 1850, to deal with police, conservancy and town improvements. In 1867 the civil station and city were amalgamated for municipal purposes, while in 1868, under the North-Western Provinces Municipalities Act, six wards were demarcated and the elective principle introduced. At the same time an octroi tax was levied on a scheduled list of commodities on imports into the city: a tax which remained the chief source of municipal income well into the next century.

In the same period the Royal Commission on the Health of the Army in India had authoritatively identified the sanitary problems in towns such as Allahabad, in a series of trenchant vignettes: 'The habits of the natives are such that, unless they are closely watched, they cover the whole neighbouring surface with filth . . . ' 'There is no such thing as subsoil drainage . . . Neither latrines nor urinals are drained. For all purposes to which drainage is applied in this country, as a means of preserving health, it is unknown in India.' 'There has . . . been no application apparently of any modern improvement, as regards either the examination of water sources or the means of collecting or distributing water for use for stations, bazaars or towns . . . the present condition of the water supply . . . is unquestionably a predisposing cause of disease, especially during the prevailing seasons of cholera, fever, dysentry and other zymotic diseases.'[11] The remedies were clear – those which Chadwick had prescribed for the equally insanitary state of British towns – improved water supply, drainage and cleansing. Once these were introduced the health of troops would necessarily improve, while freed of the burden of ill-health the natives, too, would become 'better and abler men, women and children for all the purposes of life'.[12]

The appointment of a Sanitary Commissioner to the Government of India, and of a staff of provincial commissioners provided the supervisory organisation which could press for these reforms, with authority, expertise and continuity. What is more, by their instructions, the new commissioners were required to concern themselves solely with the native population: the traditional areas of I.M.S. concern – Army cantonments and the European occupied Civil Station – remained entirely in army hands. When in 1878 the office of Superintendent-General of Vaccination was absorbed into that of Sanitary Commissioner, the Superintendents of Vaccination were also appointed Deputy Sanitary Commissioners. Their new duties were to spread vaccination, to monitor birth and death registration, to preach sanitation and to report to District Magistrates any disease-provoking conditions noted in towns met with on tour, and to suggest remedial measures. (This last duty, they were warned, must be done personally 'no subordinate officer of the department being permitted at present to inspect or report upon the sanitary condition of any town or village.')[13] I.M.S. jealousies had limited the range of work open to the new service – but with the happy effect of compelling attention to conditions in the Indian city and town. The new men, moreover, were competent professionals, who in many cases acted, as the Medical Officers of Health appointed in Britain had done, as active reformers. The village, town and city bazaar had otherwise only seen humble creatures of little standing - the vaccinators and the native doctors, products of the Agra or Calcutta Medical Colleges, who in 1855 were thought highly paid at Rs. 25 per month. The Sanitary Commissioners could command attention to their views.

But what were their views? Dr Planck sets them out for us in his first annual report as Sanitary Commissioner in 1868:

> *Typhoid Fever.* This form of fever is due to a poison of animal origin, as malarial fever is due to a poison of vegetable origin. I write in the simplest words for the advantage of non-professional readers, and the prevalence of contagious fever is always found to be in close relation to the imperfect manner in which sewage matters are removed.
> *Malaria.* Malaria [literally Bad Air] is believed to be rather heavier than pure air, and to be more prevalent, or men less able to resist its effect, in the early morning hours. It is therefore very desirable that all who can should sleep on a surface well raised above the ground.
> *Malarial Fevers.* . . . the principal causes of which are the undrained condition of the sites of towns or villages, or their dranage into excavations which neighbour upon, or often exist in, those centres of population, so as to form large collections of stagnant water.
> *Cholera.* There are two circumstances with reference to this fatal form of disease, which recent research would appear to have established.
> 1st - That the disease is most prevalent in places the air of which is tainted by fermenting or decomposing animal excrement.
> 2nd - That the germ or contagium of the disease is found in the dejections of cholera patients, and especially in the characteristic rice-water discharges.
> The proper precautions are therefore to clear away and prevent accumulations of house refuse, disinfecting any large collections of impurities before they are disturbed; 'especial cleanliness must be enforced in sewers, drains, foul ditches, sewage ponds, slaughter houses, and places where beasts are kept', and 'care must be taken that the brick work of all the wells is in good repair, and that every well from which drinking water is taken has its mouth surrounded by a low wall'.
> *Dysentry and Diarrhoea.* . . . where malaria abounds, where conservancy is neglected, where coarse and ill-cooked food is often eaten, deaths from dysentry and diarrhoea will often occur in fatal form.
> *Smallpox.* It is I think well known to all educated persons, that small-pox has . . . in all probability no other mode of communication than from one person to another . . . Therefore it fortunately happens that an efficient method of preventing its prevalence has been discovered in vaccination . . .[14]

Two things stand out from this lengthy extract - the inadequacy of Dr Planck's understanding of the mechanics of disease, and his firm grip upon many of the practical precautions to be taken to avoid contagion. His use of the phrase 'germ or contagium' reminds us that though he was writing ten years after Pasteur's work on fermentation, and eight after Lister's demonstration of the value of antiseptics, it was still to be another eight years before Koch isolated the anthrax bacteria and produced his postulate that specific diseases were caused only by specific bacteria. Not for another thirty years would the role of insects in transmitting plague, kala azar and malaria be established. But, as the Army Sanitary Committee, reviewing the excited report of the isolation in Indian cases of the same *bacillus typhosus* as had been tracked down in Europe, rather grumpily observed, given the new identification the old sanitary measures had still to be pressed home. What those measures were Dr Planck already in many cases knew when he first surveyed Allahabad, the city in which his headquarters were established, in 1868.

In 1868 the boundaries of Allahabad enclosed a great blunt wedge of land, six miles long by four miles broad, hemmed in on three sides by the Ganges and the Jamna, whose confluence made the town a sacred place of pilgrimage. The city proper, a square mile in extent, straddled the Grand Trunk Road and so north to the Jamna. South of the city, across the East Indian Railway, spread the formal grid-iron of Cannington, the new civil station, where the classical Secretariat was under construction. And round that again, in a great arc touching the Ganges to east and north, lay the Fort, remnants of the old civil station, and two new sprawling cantonments - in which were embedded Daraganj, Sheokothi and Katra, pilgrim centres and bazaar. It was a vast site, most of it standing high, with ample well-water. Since 1853 the population had grown by 35,000 to 106,000 in 1868 and it would go up by as much again by 1873. (In January of each year, at the time of the Magh Mela, the population was for some three weeks swollen by perhaps 100,000 pilgrims, a figure vastly exceeded every twelfth or Kumbh Mela year.) Of the municipal revenue of Rs. 96,000, Rs. 11,000 went on conservance, Rs. 37,000 on road maintenance, tree planting and such works as drains, culverts and latrines - in all half the total budget.

The contrast between city and civil station - north and south of the railway track - was striking. The city had two or three main streets of fair width, but from these a labyrinth of smaller and smaller lanes led off into the interior of the muhallas [the poorest districts of huts], many of which still had gates to seal off the whole block at night. There were several mandis [markets] for grain, fodder, etc, one or two considerable serais, and a townscape of mixed pucca and kutcha housing[15] given distinction only by the Mughal Khusru Bagh and Khuldabad Sarai. Houses looked traditionally inward and with space already limited were growing upwards. The civil station, laid out by the Commissioner Thornhill in 1858, was a gridiron of regular broad roads, those nearest the station already metalled, with newly planted avenues of trees, framing bungalows which stood in two, three and even ten acre compounds. Though the numbers in the civil lines were always larger than expected, since many bungalows were in multiple occupation, and the long low ranges of servants' quarters were often in part let out to artisans or labourers, while the coachhouses and compounds were leased to ekka-drivers,[16] petty shopkeepers and stall holders, the contrast in density between the European and Indian wards of the city was still extreme.

So, often enough, was the contrast in the servicing and sanitariness of the two parts of the municipality. Thus, on roads and surface drainage, Dr Planck reported in 1868 that though the city was fairly swept, and the main roads well made 'but their side-drainage is faulty; many of the by-roads are unmade, and almost impassable in the rains. . . . in a sanitary point of view the urgent requirement in this city is for the creation of an efficient surface drainage system . . . every hole or excavation now existing being at the same time filled up.' He went on, 'The Civil Station, inhabited by the European community, has been covered in all directions with very good roads; its drainage has been carefully attended to; . . . and I would urge that the same attention should now be given to the city, where men congregate and much business is done.'[17]

The metalling of the roads, scarified and then re-dressed every four to six years with rammed kunkur,[18] and their regular watering to bind the surface and lay the dust was an expensive business, made more so as the side ditches were made pucca with brick. Between 1873-4 and 1882-3 the yearly average on maintenance and petty repairs of metalled roads was over Rs. 24,000, and that exclusive of road watering, sweeping and drainage. This represented a tenth of all municipal expenditure. (The one bright spot in the roads account was provided by the road sweepings, which in the age of the horse, found ready sale as fuel for brick-kilns, while these were still permitted within municipal bounds, and thereafter were taken by the army for the improvement of their great sweep of grass lands to the south of the Civil Lines.) By far the larger part of this outlay was upon roads outside the city proper, while as is clear from the covert criticism of Devon, the Municipal Engineer, in 1900 maintenance was weighed out upon racial grounds, rather than upon those of weight of traffic or the needs of trade. As for the minor interior lanes of the city muhallas, these had to wait for attention until the devastating outbreak of plague in 1902, when an emergency grant of Rs. 15,000 for paving and Rs. 20,000 for draining those lanes 'where plague has broken out or is likely to break out', was made by the Lieutenant-Governor in person. [. . .]

The other recurring expenditure by the municipal authorities - and the most urgently necessary - was upon conservancy. No more trenchant comments were passed by the Royal Commission than upon this subject, it figured largely in every Sanitary Commissioner's report on the state of towns and villages, it obtruded itself upon everyone's attention, and if inadequate was the major cause of that 'death penalty of a city'. On the outskirts of the city, where the cover of standing crops, the ravines running down to the river, tanks or old brick fields invited it, what Dr Planck described as open-air conservancy was practised, a 'daily offering of impurity exposed on the surface of the earth'.[19] This villagers' habit always created a 'nuisance', and as the theories of bacterial infection gained acceptance, created alarm too, as a source in the dry weather of infectious dust and in the rains of polluted water. The one effective action to which alarm led, however, was the forcible removal of native villages and hamlets from within the civil station and cantonments 'to preserve the health and life of the remaining and more important section of the community'.[20] The prime example of this was the uprooting in 1901 of the village of Nimi Bagh, as an insanitary eyesore, much too close to Government House.

One gut reaction to popular but insanitary habits was thus to attempt to stamp them out by use of bye-laws, fines or compulsory acquisition powers. The more constructive response was to find alternative modes of sanitation adapted to urban conditions. The simplest answer was to define an area which might be used for defecation, and to appoint sweepers - or to allow pigs and cattle - to keep the enclosures clean. On his first Allahabad inspection Planck noted several such fenced enclosures provided on the outskirts of the city as latrines. The next stage was to build latrines and urinals of a kind more familiar to the European mind, and more appropriate to the dense building pattern of the city. Almost endless inventiveness was displayed by sanitary officials and by amateur enthusiasts in devising new types of latrine, some, such as those at Cawnpore, being quite palatial:

> These latrines are in the shape of a double row of cells with a wide passage between, cells and passage covered by one roof. The passage is the sweepers' domain, and by it impurities as they collect can be deodorized by means of earth and removed to the carts, the lower or receptacle part of being open towards the passage. The whole structure, 90 feet long by 20 feet broad, is of brickwork . . . a separate residence for sweepers is provided, so that they need never be absent.[21]

At the other extreme was the simple structure of timber clad with iron sheeting mounted on wheels so as to permit it being moved daily over another set of trenches. In between came a whole variety of latrines, some open some roofed; some of brick set in lime - or since this was attacked - set in mud mortar; some tarred, some coated with dammar[22] ; some to Dr Bellew's Bombay pattern, others in the Lucknow style, and yet others in the widely copied railway functional which became the norm in Allahabad.

By 1881 there were 34 public latrines in the city and 12 in Katra-Colonelganj, while in 1888-9 no less than 11 kutcha city latrines were rebuilt in burnt brick, stone floored, and with ashlar seats, with sweepers' houses attached, and the total had risen to 56 public latrines.

But no number of latrines would be of any use unless they were kept clean - the first need, as Dr Planck insisted, was to make them 'acceptable to the Indian people . . . unless the municipal authorities of a town are prepared to ensure this amount of cleanliness, they had better not waste their money in providing latrines at all'.[23] The ideal solution in financial no less than aesthetic terms, was to employ the water-closet system. On no point had Chadwick been more emphatic than this, arguing that the great barrier to urban conservancy, the expense and annoyance of hand labour and cartage, could only be met in great cities 'by the use of water and self-acting means of removal by improved and cheaper sewers and drains'.[24] That solution was not open to Allahabad before the twentieth century. Meanwhile reliance had to be placed upon two measures - the use of dry earth, spread over the whole latrine and thereafter freshly applied as the latrine was used to deodorise the faeces and absorb the urine, and secondly the regular and speedy removal of contents.

Private household latrines in the city and bazaars took two traditional forms - the *sandas,* or well - latrine, and the *Mugli* privy. The *sandas* was a dry well, as much as 30 to 40 feet deep, sunk in the courtyard of the house, closed with timber or stone, leaving a small hole in the centre. The *Mugli* privy as described by the Secretary of the Agra Municipality was 'the cupboard or chest form of latrine, which opens by a shutter on to the public street',[25] designed to permit its being cleaned from outside by the sweeper without his entering the house. Both merited the criticism of the Government of India: 'There is every reason for suspecting that the chief disease causes in all Indian towns are to be found within the walls which enclose the compounds and houses.'

The obvious danger from the *sandas* lay in its usually being sunk in the same courtyard as the well from which the drinking water of the family was drawn. When the first chemical analysis of wells was undertaken in Allahabad in 1891 very few were found to be uncontaminated, and many of those in the city were declared dangerous to public health. By the early 1880s official pressure was being applied to secure the discontinuance of the use of *sandas*es. Act III of 1894 then provided all the authority necessary to compel closure, and empowered the municipal authority to enter houses to inspect their sanitation and required them to undertake such inspection systematically. In 1898 the Special Health Officer for Allahabad, Surgeon O'Connor, could report that some 70 *sandas*es had been filled in on his order, though householders had objected, including some who were members of the Board. No prosecutions, however, were instituted 'for continuing a habit that seems to have been in use with them for probably over two hundred years or more'.[26]

The dangers of the *Mugli* privy were even more obvious:

> Connected with almost every house is an open square pit . . . situated either beneath the wall of the house itself or just on the margin of the public roadway; in many instances the family latrine, a small dark cupboard-like place, stands inside the house, near to the pit to which the latrine is connected by a short drain. The condition of these drains and their cesspools . . . is, in a sanitary point of view, very sad. They are approachable only with disgust, contain impurities, long since collected there, and present a state of things which calls urgently for amendment.
>
> In some instances, the family retiring place in the better sort of houses is situated near to the top of the house, and is then connected to the cess-pool below by a narrow [open] channel running down the face of the house outside.[27]

When in exercise of its powers under Act XV of 1883 the Allahabad Municipality drew up new bye-laws, it included one declaring that 'no Mughli Privies opening on any road, lane, street or public place should be sanctioned'. In

June 1889 the Working Committee instructed those members of the Board in charge of conservancy in their wards to enforce this provision. When in July 1889 Dr Ohdedar, as conservance member ordered a certain Masnad Ali to remove his 'Moghli paikhana', he was overruled by the Board, but though rights in existing property were thus respected, permission to build new or alter old houses was made conditional upon the provision of an improved latrine – 'the privy must not be a Mughli privy, but one with a good door',[28] . [. . .] according to the new specifications.

Upon what precise pattern the mainly European and Anglo-Indian occupants of the Civil Lines bungalows arranged their sanitation is not clear – thunderboxes for the bungalow, served by the sweeper, and some sort of latrine in the compound for the servants seems to have been the pattern. What is certain is that sanitation was in general appallingly neglected. Dr Planck recorded how the bungalow was served in 1868: ' . . . house conservancy is left entirely to the management of the sweepers, who as a general rule . . . throw out impurities into any convenient place not far from the house and leave them to dry up. If any man should doubt this as too horrible to believe, I think he would be wise to investigate the circumstances of his own house.'[29] As for the servants, crowded to the rare of the compounds, they were often left ill – or unprovided for. T. N. Ghose, the conservancy member for the ward, reported in 1885 that 'many of the landlords in the Civil Station, although served with notices to repair and construct latrines and supply buckets to the latrines have failed to do so'.[30] And if they did provide latrines, they seem rarely to have used the household sweeper to keep them clean. Major Chaytor White, I.M.S., reported scathingly in 1908 upon the Civil Station:

> At present in many cases no provision is made by house owners for serving private latrines in compounds. The servants often club together and pay for the survices of a sweeper or pay the bungalow sweeper something for his services. In large compounds the proprietor may let out servants' houses separately – the rent of the bungalow being thereby reduced. The consequence is that all and sundry occupy these compounds and the state of the latrines is most unsatisfactory. Under Act I of 1900 Municipal Section 91 (1) powers are given to compel house-holders to keep the latrines in order. Notices have been served times without number on various house-holders, but they are treated with contempt. The compound of the Allahabad Club on which I made a special report is in a deplorable condition: the latrine accommodation is totally inadequate . . .. Notices have been served but nothing has been done. In England when a notice to abate a nuisance is served a reasonable time is given for the improvement and if on visiting the Sanitary Inspector finds nothing has been done no second chance is given. I consider that the Municipality is far too lenient and that in future proper fines should be inflicted. It has been suggested to me that a sanitary tax of Rs.2/- a month levied per private latrine which should include the service of the latrine . . . [31]

However, when a special sub-committee of the Board reviewed Chaytor White's report, they opined that it was not advisable to undertake municipal cleaning of the private latrines in the Civil Station.

The immediate condition of latrines, public or private, depended upon close and energetic supervision of the persons in charge of them, the conservancy of the municipality as a whole upon willingness to spend money and to organise in a business-like fashion. It was a considerable task and required a small army to perform it, even in 1870/1:

	City	Rs.		*Civil Station*	Rs.
9	Jemadars	1,200	1	Head Inspector	1,440
1	Bukshee	300	1	Jemadar	120
2	Chuprassies	120	2	Chuprassies	120
			10	Beldars	480
240	Sweepers and Mehtranies	8,600	40	Drivers	2,440
			40	Assistants	1,920
50	Drivers	3,000	40	Sweepers and Mehtranies	1,680
			2	Domes	130

The force thus assembled had then to be equipped with some hundreds of vehicles - country carts for the sweepings, heavy iron filth carts for the product of the latrines and cask carts and hand barrows for the daily 30,000 gallons of sullage water and urine, together with a few donkeys to carry paniers through the innermost lanes. The disposal of the road sweepings - some 3,000 cubic feet or perhaps 100 tons a day - has already been noted. The disposal of the sewage remains to be considered.

For a while the city sewage was trenched on ground close to the latrines, and until it was discovered that it was a prime carrier of the germs of enteric fever the urine was used to water Alfred Park, but for most of the period both were carted to municipal trenching grounds outside the city. This was the system early used in Cawnpore and seems to have been modelled thereon. Shallow trenches were opened, filled with sewage, lightly covered with soil and left for a season. They could then be let, at high rents, to cultivators. The problem was to secure long-term possession of the land, which required an astute purchase, and to keep conservancy cattle strong enough to haul the immensely heavy filth carts in the hot weather and the rains over many miles of roads. Years of famine or heavy monsoon always saw conservancy approach the point of collapse.

In Calcutta the problem had been solved by building a municipal railway in 1867 which ran through the city collecting refuse from street depots and carrying it to the Square Mile in the Salt Lakes. In 1887 Allahabad investigated the possibilities of a tramway, sending plans and estimates 'to Capitalists at Home'.[32] Nothing came of this, and the tramway languished until in 1905 a syndicate applied for an electric tram and lighting licence - the tramway to double as a conservancy and passenger system. The Collector thereupon went over to Cawnpore to examine the conservancy tramway and incinerators there. He was discouraging and the electrification scheme fell through.

Meanwhile Allahabad had got a water-works, the possibility of water-borne sullage and sewage disposal had opened up, and eventually a considerable part of the city's output was pumped across the Jamna by Schone's Ejectors and used to irrigate farm land there.

The system, first in operation in 1917, had come just in time. In 1909 it was noted that the supply of sweepers, on a monthly pay of Rs.4 a man and Rs.3 a woman, was barely adequate. Two years later rising prices had created a serious problem: 'the Mohalla sweeper finds his hereditary calling no longer provides a decent livelihood. The poorer classes in the city are in consequence seriously embarrassed how to get their houses cleaned.'[33] By 1918 it was clear that the sweeper force was now 'stationary, if indeed not thinning through various outlets to cleaner labour open for untouchables in a town growing under modern conceptions of life and labour'.[34]

Since enteric and cholera were suspected to be water-borne diseases the Royal Commission had drawn particular attention to the dangers of polluted water - shallow contaminated wells, tanks into which a soiled urban surface often drained, rivers which might prove as treacherous a source as the polluted Thames. This concern was reflected in Allahabad by an attempt - abortive - to discourage the use of river water taken from below the dhobi ghat,[35] by orders for regular cleaning of public wells and by the listing of tanks in which the washing of animals was prohibited. More positively, from 1883 or a little earlier consideration was given to the possibility of providing a piped water supply. Curiously enough the main arguments adduced in favour of such a supply were that without it surface drains could not be adequately cleaned and the profits of a sewage farm would not be realised, and that supplies for road-watering would also fail. The minimum cost, Rs.$15\frac{1}{2}$ lakhs[36], was felt, however, to be prohibitive. Nevertheless in 1889 plans were submitted to government and approved, and a loan applied for. By 1890 tenders had gone out and contracts had been placed, and the following year saw pumping begun. Almost at once petitions came in from inhabitants in the city muhallas for the laying of pipes in their lanes, and individuals even offered to lay street mains at their own cost if they might then be allowed a house connection. By 1894 a million gallons a day were being supplied, to houses or street stand-pipes, and for the great Kumbh Mela of that year temporary pipes were run out for the use of the pilgrims. Since the valuation of all the houses in the municipality and their assessment to water-tax had gone off quietly and without too large arrears, the scheme might seem to have been a success. From a sanitary point of view, moreover, the result was most encouraging. The municipal analyst reported that the purity of the water supplied was very satisfactory, and as the range of the mains extended it was possible to speed up the process of filling in the polluted tanks. The analyst was also required to begin the testing of public and then of private wells. In 1881 Markham, as President of the Municipality, had blandly declared 'the quality is good, and the supply ample'.[37] As analysis proceeded more and more were found to be polluted, and notices to clean or close them were sent out – and then when re-analysis found them still tainted, power was taken under Act XV of 1883 prohibiting their use for drinking purposes.

Within a very few years, however, second thoughts about the sanitary advantages of a good piped water supply had emerged. The North-Western Provinces and Oudh Water Works Act, Act 1 of 1891, under which Allahabad operated, had taken careful note of all those legislative failures under which London had laboured in respect of its water-supply. It had firmly required the Municipal Board to supply 'pure and wholesome water sufficient for the use of the inhabitants for domestic purposes': there would be neither the scandals of cholera-bearing water such as the Lambeth Company had supplied, nor the cutting off of all supplies on Sundays which Jephson has so deplored; nor indeed would it be possible for the remissness of a landlord to imperil his tenant's water supply, since the latter could compel the former to arrange for water to be brought into the house. But what it had failed to do was to prescribe a drainage system capable of dealing with the mass of new water poured into the city. Water at the stand pipes was free, and piped to the house, was still kept deliberately cheaper than well water brought up by bullock power. The conservancy department had seized upon the opportunity to install scours from which to flush surface drains, into which sullage water was increasingly diverted. By 1893 the District Magistrate was reporting 'owing to the introduction of the filtered water supply the want of proper drainage is being felt more and more each year',[38] and two years later the municipal engineer noted, 'very few houses have proper arrangements for carrying off waste water, in consequence of which a large percentage of water percolates into walls, floors and ground near water taps and damp unhealthy areas are found all over the city'.[39] It was well that the 1891 Water Works Act had been followed by Act III of 1894, the North-Western Provinces Sewerage and Drainage Act.

Of the drainage system of Allahabad, Dr Planck had this to say in 1879: ' . . . the street drainage of the city is unsatisfactory. There is no system, and no funds exist to create one. The surface drainage, of both the city and the civil station, is delivered chiefly into deep pools or tanks'. The main sewerage artery, the Ganda nala, running from the railway station right across the city to the Jamna was reasonably described as 'a noisome ditch everywhere, with crawling fetid contents'.[40] As late as 1896 the road side drains in parts of the Civil Station were described as 'nothing better than elongated earthen cesspools'.[41] Every year some drainage black-spot was tackled – but never to an overall plan.

It was the growing problem created by the new water-supply which compelled rethinking. First the Health Officer, Surgeon-Captain Davidson, demanded a large-scale survey and mapping of the city, forward planning of building and drainage, and tight control of both. He was told that with Rs.96,000 annual interest payments on the water-works loan no more could be done. Instead in 1900 a list of no less than 58 minor drainage works was drawn up, to be tackled piecemeal as funds could be spared. Under the pressure of the municipal engineer, however, a basic survey of levels was begun in 1904, and this was enlarged at the instance of the Sanitary Engineer to Government into a full drainage system. After reference to Mr Lane Brown, M.I.C.E., who added his professional weight – and fees – to the scheme, the project finally got under way in 1913: a triumph for the influence of the expert, a heavy burden on municipal finance.

The last sanitary problem to be successfully tackled in London was that of overcrowding, whether of persons or buildings. The only early triumph there had been in the control of lodging-houses, tackled by an Act of 1851. But that act became the model for Bengal Act IV of 1871 to control lodging-houses for pilgrims at Puri, and upon that in turn was built Act I of 1892, dealing with lodging-houses in the North-Western Provinces and Oudh. Initially this was applied to Benares and Hardwar, but not to Allahabad – an odd omission when a Kumbh Mela was due, and when Planck had drawn so compelling a warning from the last great Allahabad Kumbh in 1882:

> The houses generally overcrowded, many of them quite full of people lying, at night time, in pretty close order wherever shelter from the sky could be obtained. Rooms, verandahs, out-houses, temporary shelters of mats . . . all accommodating, to their utmost, their *quota* of people.
>
> Surface cleanliness neglected, the drainage-ways choked with sweepings, chiefly large leaves in which food had been purchased – moistened with much refuse water used for cooking or washing, with great suspicion of urine admixture. Attached private retiring places, generally in one corner of the courtyard, overflowing with impurities . . . A condition of things always deplored by the householder, but excused upon the plea that sweepers could not, at this time be found to effect the usual business of cleanliness.
>
> The water of the wells, stirred up by ceaseless drawing, slightly discoloured and of unpleasant odour . . .

All of which was but preparation for the last scene, at the station, of the struggling and then weary crowd of pilgrims, from which 'in every day of its existence, cholera cases were carried away . . . '[42]

However, what government had not prescribed, the municipality took to itself. Despite the objections of [. . .] Daraganj dharmsalas[43] Lodging-House inspectors were appointed and the law imposed, a recognition that once a year the municipality had a special duty larger than that owed to its own citizens.

To the overcrowded conditions under which many of those citizens lived it chose, indeed, to turn a selectively blind eye. If the District Magistrate McNair pointed to the crowded Atarsuiya or Dundipur areas of the city, Brownrigg, the Commissioner, splendidly replied:

> I do not think that the Allahabad City can be said to be seriously congested. There may be some overcrowding in the two Mohallas . . . but I doubt if it is of such a nature as to require heroic measures. Every large town is bound to have some parts which are more or less congested.
>
> As to providing model 'bastis' for the poor, I do not believe for a moment that any plan of this kind would meet with success. It seems to me altogether utopian. What many persons, though, in Allahabad really would like is cheap houses, detached residences, where Government servants could live . . . something after the fashion of Lukerganj.[44]

The outcome of the Brownrigg attitude took two forms. Insanitary overcrowding in Rajapur was dealt with by evicting the inhabitants of

Figure 32.2 Allahabad: sewered, paved and lit corner of Bharti Bhavan Muhalla

'this little pesthole from the heart of the civil lines'.[45] And to the model settlement for Government Press employees, Lukerganj, was added the early Mumfordganj and the George Town Extension which provided for the senior clerks of the Secretariat and for professional men retreating from the dangers of plague in the city. The *Indian People* was happy to applaud this action to deal with 'this most pressing need of Allahabad . . . congestion in the Civil Lines'.[46]

But attention to the problems of the city could not long be withheld. In December 1883 Austen Chamberlain had re-opened the question of slum clearance and working class housing in Britain with a sharp attack on the Torrens and Cross Housing Acts, 'tainted and paralysed by the incurable timidity with which Parliament [deals] with the sacred rights of property'. *The Times* joined in the attack on overcrowding in 1884 and the Royal Commission on the housing of the working classes was appointed whose work was to issue in Housing of the Working Classes Acts in 1886 and 1890. In Bengal an amending Act VI of 1881 gave the Commissioners of Calcutta power to acquire bustees[47] in order to improve them, while far more extensive legislation, Act II of 1888 and Act III of 1889 imposed improvement of bustees upon owners, or in default upon the municipality and provided for driving roads and lanes of standard width through such areas. Add to this the example of the Bombay Improvement Trust's work, and nearer to hand, that of the Aminabad improvement in Lucknow, and some action in Allahabad on similar lines became inevitable.

The final impetus was given, perhaps, by the Sanitary Conference at Naini Tal in September 1908. In January 1909 a trust deed was executed in favour of the municipal board at Allahabad, empowering an Improvement Trust to open up the most crowded areas of the city by making new roads, and to provide sanitary houses for the displaced population. An initial grant of Rs.125,000 was made by government to set going what was expected to be a self-supporting operation. A whole series of operations followed, under the chairmanship of Professor Stanley Jevons of Allahabad University, most notably the driving through of Hewett Road, Crosthwaite Road, Sheo Charan Lal Road and Zero Road, relieving the pressure on the Grand Trunk Road, undoing in part the ills caused by the presence of three railway systems driven into the heart of the city, and slicing through the heaviest concentrations of working class population. These measures, and those later initiated by the Allahabad Improvement Trust refashioned in 1919 by the United Provinces Town Improvement Act VII of 1919, were on the whole financially successful - the Hewett Road frontages sold for sums that astonished everyone. They did do something to cope with the traffic of a city which had much more than doubled in forty years. They did let in that light and air on which sanitarians placed so much stress - though in a crude way, which as was later admitted would scarcely have secured the approval of the generation of 'organic' town planners such as Patrick Geddes.[48] What they did not do, of course, was to provide satisfactorily for the slum dwellers who had lost their homes. The new model townships - Bai ka Bagh and its successors - ended up in middle-class hands, as Collector Hopkinson had foreseen.

Notes

1 *Eighteenth Annual Report of the Sanitary Commissioner of the North-Western Provinces, 1885*, pp. 60-1 (hereafter Sanitary Report NWP).

2 M. W. Flinn (ed.), *Report on the Sanitary Condition of the Labouring Population of Great Britain*, by Edwin Chadwick, 1842, Edinburgh, 1965, p. 8.

3 Sir George Newman, 'The Health of the People', in H. J. Laski, W. I. Jennings and W. A. Robson (eds), *A Century of Municipal Progress, 1835-1935*, London, 1935, p. 159.

4 Flinn, p. 423.

5 Flinn, pp. 424-5.

6 Newman, p. 160.

7 *Royal Commission on the Health of the Army in India*, 1863, p. 77 (hereafter *Health of the Army*).

8 The Magh Mela is the annual Pilgrim Fair at the confluence of Ganges and Jumna at Allahabad in January and February. Every twelfth year there occurs the greater and more auspicious Kumbh Mela.

9 Fanny Parkes, *Wanderings of a Pilgrim in Search of the Picturesque*, London, 1850, vol. 1. pp. 72-3 and 206.

10 Magistrate to Commissioner, 28 June 1861, Commissioners' Records, Basta 78, File 4, Serial 116. In Calcutta the house-tax was the mainstay of municipal finance.

11 *Health of the Army*, pp. 82-94.

12 *Ibid.*, p. 81.

13 Eleventh Sanitary Report NWP, 1878.

14 First Sanitary Report NWP, 1868.

15 That is housing of fired or sun dried brick and pise. The bricks were either made on site or in brick-fields west of the city. Brick was the building material of the civil station too, though classically clad in stucco.

16 ['Ekka' = a small, Indian, one-horse vehicle].

17 First Sanitary Report NWP, 1868, para. 33.

18 [A coarse variety of limestone, commonly occurring in India].

19 First Sanitary Report NWP, 1868, para. 102.

20 *Ibid.*, para. 36.

21 First Sanitary Report NWP, 1868, para. 393.

22 [An oriental resin taken from trees].

23 First Sanitary Report NWP, 1868, para. 389.

24 Flinn, p. 117.

25 Agra Municipality to Government NWP, 15 March 1869, NWP General (Sanitation) Proceedings, 17 April 1869, No. 2.

26 *Report on the Administration of the Allahabad Municipal Board*, 1898-9.

27 Sanitary Commissioner to Government NWP, 6 October 1868, NWP, General (Sanitation) Proceedings, 17 April 1869, No. 2.

28 Working Committee Proceedings, Allahabad Municipal Board, 19 May 1892.

29 First Sanitary Report NWP, 1868, para. 103.

30 Working Committee Proceedings, 12 November 1885.

31 Sanitary Commissioner to Municipal Board, 17 February 1908, CR, Dept. XXIII, File 63, Serial 3.

32 Report, Administration of the Allahabad Municipal Board, 1886-7.

33 *Ibid.*, 1910-11, p. 10.

34 *Ibid.*, 1917-18, p. 12.

35 ['Dhobi' = native washerman; 'ghat' = steps or passage leading to the riverside]

36 ['Lakh' = 100,000].

37 Report, Administration of the Allahabad Municipal Board, 1880-1.

38 *Ibid.*, 1897-8.

39 *Ibid.*, 1899-1900.

40 Twelfth Sanitary Report NWP 1879, para. 124; Thirteenth Sanitary Report 1880, para. 97.

41 Report, Administration of Allahabad Municipal Board, 1895-6.

42 Fourteenth Sanitary Report NWP 1882, paras. 197 and 208.

43 [Buildings devoted to religious or charitable purposes - especially rest-houses for travellers].

44 Commissioner Brownrigg to UP Govt., 3 August 1908, Commissioners' Records, Dept XXIII, File 98.

45 Commissioner to Collector, 9 September 1911, *ibid.*
46 *The Indian People*, 19 January 1908.
47 ['Bustee' = village of humble, Indian huts.]
48 The Improvement Trust Sub-Committee of the Municipal Board appointed in 1908 discussed the opening up of the city and construction of parks and open spaces with Professor Geddes and the map of proposals they prepared 'was drawn with his general approval'.

ACKNOWLEDGEMENTS

PART 1

EVANS, C.
'Industry and urban form in eighteenth-century Merthyr' in Clark, P. and Corfield, P. (eds) *Industry and Urbanisation in Eighteenth-century England*; 1994, Leicester University Press. Every Effort was made to trace copyright holders.

GRANT, R. K.
'Merthyr Tydfil in the mid-nineteenth century: the struggle for public health' in *Welsh History Review*; 1989. Reprinted by kind permission of the University of Wales Press.

YOUNG, C. F.
'Increasing fire-risks in cities in an age of industrialization' in Young, C. F. *Fires, Fire Engines and Fire Brigades*; 1866, London. Every effort was made to trace the copyright holders.

KELLETT, J. R.
The Impact of Railways on Victorian Cities; 1969, Routledge and Kegan Paul, London. Reprinted by kind permission of Routledge and Kegan Paul.

THOMPSON, F. M. L.
'The case against the historical inevitability of suburbs in Thompson, F. M. L. (ed.) *The Rise of Suburbia*; 1982, Leicester University Press. Every effort was made to trace the copyright holders.

VIGIER, F.
Change and Apathy: Liverpool and Manchester during the Industrial Revolution; 1970, MIT Press, Massachusetts. Reproduced by kind permission of MIT Press.

HASSAN, J. A. and WILSON, E. R.
'The Langdale water scheme 1848–1884' in *Industrial Archaeology*; 1979. Reprinted by kind permission of The Association for Industrial Archaeology.

DEVINE, T. M.
'Urban growth in industrializing Scotland, 1750–1840' in Devine, T. M. and Mitchison, R. (eds) *People and Society in Scotland: I, 1760–1830*; 1988, John Donald Publishers, Edinburgh. Reprinted by kind permission of John Donald Publishers.

RIDDELL, J. F.
Clyde Navigation: A History of the Development and Deepening of the River Clyde; 1979, John Donald Publishers, Edinburgh. Reprinted by kind permission of John Donald Publishers.

RIDDELL, J. F.
'Glasgow and the Clyde' in Reed, P. (ed.) *Glasgow: The Forming of the City*; 1993, Edinburgh University Press, Edinburgh. Reprinted by kind permission of Edinburgh University Press.

RANKINE, W. J. M.
'Letter to Glasgow Town Council: Proposal for Glasgow's water supply from Loch Katrine' in *To Commemorate the Public Services of the late Robert Stewart, Glasgow*; 1868. A privately circulated volume now in the Mitchell Library, Glasgow.

PORTER, R.
London: A Social History; 1994, Penguin, London. Reprinted by kind permission of Penguin.

WHETHAM, E. H.
'The London milk trade, 1860–1900' in *Economic History Review*, 17; 1964–5. Every effort was made to contact the copyright holders.

TAYLOR, D.
'London's milk supply, 1850–1900: a reinterpretation' in *Agricultural History*, 45; 1971, University of California Press, Davis. Reprinted by kind permission of the University of California Press.

ATKINS, J.
'The growth of London's railway milk trade, c. 1845–1914' in *Journal of Transport History*, 4; 1978, Manchester University Press, Manchester. Reprinted by kind permission of Manchester University Press.

RATCLIFFE, B.
'Manufacturing the metropolis: the dynamism and dynamics of Parisian industry at the mid-nineteenth century' in *Journal of European Economic History*, 23; 1994. Reprinted by kind permission of the editor of the Journal of European Economic History.

PINKNEY, D.
Napoleon III and the Rebuilding of Paris © 1952. Reprinted by kind permission of Princeton University Press.

PART 2

SIMPSON, M.
'Urban transport and the development of Glasgow's West End, 1830–1914' in *Journal of Transport History*, 2nd Series; 1972. Every effort was made to contact the copyright holders.

BUTTERFIELD, R.
'The industrial archaeology of the Twentieth Century: the Shredded Wheat Factory at Welwyn Garden City' in *Industrial Archaeology Review*, 16; 1994. Reproduced by kind permission of the Association for Industrial Archaeology.

GLENDINNING, M. and MUTHESIUS, S.
Tower Block: Modern Public Housing in England, Scotland, Wales and Northern Ireland; 1994, Yale University Press. Reproduced by kind permission of Yale University Press.

WOLFE-BARRY, J.
'Solving late Victorian London's traffic problems' in *Journal of the Society of Arts*, 47; 1898–9. PUBLIC DOMAIN.

LAWRENCE, J.
'Steel frame architecture versus the London Building Regulations: Selfridge's, The Ritz and American technology' in *Construction History* 6; 1990. Reproduced by kind permission of The Chartered Institute of Building.

EVENSON, N.
Paris, a century of change, 1878–1978; 1979, Yale University Press, New Haven. Reprinted by kind permission of Yale University Press.

KOLLHOFF, H.
'The metropolis as a construction: engineering structures in Berlin 1871–1914' in Kleihues, J. P. and Rathgeber, C. (eds) *Berlin/New York: like and unlike. Essays on Architecture and Art from 1870 to the Present*; 1993, Rizzoli,

New York. Reprinted by kind permission of Rizzoli International Publications, Inc.

HUGHES, T.
'The city as creator and creation' in Kleihues, J. P. and Rathgeber, C. (eds) *Berlin/New York: like and unlike. Essays on Architecture and Art from 1870 to the Present*; 1993, Rizzoli, New York. Reprinted by kind permission of Rizzoli International Publications, Inc.

WHITE, R. B.
Prefabrication: a history of its development in Great Britain; 1965, HMSO, London. Crown copyright is reproduced with the permission of the Controller of Her Majesty's Stationery Office.

DIEFENDORF, J. M.
'Urban reconstruction and traffic planning in postwar Germany' in *Journal of Urban History* 15; 1989. Reprinted by kind permission of Sage Publications, Inc.

PART 3

BATER, J. H.
'The development of public transportation in St Petersburg, 1860–1914' in *Journal of Transport History*, New Series 2; 1973.

ROBBINS, M.
'London Underground and Moscow Metro' in *Journal of Transport History*, 18; 1997.

RUBLE, B.
'From Khrushcheby to Korobki, in Brumfield, W. C. and Ruble, B. (eds) *Russian Housing in the Modern Age*; 1993, Woodrow Wilson Press and Cambridge University Press. Reproduced by kind permission of the author.

HARRISON, J.
'Allahabad: a sanitary history' in Ballhatchet, K. and Harrison, J. (eds) *The City in South Asia*; 1980, Curzon Press, London. Reproduced by kind permission of Curzon Press.

INDEX